T0334153

PROCESSING AND PROPERTIES OF STRUCTURAL NANOMATERIALS

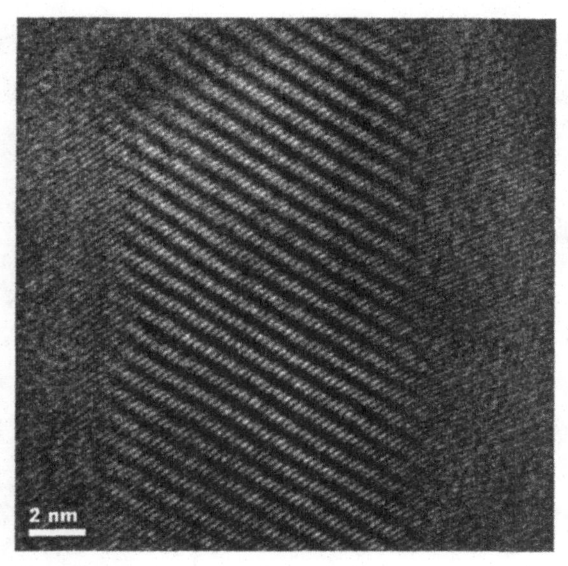

Cover Image:

The cover is a high-resolution electron microscopy (HREM) image of a nanocrystalline $Al_{93}Fe_3Ti_2Cr_2$ alloy taken at the University of Connecticut, showing moiré fringes of two nanoscale Al grains free of dislocations.

PROCESSING AND PROPERTIES OF STRUCTURAL NANOMATERIALS

Proceedings of Symposia sponsored by
the Powder Materials Committee of
the Materials Processing and Manufacturing Division (MPMD) and
the Mechanical Behavior of Materials Committee (Jt. ASM-MSCTS) of
the Structural Materials Division (SMD) of
TMS (The Minerals, Metals & Materials Society)

Held at the Materials Science & Technology 2003 Meeting in
Chicago, Illinois, USA
November 9-12, 2003

Edited by

Leon L. Shaw
C. Suryanarayana
Rajiv S. Mishra

A Publication of

A Publication of **TMS (The Minerals, Metals & Materials Society)**
184 Thorn Hill Road
Warrendale, Pennsylvania 15086-7528
(724) 776-9000

Visit the TMS web site at
http://www.tms.org

Statements of fact and opinion are the responsibility of the authors alone and do not imply an opinion on the part of the officers, staff, or members of TMS, The Minerals, Metals, and Materials Society. TMS assumes no responsibility for the statements and opinions advanced by the contributors to its publications or by the speakers at its programs. Registered names and trademarks, etc., used in this publication, even without specific indication thereof, are not be considered unprotected by the law.

No part of this book may be reproduced, stored in a retrieval system, or transmitted in any form or by any means, electronic, mechanical, photocopying, microfilming, recording, or otherwise, without written permission from the publisher.

Library of Congress Catalog Number 2003106811
ISBN Number 0-87339-558-1

Authorization to photocopy for internal or personal use beyond the limits of Sections 107 and 108 of the U.S. Copyright Law is granted by TMS, provided that the base fee of $7.00 per copy is paid directly to the Copyright Clearance Center, Inc., 222 Rosewood Drive, Danvers, MA 01923 USA, www.copyright.com. Prior to photocopying items for educational classroom use, please contact the Copyright Clearance Center, Inc.

For those organizations that have been granted a photocopy license by the Copyright Clearance Center, a separate system of payment has been arranged.

This consent does not extend to copying items for general distribution or for advertising or promotional purposes or to republishing items whole or in part in any work in any format.

Please direct republication or special copying permission requests to the Copyright Clearance Center, Inc., 222 Rosewood Drive, Danvers, MA 01923 USA; (978) 750-8400; www.copyright.com.

TMS

Copyright 2003, The Minerals, Metals, and Materials Society. All rights reserved.

If you are interested in purchasing a copy of this book, or if you would like to receive the latest TMS publications catalog, please telephone (800) 759-4867 (U.S. only) or (724) 776-9000, EXT. 270.

TABLE OF CONTENTS

Processing and Microsctructure Development

PREFACE

Nanocrystalline (nc) materials have enormous potential to provide structural materials in the future with significant property improvements over today's conventional coarse-grained counterparts. Many interesting properties including superior hardness, strength and wear resistance as well as exceptional levels of superplasticity (900% elongation) at a temperature as low as $0.36T_m$ (where T_m is the absolute melting temperature) have been reported. The recent progress in engineering nanomaterials with a bimodal grain size distribution has demonstrated nc materials that possess superior tensile strength coupled with impressive ductility (such as 30% uniform tensile elongation). In order to provide a forum for scientists from universities, research laboratories and industry to present and discuss these most recent exciting results, the Powder Materials Committee and the Mechanical Behavior of Materials Committee of the Minerals, Metals and Materials Society have co-sponsored a symposium entitled "Processing and Properties of Structural Nanomaterials".

The papers presented herein are the output of this symposium focusing on recent advancements in fundamental understandings and technological applications related to processing and properties of structural nanomaterials. The symposium was held in Chicago, Illinois, November 9-12, 2003. The topics covered in this symposium included synthesis/processing, characterization of structure and mechanical properties, processing/structure/property relationships, thermal stability, phase transformations, theory, modeling, performance and applications. A total of 54 invited and contributed papers were presented in the symposium. The papers submitted to this symposium have been organized into three categories:

I. Modeling and Simulation
II. Structure and Property Relationships
III. Processing and Microstructure Development

27 papers are published in this symposium proceedings.

We would like to take this opportunity to thank the authors for presenting the papers and session chairpersons for organizing the sessions and leading the stimulating technical discussions. We would also like to thank those who have helped reviewing the papers published in this proceedings.

Leon L. Shaw
University of Connecticut

C. Suryanarayana
University of Central Florida

Rajiv S. Mishra
University of Missouri - Rolla

MODELING AND SIMULATION

Processing and Properties of Structural Nanomaterials
Edited by Leon L. Shaw, C. Suryanarayana and Rajiv S. Mishra
TMS (The Minerals, Metals & Materials Society), 2003

Are Deformation Mechanisms Different in Nanocrystalline Metals? Experiments and Atomistic Computer Simulations.

H. Van Swygenhoven, Z. Budrovich, P. M. Derlet & A. Hasnaoui

Paul Scherrer Insitute, NUM/ASQ, Villigen CH-5232, Switzerland

Keywords: Nanocrystalline, Mechanical Properties, Simulation

Abstract

Molecular dynamics computer simulations of fully 3D-nc metals with mean grain sizes up to 30 nm show that from a certain grain size on, there is a dislocation activity within the grains. The picture, however, is different from what is known from coarse-grained materials: dislocations are emitted from the grain boundaries (GB) and absorbed in the opposite GBs. The atomic mechanism behind this emission process observed in atomistic simulations will be discussed. Experimental results on ED-nc-Ni are also discussed in the framework of this new type of dislocation activity from GBs.

Introduction

The mechanical behaviour of fully dense nc metals is mainly characterized by a high yield stress and a limited tensile elongation. From the perspective of the Hall-Petch relation, which relates the yield stress to the inverse square root of the grain size, such a high yield stress trend with respect to reducing grain sizes is expected. However the Hall-Petch relation, which usually is explained on the basis of dislocation pile-ups at grain boundaries [1, 2], must break down at grain sizes for which pile-up can no longer be supported [3] and this calls into question the validity of such a plasticity mechanism at the nano-scale. Furthermore, the operation of dislocation sources is grain size dependent, in that there is a critical length scale below which sources can no longer operate [4] since the stress to bow out a dislocation approaches the theoretical shear stress. For fcc metals the critical grain size is believed to lie between 20-40nm. Thus, it is not only the existence of a dislocation pile-up process, but also the creation of dislocations which is called into question at these grain sizes. Alternatively one could also evoke some type of grain boundary (GB) accommodation mechanism(s) such as GB sliding or Coble creep, as possible deformation mechanisms in nc structures. Here observed strain rates are proportional to the GB diffusion coefficient divided by a higher power of the grain diameter. The possibility for having higher effective diffusion coefficients in nanosized GBs [5], in addition to the nanosized grain diameter, raised the hope that such diffusion based mechanisms could become active at room temperature.

There are, however, some experimental results suggesting that trying to understand the deformation mechanism needs more than extrapolation of equations derived for coarse grained material. For example, the rapid decrease in the slope of an nc stress-strain curve before failure is usually interpreted in terms of a lack of work hardening leading to plastic instability. However some deformation tests under both tensile and compression conditions showed a considerable elongation before failure after reaching a zero derivative in the stress-strain curve and necking is not always observed [6-8]. The experiments are however performed at different strain rates and different microstructures which makes comparison difficult, especially when rate sensitive deformation processes are activated. On the other hand, other deformation techniques, such as grinding and cold rolling evidence considerable amounts of plastic deformation [9, 10]. Therefore the idea that failure might be a result of internal flaws such as porosity, and impurities is still under discussion [11, 12]. The importance of voids in nc materials [13] has recently been demonstrated in in-situ deformation experiments [14]. Ductile deformation mechanism are also evidenced by the dimple-like features on the fracture surface [15]. Such dimples have been recently explained as the result of interplay between local shear planes resulting from GB sliding and the presence of GBs that do not participate in the sliding mechanism [16].

Molecular dynamics (MD) computer simulations can aid and complement experimental results and provide insight into the guidance of new experiments [17, 18, 19]. Although it is very well known that this technique suffers from the restriction of size and length scales, it usually brings a different perspective than that common to experiments, i.e. the atomic scale. MD simulations have suggested the existence of a dislocation mechanism in nc fcc metals different from the conventional dislocation-dislocation interaction mechanism. The simulations highlight the role of the GBs in the deformation mechanism of nc metals by means of two processes: the first can be identified as GB sliding triggered by atomic shuffling and stress assisted free volume activity [20], both of which are intimately linked to GB and triple junction (TJ) migration [21]. The second is dislocation activity where the GBs are the source and sink for partial dislocations, promoting GB relaxation and reorganization of the remaining GB dislocation network which prevents the emission of a trailing partial [22, 23]. Partial dislocation activity is observed in grain sizes above 10nm, where the critical grain size depends on the stacking fault energy of the material. Simulations suggest that unrelaxed GBs containing many extrinsic GB dislocations would favour the emission of these partial lattice dislocations. The observation of only partials even at grain sizes of 30nm is still an open question: full dislocations have been observed in Al grains with similar or higher grain sizes, but the sample geometry is a 2D columnar structure [24, 25] where the process of nucleation and emission is expected to be different [26]. In fully 3D structures it is highly probable that the inherent time restriction of the MD technique is the reason that only partials are observed [18,19]

In the following paragraphs we discuss the deformation mechanism that are suggested by MD simulations and discuss some new experimental results in the framework of these mechanism.

Mechanical Properties Derived from Simulation

Fully 3D nc Ni samples are synthesized using the Voronoi construction in which fcc crystals are constructed at random positions and with crystallographic orientations resulting in a 3D GB network containing a large range of GB misorientations. For a given simulation cell size, the number of fcc crystallites then determines the mean grain size. To simulate a bulk nc system 3D periodic boundary conditions are employed. This initial configuration is then relaxed to a minimum enthalpy using molecular statics, followed by 100ps of molecular dynamics at 300K to allow the nc structure to achieve a more equilibrium configuration. Nc samples are deformed by applying a uniaxial tensile stress within the Parrinello-Rahman framework [27]. A Second-Moment Tight-binding potential [28] was used to describe the metallic bond for model Ni. For more details about the sample synthesis and simulation conditions we refer to [29, 30].

All simulations indicate that below a critical grain size, plasticity is dominated by inter-grain deformation mechanisms, and with increasing grain size (10-12nm) a dislocation based intra-grain deformation mechanism is activated [17, 20, 22-24, 31, 32]: dislocations are emitted from GBs and only partial dislocations are observed in fully 3D GB networks with grain sizes below 30 nm. Such dislocation activity is fundamentally different from that seen in the polycrystalline regime, where plasticity is governed by dislocation-dislocation interactions.

Inter-grain deformation and collective processes

In all samples with mean grain sizes up to 20nm, grain boundary sliding is observed as being the main contribution to the observed plasticity. Careful analysis of the GB structure during sliding under constant tensile load shows that sliding includes a significant amount of discrete atomic activity, either through uncorrelated shuffling of individual atoms or, in some cases, through shuffling involving several atoms acting with a degree of correlation [22]. In all cases, the excess free volume present in the disordered regions plays an important role. In addition to the shuffling, we have observed hopping sequences involving several GB atoms. This type of atomic activity may be regarded as stress assisted free volume migration. Together with the uncorrelated atomic shuffling they constitute the rate controlling process responsible for the GBS. A detailed description of these processes is given in [22]. Such sliding is accompanied by stress build up across neighbouring grains which is accommodated by GB and triple junction migration involving further discrete atomic activity.

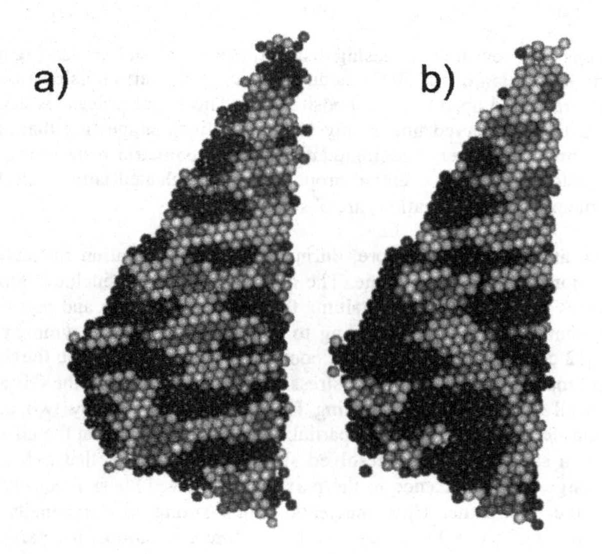

Figure 1: Atomic section of grain boundary 1-7. In a) and b) the atoms are coloured according to their local hydrostatic pressure. a) represents the configuration at elastic loading and b) at ~2% plastic strain

More recently, collective processes such as cooperative grain boundary sliding via the formation of shear planes spanning several grains have been observed [16, 21]. For simulation of such phenomena, a sample with 125 grains and a mean grain size of 5nm was deformed. A large number of grains was necessary in order to minimize the effects imposed by the periodicity used to simulate bulk conditions.

The small grain size is chosen to reduce the total number of atoms in the sample (to 1.2 million) so that longer deformation times are possible at acceptable strain rates. In order to increase grain boundary activity the deformation was done at 800K. The underlying mechanisms that were observed for the formation of shear planes are (1) pure GBS induced migration of parallel and perpendicular GBs to form a single shear interface (2) Coalescence of neighbouring grains that form a low angle GB facilitated by the propagation of Shockley partials. (3) Continuity of the shear plane by intragranular slip. A detailed description of the processes is given in [21].

Fig. 1a shows the local hydrostatic pressure [18, 19, 33] distribution in GB 1-7 just after elastic loading in a viewing direction perpendicular to the GB plane and fig.1b shows the pressure after uniaxial tensile deformation for 0.5 nsec corresponding to a plastic strain of about 2%. After elastic load, the GB clearly contains regions of high compressive pressure and regions of high dilative pressure that are aligned in regular arrays. The pressure varies from -2GPa (blue) to 2GPa (red). At 0.5 nsec of deformation the regions of high compressive stress reduce and whole areas in the GB became dilative. A corresponding increase in order in terms of the pressure distribution is also evident. From temporal analyses of this atomic configuration, GBS and GB migration were observed. Such activity was facilitated by discrete atomic events involving shuffling and free volume migration.

Intra-grain deformation

All atomistic simulations have revealed increasing dislocation activity with increasing grain size [17, 22-25, 31]. Below 30 nm, simulation of full 3D structures reveals only partial dislocations emited at GBs, travelling through the grain, and absorbed in opposite GBs. The entire process is accompanied by the same type of atomic activity observed under only GBS conditions suggesting that at the larger grain sizes, where sliding cannot be entirely accommodated due to geometric restrictions, stress across the entire grain builds up allowing for the eventual propagation of nucleated lattice partial dislocations. In 2D columnar microstructures, full dislocations are observed [24, 25].

Figs. 2a and 2b show atomic sections before, during and after dislocation nucleation. The viewing direction is along the normal of this slip plane. The atomic section now includes atoms from the two adjacent (red) hcp planes within the grain containing the partial dislocation and part of the surrounding grains. In fig 2a the atoms are coloured according to their local crystalline symmetry [34] (grey=fcc, red=hcp, green=other 12 coordinates, blue=non 12 coordinated), whilst in fig 2b the atoms are coloured according to their local maximum resolved shear stress [18, 19, 33]. In fig 2a, the GB and TJ regions are easily identified, as well as the growing stacking fault defect indicated by two adjacent (red) hcp coordinated (111) planes left behind the emitted partial. Fig. 2b displays that at the GBD and the point of nucleation there exists a strong maximum resolved shear inhomogeneity that extends into the grain. Infact, there exists a large local difference in the maximum resolved shear precisely at the two (111) planes which define the slip plane. Upon nucleation, this strong inhomogeneity is removed and coincides with the removal of a GBD. One can see that before emission of the partial dislocation, the associated GB contains strong variations in maximum resolved shear stress (and also local hydrostatic pressure; not shown) , which upon dislocation nucleation and propagation across the grain are considerably relieved resulting in a more ordered and faceted GB structure.

That GBs act as sinks for lattice dislocations under the action of applied stress is a well established property, but it is also known that the opposite mechanism, i.e. emission of lattice dislocations from GBs is also possible [35, 36]. There is however no detailed understanding of the emission process and of the local changes in GB structure accompanying the emission. An important observation from simulation has been that the precursor to the nucleation of lattice dislocations in the regions of misfit involves similar atomic activity as that seen for GBS. Indeed, results show that local atomic shuffling often involving stress assisted diffusion of free volume from a nearby TJ to the region of misfit surrounding a GBD, allows the "formation" of the Burgers vector of a partial lattice dislocation resulting in a dissociation of the GBD. This was the first time a direct link between the GBS process, its associated atomic scale activity, and the nucleation of an initial partial lattice dislocation from the GBD, was observed.

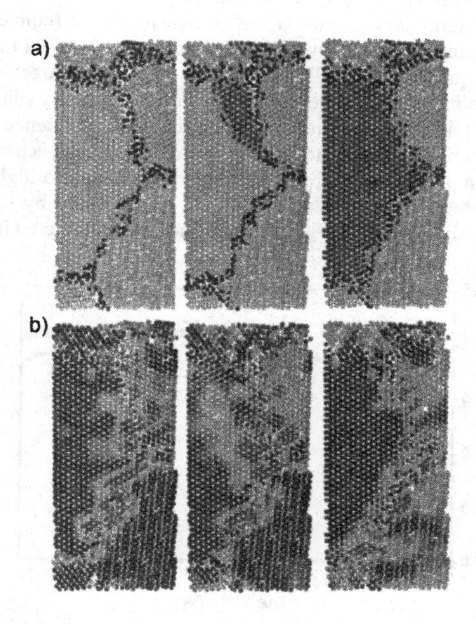

Figure 2: Atomic section for grain boundary 12-13. a) the atoms are coloured according to their local crystalline symmetry and in b) according to their local maximum resolved shear stress. Within each a) and b) are shown the configurations for three different plastic strains with a nucleating and propagating partial dislocation across a grain.

Mechanical Properties Derived from Experiment

Recent experimental data provide evidence or at least indicate the possibility of the deformation mechanisms suggested by simulation. For example it is often reported that no major dislocation debris is observed during post-mortem TEM analysis of samples deformed under tensile conditions [15, 37]. This is entirely compatible with the results of simulations: that slip activity is derived from nucleation and absorption of dislocations at grain boundaries. In-situ deformation measurements with TEM, demonstrate dislocation activity at the crack tip, but recent experiments show that other dislocation sources such as surface defects are possible [14] and that when no porosity is present, continuous thinning of the sample can be observed in-situ, suggesting significant slip transfer. However, when indentation, grinding or cold rolling is performed on nc systems, deformation twinning is observed post mortem [9, 10]. These deformation modes are not only less sensitive to flaws resulting from pores, cracks or other surface defects, but they also incorporate much higher shear stresses that might be more difficult to accommodate by only intergrain mechanisms.

In tensile and compressive testing the derivative of the stress-strain curve provides information on a possible work hardening mechanism. Fig. 3 shows true stress-true strain curves for electrodeposited nc-Ni with a mean grain size of 26 nm deformed at 6×10^{-3}/sec and at a lower strain rate of 6×10^{-5}/sec. For detailed analysis of the structure and the procedure to make the dog bones we refer to [8, 38]. At both strain rates, the stress-strain curves show a fast decreasing slope (conventionally interpreted as a fast decrease in work hardening) followed by a plateau and consequently a negative slope. The lack of work hardening infers the presence of full dislocations from GBs together with a GB relaxation mechanism. For coarse grain materials, the negative slope region normally can be associated with plastic instability; however inspection of the fractured dog bone reveals no evidence of necking and a dimple-like fracture surface. This indicates that softening occurs and that the mechanism behind the softening involves GB accommodation processes and promotes deformation since a higher elongation to failure is reached at the lower strain rate. The softening could also be explained by the presence of local shear planes facilitating collective sliding of a few grains as has been demonstrated in simulations [21].

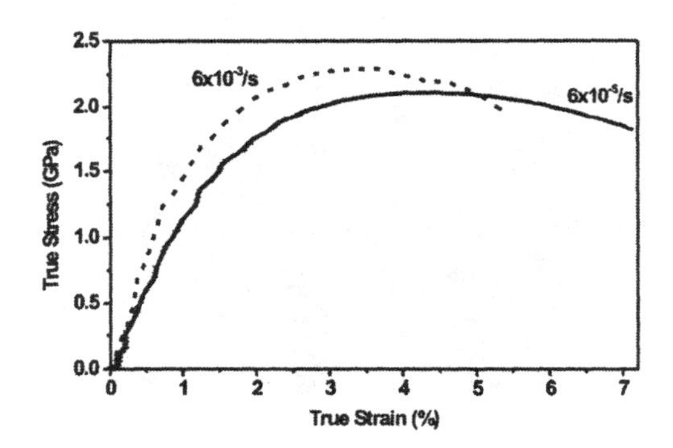

Figure 3: Experimental true stress/true strain curve of nc-Ni at two different strain rates.

Discussion and Conclusion

Atomistic simulations on nc fcc metals have shown that the GB and GB network structure play a crucial role in plasticity for both the inter-grain and intra-grain deformation mechanisms. The major discovery of atomistic simulation has been that in the high-strain rate regime, there is a transition with increasing grain size, from an inter-grain plastic deformation mechanism to a mixture of inter- and intra-grain mechanisms where the latter takes the form of dislocation nucleation/annihilation at GBs and propagation through grains. The simulations suggest three possible intra-grain deformation processes, all based on partial dislocation activity: (1) partial dislocations, (2) full dislocations where a trailing partial follows after some time in the same slip plane and, (3) deformation twinning where a trailing partial follows on an adjacent plane. Note that the first and the last mechanism leaves behind some kind of dislocation debris, while in the case of full dislocations nothing can be observed post-mortem.

In summary, the simulations together with some of the experimental results suggest that the main answer to the ductility problem in nc-metals has to be found in the GB structures, having to be designed in such a way that they allow emission of dislocations from GBs and accommodation of local stress by means of local atomic structural changes. In such deformation mechanisms the decrease in the true stress-strain slope from a positive to a zero value and even later to a negative value may in fact not be related to conventional work hardening, but should rather be seen as the result of an evolving GB network accommodating the applied stress by both an inter- and intra-grain deformation mechanisms. A lack of a work hardening mechanism is entirely compatible with the lack of accumulation of dislocation debris, as suggested by both simulation and experiment. The softening could result from the formation of local shear planes as is suggested by simulations and witnessed by the dimple features on the fracture surface.

References

1. E. O. Hall, Proc. Phys. Soc. London B64 (1951) 747.
2. N. J. Petch, J. Iron, Steel Inst. London 74 (1953) 25.
3. J. R. Weertman, (2002) Mechanical behaviour of nanocrystalline metals, Nanostructured Materials: Processing, Properties, and Potential Applications, William Andrew Publishing, Norwich.
4. M. Legros, B. R. Elliott, M. N. Rittner, J. R. Weertman, & K. J. Hemker, Phil Mag A 80 (2000) 1017.
5. B. Cai, Q. P. Kong, L. Lu & K. Lu, Mat. Sci. Eng. A 286 (2000) 188.
6. Y. Champion, C Langlois, S. Gueren-Mailly, P. Langlois, J-L. Bonnentien & M. J. Hytch, Science 300 (2003) 310.
7. D. Jia, Y. M. Wang, K. T. Ramesh, E. Ma, Y. T. Zhu & R. Z. Valiev, App. Phys. Lett. 79 (2001) 611.
8. Z. Budrovic, H. Van Swygenhoven, P. M. Derlet, S. Van Petegem & B. Schmitt, submitted (2003).
9. M. Chen, E. Ma, K. J. Hemker, H. Sheng, Y. Wang & X. Cheng , Published online April 24, 2003; 10.1126/science.1083727 (Science Express Reports)
10. J. Markmann, P. Bunzel, Hoesner, K. W. Liu, K. A. Padmanabhan, R. Birringer, H. Gleiter & J. Weissmueller, Scripta Mater. (2003) in press.
11. P. G. Sanders, C. J. Youngdahl & J. R. Weertman, Mater. Sci. Eng. A 234–236 (1997) 77.
12. Y. M. Wang, E. Ma & M. W. Chen, Appl. Phys. Lett. 80 (2002) 2395.
13. S. Van Petegem, F. Dalla Torre, D. Segers & H. Van Swygenhoven, Scripta Mater. 48 (2003) 17.
14. R. C. Hugo, H. Kung, J. R. Weertman, R. Mitra, J. A. Knapp & D. M. Follstaedt. Acta Mater. 51 (2003) 1937.
15. K. S. Kumar, S. Suresh, M. F. Chisholm, J. A. Horton & P. Wang, Acta Mater. 51 (2003) 387.
16. A. Hasnaoui, H. Van Swygenhoven & P. M. Derlet, Science 300 (2003) 1550.
17. H. Van Swygenhoven, Science 296 (2002) 66.
18. P.M. Derlet, A. Hasnaoui & H. Van Swygenhoven, Scripta Mater. 49 (2003) 629.
19. H. Van Swygenhoven, P. M Derlet, Z. Budrovic & A. Hasnaoui, Zeitschrift fur Metallkunde, in press (2003)
20. H. Van Swygenhoven & P. M. Derlet, Phys. Rev. B 64 (2001) 224105.
21. A. Hasnaoui, H. Van Swygenhoven & P. M. Derlet, Phys. Rev. B 66 (2002) 184112.
22. H. Van Swygenhoven, P. M. Derlet & A. Hasnaoui, Phys. Rev. B 66 (2002) 024101.
23. P. M. Derlet, H. Van Swygenhoven & A. Hasnaoui, Phil. Mag. A (2003) in press.
24. V. Yamakov, D. Wolf, M. Salazar, S. R. Phillpot & H. Gleiter, Acta Mater 49 (2001) 2713.
25. V. Yamakov, D. Wolf, S. R. Phillpot, A. K. Mukherjee & H. Gleiter, Nature Materials 1 (2002) 1.
26. P. M. Derlet & H. Van Swygenhoven, Scripta Mater. 47 (2002) 719.
27. M. Parrinello & A. Rahman, J. Appl. Phys. 52 (1981) 7182.

28. F. Cleri & V. Rosato, Phys. Rev. B 48 (1993) 22.
29. H. Van Swygenhoven, D. Farkas & A. Caro, Phys Rev B 62 (2000) 831.
30. P. M. Derlet & H.Van Swygenhoven, Phys Rev B 67 (2002) 014202.
31. Van Swygenhoven H, Spacer M & Caro A. Acta Mater 47 (199983 117.
32. J. Schiøtz, F. D. Di Tolla & K. W. Jacobsen, Nature (London) 391 (1998) 561.
33. J. Cormier, J. M. Rickman & T. J. Delph, J Appl Phys 89 (2001) 99.
34. D. J. Honeycutt & H. C. Andersen, J Phys Chem 91 (1987) 4950.
35. H. Gleiter, Progr. Mater. Sci 25 (1981) 125.
36. R. Z. Valiev, V. Yu. Gertsman & O.A. Kaibyshev, Phys. Stat. Sol. (a) 97 (1986) 11.
37. S. X. McFadden, A. V. Sergueeva, T. Kruml, J-L. Martin & A. K. Mukherjee, MRS Symposium Series Vol. 634 (2000) B1.3.
38. F. Dalla Torre, H. Van Swygenhoven & M. Victoria, Acta Mater. 50 (2002) 3957

Processing and Properties of Structural Nanomaterials
Edited by Leon L. Shaw, C. Suryanarayana and Rajiv S. Mishra
TMS (The Minerals, Metals & Materials Society), 2003

Yield Stress of Nanocrystalline Materials: Role of Grain Size Distribution

Chandra S. Pande and Robert A. Masumura

Naval Research Laboratory
Materials Science and Technology Division
Washington DC, D.C. 20375

Keywords: Hall-Petch, nanocrystalline materials, grain size distribution

Abstract

Models of strengthening by nanocrystalline processing are briefly considered. Experimental results indicate that as grain size becomes smaller and smaller the strengthening by grain refinement departs from that predicted by Hall-Petch relation and further grain refinement could in fact reduce the strength. We consider the reason for this "inverse" Hall-Petch behavior. It is found that a model based on Coble creep (with a threshold stress) for finer grains and conventional Hall-Petch strengthening for larger grains appears to be most successful in explaining experimental results but only when a grain size distribution occurring in most specimens is incorporated into the analysis. It is shown mathematically that the inverse Hall-Petch region is strongly dependent on the width of the grain size distribution.

Introduction

The classic Hall-Petch law [1,2] describes the relationship between yield stress τ and grain size d, *viz.*,

$$\tau = \tau_o + K\, d^{-1/2} \tag{1}$$

where τ_o is the friction stress and K is a material constant. Masumura *et al.* [3] by plotting a large amount of available data in a Hall-Petch plot showed that yield stress-grain size exponent for relatively large grains appears to be very close to -0.5 and generally this trend continues until the very fine grain regime (~100 nm) is reached. With the advent of nanocrystalline materials whose grain sizes are of nanometer (nm) dimensions, the applicability and validity of Eq. (1) becomes of interest. Sanders *et al.* [4] and others have shown that there is indeed a peak in Hall Petch plot and if the grain sizes are reduced further softening of the material may occur ("inverse" Hall-Petch [5-8]). Chokshi *et al.* [6] postulated that, at sufficiently small grain sizes, the Hall-Petch model based upon dislocations [9-11] may not be operative. However in this region a new mechanism of deformation may be operative. In this region, they [6] have proposed room temperature Coble creep as the mechanism to explain their results. Certainly, there is an order of magnitude agreement and the trend is correct, however, the functional dependence of □ on d they obtain is incorrect as pointed out by Neih and Wadsworth [12]. Conventional Coble creep demands that $\tau \sim d^3 = [d^{-1/2}]^{-6}$, i.e., the τ vs. $d^{-1/2}$ curve falls very steeply as $d^{-1/2}$ increases. This is not found experimentally [6]. When we fit the data of Chokshi *et al.* to a Coble type equation, *viz.*,

$$\tau = \gamma + \alpha\, d^3 . \tag{2}$$

We find that the fit requires a large value of $\gamma = 360$ MPa which suggests that the threshold stress for Coble creep is of the form Gb/d [13] since this would have the correct magnitude for a 10 nm grain size. The origin of this threshold may be related to vacancies that are created and destroyed on dislocations climbing along grain boundaries [3]. The dislocations are pinned at grain boundary nodes and require a stress of Gb/d to climb. Chokshi *et al.* [6] showed that their data, however, fit better the relation

$$\tau = \beta - K' d^{1/2}, \tag{3}$$

with $\beta = 937$ MPa and $K' = 0.027$ MPa $\sqrt{m}$, instead of Eq. (2). Eq. (3) cannot be related simply to any known mechanism.

Masumura *et al.* [3] provided such a model also based on Coble Creep for smaller grains and obtained an analytical expression for τ as a function of the inverse square root of d in a simple and approximate manner that could be compared with experimental data over a wide range of grain sizes. Specifically, their model assumes conventional Hall-Petch strengthening for larger grains and Coble creep with a threshold stress for smaller grains. It assumes that in a material with a distribution of grain sizes, larger grains deform by a dislocation glide process up to a grain size d^* and the rest by vacancy transport. As the average grain size decreases, the fraction deforming by glide decreases and the overall response changes from strengthening to softening. The form of the yield stress against grain size curve depends on the relative values of Hall-Petch slope k, the conventional Coble constant B, the threshold constant A and the width of the grain size distribution σ. The resulting expression was shown to be in reasonable agreement with experimental data of the yield stress of NiP of McMahon and Erb [14], The value for d^* = 5.5 nm as determined from this analysis is in agreement with the original hardness data where the hardness (or stress) begins to decrease with decreasing grain size at a grain size of 5-6 nm.

Unfortunately their result for yield stress is expressed in terms of complicated error functions. To bring into focus the role of the grain size distribution (especially the width of the distribution) in determining the in Hall-Petch plot we derive result of this model in a simple and approximate fashion below.

Derivation

We define the shear stress arising from several contributions and give its dependence on grain size, d, as

$$\tau_d = \tau - \tau_o = K d^{-1/2} + k_1 + B_o\, d^{-1} + B d^3, \tag{4}$$

where the first term on the right hand side is valid over large grains and the other three over small grains with k_1, K, B_o, and B being constants. In Ref. [3], k_1 was taken to be zero. At some critical grain size d^*, the switch from larger grain effects (Hall-Petch) to other phenomena such as Coble creep (grain boundary vacancy) and/or tri-junction mobility (grain boundary dislocation) at smaller grain sizes. See Ref. [15] for a recent discussion of Coble creep. Thus at d^*

$$K(d^*)^{-1/2} = k_1 + B_o\,(d^*)^{-1} + B(d^*)^3. \tag{5}$$

So far, we have assumed that all the grains of the same size. In practice the grain sizes are log-normally distributed both for large grains [16] and nanocrystalline materials [17]. The contribution for $d > d^*$ where the Hall-Petch effect dominates can be given as

$$I_1 = \int_{v^*}^{\infty} \frac{K}{\sqrt{d}}\, v\, f(v)\,dv \tag{6}$$

and for $d < d^*$, the smaller grain size is,

$$I_2 = \int_0^{v^*} \left(k_1 + B_o\, d^{-1} + B d^3\right) v\, f(v)\,dv, \tag{7}$$

where v is the volume and is given as $v = \gamma d^3$ with γ a shape factor that is set equal to one for simplicity.

The grain size distribution $f(v)$ is the log-normal distribution[18],

$$f(v)=\frac{1}{\sqrt{2\pi}\ \sigma\, v}\exp[-\frac{1}{2\sigma^2}\left(\ell n[\frac{v}{v_{med}}]\right)^2] \tag{8}$$

where σ is the standard deviation in $\ell n(v)$ and v_{med} is the median of the distribution. The averaging process is described in [3] and [18].

Eqs. (6) and (7) can be re-written to obtain a useful form. First, I_1 is re-cast and evaluated explicitly as

$$I_1=\int_0^{\infty}\frac{K}{\sqrt{d}}v\, f(v)dv=\exp[\frac{25\sigma^2}{72}]K(v_{med})^{5/6}. \tag{9}$$

Secondly, Eq. (7) is modified as

$$I_2=-\int_0^{v^*}\left(\frac{K}{\sqrt{d}}-k_1-B_o\, d^{-1}-Bd^3\right)v\, f(v)dv=-\int_0^{v^*}G(v)dv. \tag{10}$$

Expanding $G(v)$ around v^* to reduce the complexity of I_2,

$$I_2=-\int_0^{v^*}G(v)\,dv\cong-\int_0^{v^*}\left\{G(v^*)+\Delta v\frac{\partial G(v^*)}{\partial v}\right\}dv \text{ and } \Delta v=v-v^*, \tag{11}$$

but from Eq. (5), $G(v^*)=0$, and hence

$$I_2=-\left(\frac{\partial G(v^*)}{\partial v}\right)\int_0^{v^*}(v-v^*)dv=\left(\frac{\partial G(v^*)}{\partial v}\right)\frac{(v^*)^2}{2}. \tag{12}$$

The yield stress can be simplified by re-formulating I_1 and I_2 in terms of a scaled grain size parameter normalized to the mean of the distribution,

$$\tau_{norm} = I_1 + I_2 = \xi + \lambda \xi^3 \frac{\exp[-\frac{\sigma^2}{8} - \frac{18(\ell n[\xi])^2}{\sigma^2}]}{\sigma} \text{ and } \xi = \sqrt{\frac{d^*}{\bar{d}}}, \tag{13}$$

where λ is

$$\lambda = \frac{-2B_0 + K\sqrt{d^*} + 6B(d^*)^4}{12\sqrt{2\pi}\ d^*}. \tag{14}$$

As mentioned before, the experimental data of yield stress for nanocrystalline materials often exhibit a maximum leading to an "inverse" Hall-Petch effect. To simplify the analysis and make amenable to comparison for experimental data, τ_{norm} is modified in the following manner:

The peak of the experimental yield stress data occurs at d_{peak} should be near to the cross-over grain size, d^*, but may not coincide with it. Thus the grain size is scaled to d_{peak},

$$\omega = \sqrt{\frac{d_{peak}}{\bar{d}}} = \frac{\xi}{p}, \ \xi = \sqrt{\frac{d^*}{\bar{d}}} \text{ and } p = \sqrt{\frac{d^*}{d_{peak}}}. \tag{15}$$

At $w = 1$, the scaled normalized peak in the experimental data, we determine value of λ and let $\sigma = 3s$ where s is the standard deviation of in ℓn (grain size) to obtain the simplified expression that relates the experimental yield stress,

$$\tau_n = \frac{\tau_{norm}}{p} = \omega - \frac{s^2 w^3 \exp\left[\frac{2\{(\ell n[p])^2 - (\ell n[p\omega])^2\}}{s^2}\right]}{3s^2 - 4\ell n[p]}. \tag{16}$$

Generally, $p < 1$, *i.e.*, the peak in the experimental data occurs at a larger value than the cross-over or switch from Hall-Petch to Coble creep and tri-junction mobility [3]. Since in this calculation I_1 is determined explicitly and I_2 is most accurate in the region of the peak, we expect Eq. (16) to be applicable for most of the region of Hall-Petch plot except when $\bar{d} \to 0$. The effect of the width of the distribution strongly affects the yield stress response as seen in Fig. 1. In that figure, we have selected $p = 0.9$ or $d_{peak} = 0.81\ d^*$.

Conclusions

From Fig. 1 it is seen that the width of the grain size distribution as characterized by s has a significant effect on the shape of the Hall-Petch plot in the nanocrysatalline range. Experimentally (see for example [3] and [4]) the shape of the plot in this region has been shown to behave in a variety of fashion with no clear indication why it is so. While purity of the material, nature and cleanliness of the grain boundaries, oxides in the system etc are all possible source of scatter, our results indicate that the nature of the Hall-Petch plot may also depend sensitively on the width of the distribution.

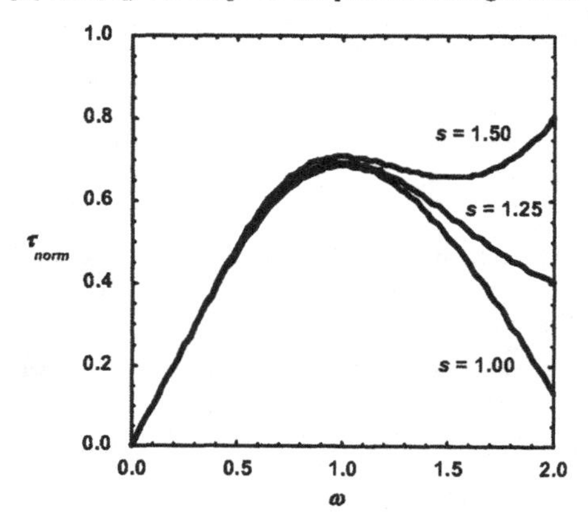

Figure 1. Width distribution effect on Hall-Petch Plot. See text for definitions of s, ω and τ_{norm}.

Another point worth noting is that the actual mechanism of the inverse Hall-Petch effect may not have much effect in changing this conclusion. Since the mechanism as given in Eq. (4) is mainly used to define the parameter d^* and is not used in a significant fashion in deriving the final result.

There are at present no experimental results where the width of the distribution has been carefully measured. Further experiments are needed to verify the predictions.

References

[1] E. O. Hall, Proc. *Phys. Soc. London,* 1951, B64 , 747.

[2] N. J. Petch, *J. Iron Steel Inst.*, 1953, 174, 25.

[3] R. A. Masumura, P. M. Hazzledine and C. S. Pande, *Acta Mat.*, 1998, 2, 4527.

[4] P. G. Sanders, J. A. Eastman and J. R. Weertman, *Acta Mat.*, 1997, 45, 4019.

[5] G.E. Fougere, L. Riester, M. Ferber, J. R. Weertman, and R. W. Siegel, *Nanostruct. Mater.*, 1995, 5, 127.

[6] A. H. Chokshi, A. Rosen, J. Karch and H. Gleiter, *Scripta Metall.*, 1989, 23, 1679.

[7] K. S. Kumar, M. S. Dipietro, M. S. Duesberry, N. Louat and V. Provenzano, unpublished research, 1994.

[8] Z. Valiev, E. V. Kozolv, Yu. F. Ivanov J. Lian, A. A. Nazarov and B. Bandelet, *Acta Mat.,* 1994, 42, 2467.

[9] J. C M. Li and G .C. T. Liu, *Phil. Mag.*, 1967, 15, 1059.

[10] J. C. M. Li, *Trans. TMS-AIME*, 1963, 227, 247.

[11] G. A. Malygin, *Phys. Solid State,* 1995, 37, 1248.

[12] T. G. Nieh and J. Wadsworth, *Scripta Met.*, 1991,25, 955.

[13] B. Burton, *Mat. Sci. Engr.*, 1972, 10, 9.

[14] G. McMahon and U. Erb, *Microstructural Sci.*, 1989, 17, 447.

[15] V. Y. Gertsman, M. Hoffmann, H. Gleiter and R. Birringer, *Acta Met.*, 1994, 42, 3539.

[16] S. K. Kurtz and F.M.A. Carpay, *J. Appl. Phys.*, 1980, 51, 5725.

[17] F. Dalla Torre, H. Van Swygenhoven and M. Victoria, *Acta Mat.*, 2002, 50, 3957.

[18] K. J. Kurzydlowski, *Scripta Met.*, 1990, 24, 879.

Processing and Properties of Structural Nanomaterials
Edited by Leon L. Shaw, C. Suryanarayana and Rajiv S. Mishra
TMS (The Minerals, Metals & Materials Society), 2003

Effect of grain size distribution on the mechanical behavior of ultrafine-grain metals

S. Cheng and W.W. Milligan

Michigan Technological University
Department of Materials Science and Engineering
Houghton, MI 49931, USA

Keywords: Grain size distribution, Hall-Petch relationship, strain hardening, nanostructured metals

Abstract

This paper explores the effect of the grain size distribution on the uniaxial yielding behavior of ultrafine-grain metals. Simple modeling based on the Hall-Petch relationship reveals that apparent strain hardening behavior may be due to a widely-distributed grain size. Yield strength and other properties may also be affected.

Introduction

The grains in conventional large grained metals are usually relatively uniform, due to the application of methods such as hot forging or rolling, in which large grains are broken into smaller ones by working and recrystallization. But in nanostructured or ultrafine-grain metals, the coarsening of initially fine grains is generally conducted by heat treatment. Some grains may preferentially grow to very large sizes compared to others. Consequently, the grain size may have a much wider distribution than conventional large-grain metals; for example, the large grains may reach over 1 μm, while the smallest ones remain under 100 nm. Figure 1 shows the microstructure of ultrafine-grain iron, in which some large grains are more than 10 times greater than the small ones. The material was produced by ball milling and rapid forging, with the details reported in [1]. Metals with such a large grain size distribution may be very different from their conventional counterparts in terms of mechanical response and other properties.

Figure 1 Microstructure of iron with a wide grain size distribution

In studies of the mechanical behavior of nanostructured or ultrafine grained metals, researchers usually report their results based on the average grain size that is calculated from the X-ray diffraction or counted from TEM statistics, but the effect of the grain size distribution is often overlooked. A close look will be taken at the effect of grain size distribution in this study.

A simple model based on the Hall-Petch relationship

In order to study the effect that grain size distribution may have on the mechanical behavior, a simple model based on the Hall-Petch relationship is developed to compare the properties of metals with different grain size distributions but with the same average grain size. For simplicity, the grain size distribution is assumed to be Gaussian:

$$f(x) = \frac{e^{-\frac{(x-\bar{d})^2}{2\cdot\delta^2}}}{\delta\cdot\sqrt{2\cdot\pi}} \qquad (1)$$

Where $\bar{d}$ is the average grain size, and δ is the standard deviation of the grain size. As an example, Figure 2 shows the grain size distributions for two metals, both with an average grain size of 500 nm, but with standard deviations of 20 and 120 nm.

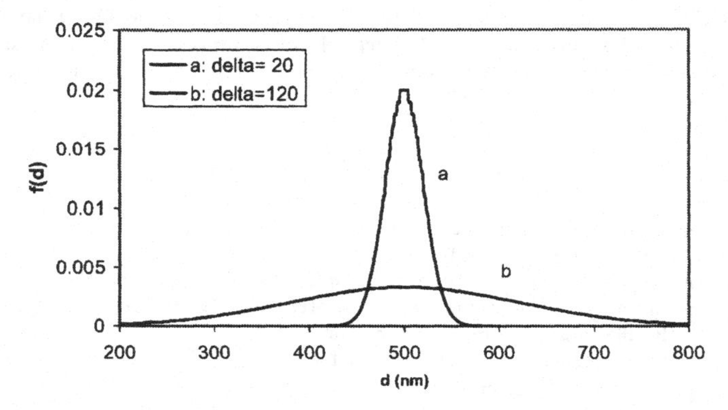

Figure 2 Grain size distributions for two metals with an average grain size of 500 nm, with different standard deviations.

It is well known that the yield strength of ductile metals is related to the average grain size by the Hall-Petch relationship: $\sigma_y = \sigma_0 + k_y d^{-1/2}$, i.e., metals with a different grain size will have a different yield strength. But this equation *per se* does not consider the effect that grain size distribution may have on the yield behavior. To study the effect that the grain size distribution may have, two basic assumptions are made: 1. the yield strength of the grains is defined by Hall-Petch relationship, i.e., grains with different size possess different yield stress, but the overall yield strength of the material is the average of the contributions from all grains. 2. the material lacks strain hardening, i.e., the strain-stress curve is elastic-perfectly plastic, as shown in Figure 3. This lack of strain hardening is appropriate for nanostructured and ultrafine-grain metals, which often exhibit elastic-perfectly plastic response [2, 3].

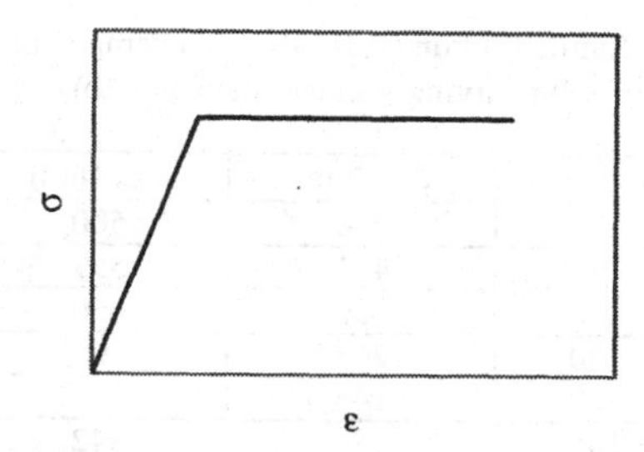

Figure 3 Schematic of elastic- perfectly plastic behavior, which is typical of nanostructured and ultrafine-grain metals

When the grain size has a distribution, some grains have larger size than the average value. According to Hall-Petch relationship, these grains have lower yield strength than rest of the grains. If the applied stress is homogenously applied to the grains, those with large grain size will yield first while others are still under elastic loading. Once the grains have yielded, the flow stress of these grains is equal to their yield strength according to the second assumption that the material has no strain hardening, so the stress remains at the yield level for that grain as deformation proceeds. With the increase of the applied stress, more and more grains yield so long as their yield strength is reached.

To determine the overall stress, a rule-of-mixtures approach is used as a first approximation. The overall stress is therefore the sum of the contributions from all of the grains, which can be calculated from:

$$\sigma = E\varepsilon\left(\int_{d_{\min}}^{x'} f(x)dx\right) + \int_{x'}^{d_{\max}} (\sigma_0 + k_y x^{-1/2}) f(x)dx \qquad (2)$$

Where E is the Young's modulus, ε is the strain, x is the grain size, d_{min}, d_{max} are the minimum and maximum grain sizes in the metal; f(x) is the Gaussian distribution function (Equation 1), and for simplicity we limit our range of grain sizes to $\pm$ 3δ, i.e.:

$$\int_{d_{\min}}^{d_{\max}} f(x)dx = 0.999 \qquad (3)$$

On the right-hand side of Equation 2, the first term is the stress in the small grains that are in the elastic state at strain ε (multiplied by their volume fraction), while the second term is the sum of the stresses in the large grains that have already yielded. The integration limit, x', corresponds to the largest grain size which has not yet reached its yield point at the strain ε. Since the elastic stress is almost exactly equal to the yield stress in these grains (which are just about to yield), the integration limit can be determined as a function of strain from the following

$$E\varepsilon = \sigma_0 + k_y x^{-1/2} \quad \text{[at yield]} \qquad (4)$$

The minimum and maximum grain sizes are determined by Equations 1 and 3, with typical examples shown in Table 1 for a 500 nm average grain size.

Table 1 Limiting grain sizes (500 nm average grain size with varying standard deviation, δ)

δ (nm)	d_{min} (nm)	d_{max} (nm)
0	500	500
20	441	559
50	350	650
100	205	795
120	146	854
150	8	942

If we take ultrafine-grain iron as an example, typical Hall-Petch constants from experimental studies are as follows:

$$\sigma_y = \sigma_0 + kd^{-1/2} = 517 + 15250 \cdot d^{-1/2} \qquad (5)$$

Stress-strain curves can be constructed numerically via Equation 2, while using Equation 4 to determine x' at each level of strain. The 0.2% yield strength and other parameters can be obtained from the predicted curves.

Model predictions

Typical strain-stress curves with different grain size distributions, but the same average grain size, are shown in Figure 4. Offset yield strengths determined from these types of curves are presented in Table 2.

Table 2 Yield strength at 0.2% offset (500 nm average grain size with varying standard deviation, δ)

δ (nm)	$\sigma_{0.2}$ (MPa)
0	1199
20	1199
50	1199
100	1206
120	1210
150	1223

From Figure 4, it is clear the yield behavior changes with the grain size distribution. Some "strain hardening" appears as grains deviate from their average size. In fact, this apparent strain hardening is simply the result of progressive yielding in grains of different sizes.

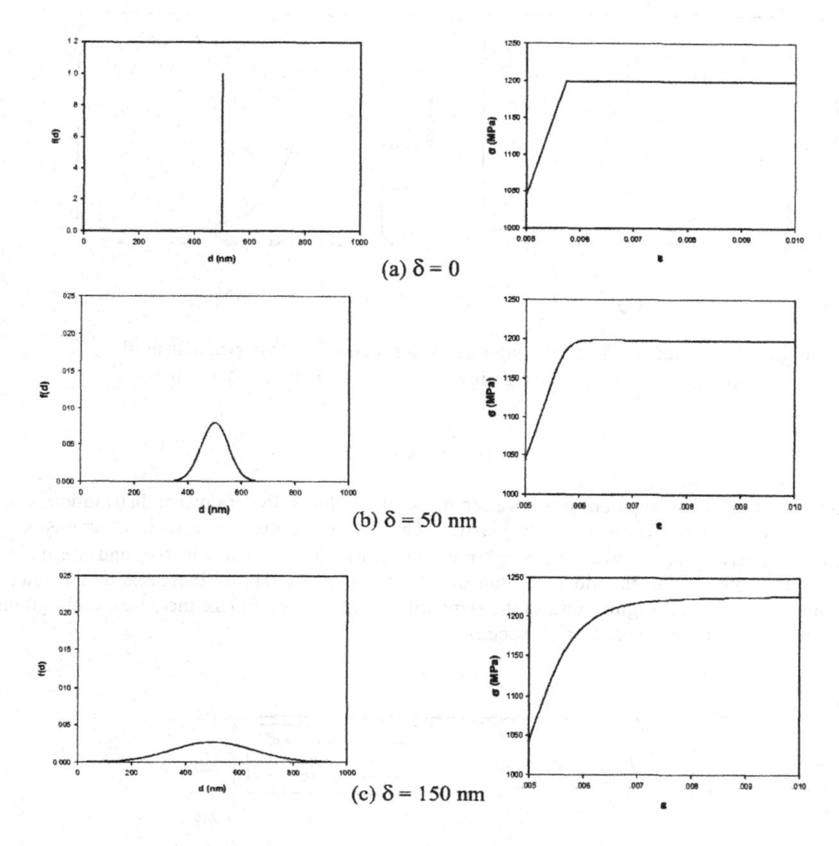

Figure 4 The yield behavior of iron with an average grain size of 500 nm but with different grain size distributions: (a) Ideally uniform grain size; (b) grain size distribution with $\delta = 50$ nm; (c) grain size distribution with $\delta = 150$ nm.

The apparent strain hardening can be quantified by taking the derivative of the stress-strain curve at each point. This results in a delta-function for the standard deviation of zero, while typical curves are shown in Figure 5 for larger standard deviations. Two major observations may be made:

1. With an increase of the standard deviation in grain size, the stress increase becomes more gradual, and it takes a larger strain before the maximum stress is reached. Yield occurs at $\varepsilon = 0.00574$ in the perfectly elastic-plastic case, but the yielding transition continues up to a strain of 0.01 in the case of $\delta = 150$ nm.

2. The yield point $\sigma_{0.2}$ increases slightly with the increase of standard deviation of grain size distribution, as shown in Table 1. The flow stress for metals with wider grain size distribution surpasses the one with ideally elastic-plastic behavior.

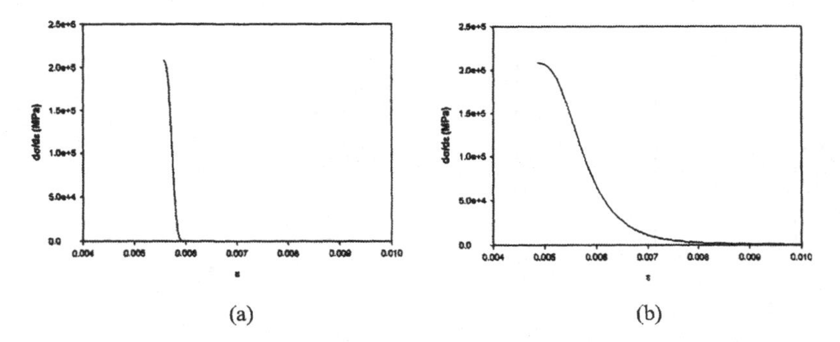

Figure 5 Apparent strain hardening rates as a function of grain size distribution, iron with 500 nm average grain size. (a) $\delta = 20$ nm; (b) $\delta = 150$ nm

Discussion

According to this model, the apparent strain hardening is purely due to the grain size distribution, since all grains were assumed to be elastic-perfectly plastic. The yielding sequence from large grains to small grains leads to experimental "strain hardening" measurements. The increase in $\sigma_{0.2}$ and the maximum stress can be understood with the aid of Figure 6. The yield stress rapidly increases as the grain size decreases into the nanometer regime, and as the standard deviation of grain size increases, more grains are likely to be on the small (strong) end of the spectrum.

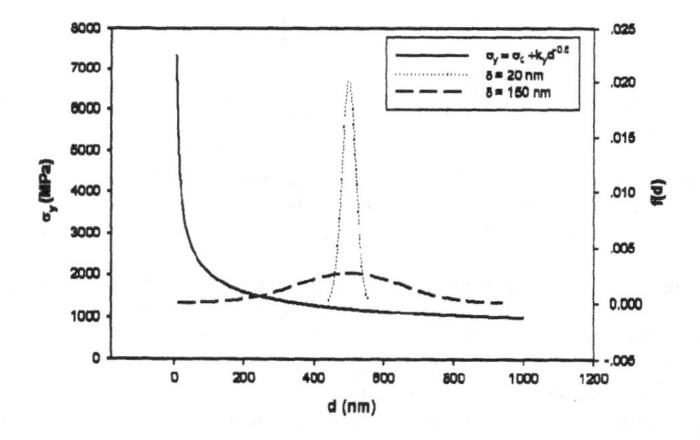

Figure 6 Hall-Petch relationship with superimposed grain size distributions, showing that more grains are in the finer (stronger) regime with large δ.

The gradual transition from elastic to perfectly-plastic is often seen in nanostructured metals [4-8]. In the case of iron, the constitutive behavior is complicated by shear banding, but the effect can be demonstrated in ultrafine-grain aluminum alloys. Figure 7 shows a set of stress strain curves from an ultrafine-grain Al-8Ti-2Ni alloy which was processed by cryomilling and extrusion. The different curves correspond to different thermal exposures which were used to coarsen the structure after manufacture. The material had a wide grain size distribution, with mean values in the 200-600 nm range (depending on heat treatment), and standard deviations in the 150 nm range. It is evident that the shapes of the curves are very similar to those predicted by our model; there is a gradual transition from elastic to perfectly plastic at about 2% strain.

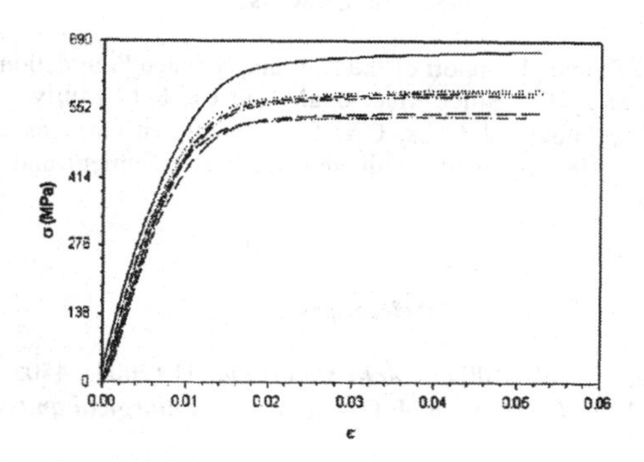

Figure 7 Stress-Strain curves for an ultrafine-grain aluminum alloy. See text for explanation.

The effect of grain size distribution has been acknowledged by other investigators. Ebrahimi et al. [9] have investigated the nanocrystalline nickel with various grain size distributions. It was found that the strength of nickel fell within the scatter band of the general Hall-Petch curve, but large variations were observed in yield strength, strain hardening rate and tensile elongation associated with a relatively small change in the average grain size. They attributed the distinct yield strength and strain hardening rate to the changes of the grain size distribution, although the scatter in the elongation was assumed due to the formation of nodules and the presence of voids [9]. Modeling performed by Mitra et al. [10] has suggested that a significant decrease of yield strength occurs as the standard deviation of grain size distribution increases. Their experimental work, however, showed that the grains followed a log-normal distribution. Recently Wang et al. [11] investigated ultrafine-grain copper with a bimodal grain size distribution. Some extremely large grains were purposely introduced by secondary recrystallization. Their results indicated that the large grains worked as "strain reservoirs"; as a consequence, elongation was substantially improved while retaining a high yield stress. They believe that the large but "soft" grains provided strain accommodation, thus enhanced the strain hardening rate, which is critical to combat the strain localization.

The current model, although simple, has shown that the grain size distribution can have an effect on the mechanical behavior of nanostructured metals. More realistic assumptions about grain size distributions and the strain hardening behavior of individual grains are necessary to further explore the effects of grain size distribution on mechanical properties.

Conclusions

A very simple model has been developed to predict the effect of grain size distribution on mechanical behavior of nanostructured metals. The model predicts that, at constant mean grain size, the distribution affects both the apparent "strain hardening" early in the stress-strain curve and the offset yield strength. The apparent strain hardening may be simply the result of different size grains yielding at successively higher stresses.

Acknowledgements

We gratefully acknowledge the financial support of the National Science Foundation through Grant DMR-99-72931, which was monitored by Dr. Bruce MacDonald and Dr. K.L. Murty. We thank Dr. Patrick Berbon of Rockwell Scientific, Thousand Oaks, CA, for valuable discussions and for providing the ultrafine-grain aluminum alloy. Jeff Spencer conducted the heat treatment and tensile testing on the ultrafine-grain aluminum alloy.

References

[1] S. Cheng, J.A. Spencer and W.W. Milligan, *Acta Materialia*, 51 (2003), 4505.

[2] J.E. Carsley, A. Fisher, W.W. Milligan and E.C. Aifantis, *Metallurgical and Materials Transactions A*, 29A (1998), 2261.

[3] W.W. Milligan, "Mechanical Behavior of Bulk Nanocrystalline and Ultrafine-Grain Metals", in *Encyclopedia of Comprehensive Structural Integrity*, eds. B. Karihaloo, R.O. Ritchie and I. Milne, (Oxford, UK, Elsevier, 2003), Chapter 8.15.

[4] P.G. Sanders, J.A. Eastman and J.R. Weertman, *Acta Materialia*, 45 (1997), 4019.

[5] S.M.L. Sastry, R.S. Iyer, C.A. Frey, B.E. Waller, and W.W. Buhro, *Materials Science and Engineering A*, 264 (1999), 210.

[6] R.Z. Valiev, N.A. Krasilnikov and N.K. Tsenev, *Materials Science and Engineering A*, A137 (1991), 35.

[7] R.Z. Valiev, N.A. Krasilnikov and R.R. Mulyukov, *Materials Science and Engineering A*, A168 (1993), 141.

[8] S.L. Semiatin, K.V Jata, M.D. Uchic, P.B. Berbon, D.E. Matejczyk and C.C. Bampton CC, *Scripta Materialia,* 44 (2001), 395.

[9] F. Ebrahimi, Z. Ahmed and K.L.Morgan, in *Structure and Mechanical Properties of Nanophase Materials—Theory and Computer Simulations vs. Experiment*, (Warrendale, PA: Materials Research Society, 2001), B2.7.1.

[10] R. Mitra, T. Ungar, T. Morita, P.G. Sanders and J.R. Weertman, in *Advanced Materials for the 21th Century: the 1999 Julia R Weertman Symposium*, (Warrendale, PA: The Minerals, Metals and Materials Society, 1999), 553.

[11] Y.M. Wang, M.W. Chen, F.H. Zhou and E. Ma, *Letters to Nature*, 419 (2002), 912.

Processing and Properties of Structural Nanomaterials
Edited by Leon L. Shaw, C. Suryanarayana and Rajiv S. Mishra
TMS (The Minerals, Metals & Materials Society), 2003

Strength Design Maps for Nanolayered Composites

Adrienne V. Lamm and Peter M. Anderson
Department of Materials Science and Engineering
The Ohio State University
2041 College Road
Columbus, OH 43210-1179 USA

Abstract

Nanoscale multilayered materials exhibit extraordinary strength, due to the ability of these systems to confine slip to individual layers. This paper adopts a premise suggested by embedded atom simulations in copper-niobium multilayers that the tensile yield strength is determined approximately by the critical applied stress to eliminate the compressive bi-axial stress in alternating layers of the composite. Yield strength maps are constructed for nanoscale multilayered composites consisting of alternating layer phases in which the volume fraction of the phases, bi-layer thickness, and ratio of in-plane stress-free lattice parameters of the two phases are regarded as variables. Also, the degree of coherency of the interfaces and the internal bi-axial stress state in each layer prior to loading are predicted as a function of these variables. The results suggest that decreasing bi-layer thickness is a limited approach to increasing the biaxial yield strength of nanolayered composites.

Introduction

Multilayered thin films are a class of engineered layered composite materials with individual layer thicknesses on the order of one to several hundred nanometers. The driving force to reduce interfacial energy can cause large internal stress [1], generate deviations from bulk elastic moduli [2, 3], induce phase changes [4], suppress plasticity [5-8], generate significant anelasticity [9, 10], and even promote unique rolling textures [11]. Hardness measurements for some A/B metallic systems indicate that a critical bi-layer period exists below which the resistance to plasticity reaches a plateau or even decreases [12-14]. Tensile testing and fractography of Ni/Ni_3Al multilayers, for example, show that multilayers can be macroscopically brittle, but ductile near the fracture surface [15]. Also, heat treating these Ni/Ni_3Al multilayers at 673K increases hardness, presumably by reducing interfacial dislocation content formed during the growth of the multilayers [21].

The purpose of this manuscript is to determine the role of microstructural parameters such as bi-layer thickness, volume fraction, modulus mismatch, and lattice parameter mismatch on the macroscopic yield strength of two-phase metallic multilayer thin films. This will be accomplished by combining mechanical equilibrium, the critical conditions for confined layer slip (CLS) in which dislocations propagate only in individual layers, and the critical condition for bulk yield whereby extensive co-deformation occurs. Recent embedded atom modeling [3] suggests that for Cu/Ni multilayered thin films with coherent and semi-coherent interfaces, extensive co-deformation across layers occurs when the applied stress is sufficient to eliminate the large compressive stress (~2GPa for the coherent case) in the alternating Cu layers. Further, heat treating Ni(Al)/Ni_3Al multilayers indicates that hardness scales with the magnitude of internal bi-axial stress state [21]. Thus, dislocations in these cases appear to be trapped in individual layers chiefly by the alternating tensile-compressive stress state rather than by the structural resistance provided by the interface. This premise is adopted in the present analysis. The resulting maps predict that decreasing bi-layer thickness (Λ) is a limited approach to increasing multilayer yield strength, valid only down to a critical Λ. Below that, decreasing the volume fraction of the compressively stressed phase is most effective.

Model Development: Interface Structure and Macroscopic Yield

Assume that two-phase multilayered thin films may be modeled with layers of alternating phases 1 and 2 with in-plane dimension $l = l_1 = l_2$ and bi-layer thickness $\Lambda = h_1 + h_2$ as shown in **Fig. 1**. The layers have been deformed from a reference stress-free dimension l_k^o (k = 1 or 2) via a bi-axial stress state σ_k (k = 1 or 2), which induces elastic (e_k) and possibly plastic (ε_k^p) in-plane strains. The corresponding compatibility conditions to reflect this are [22]

$$\varepsilon_1 - \varepsilon_2 = \ln\left(\frac{l_2^o}{l_1^o}\right) \quad \text{and} \quad \varepsilon_k = e_k + \varepsilon_k^p \tag{1}$$

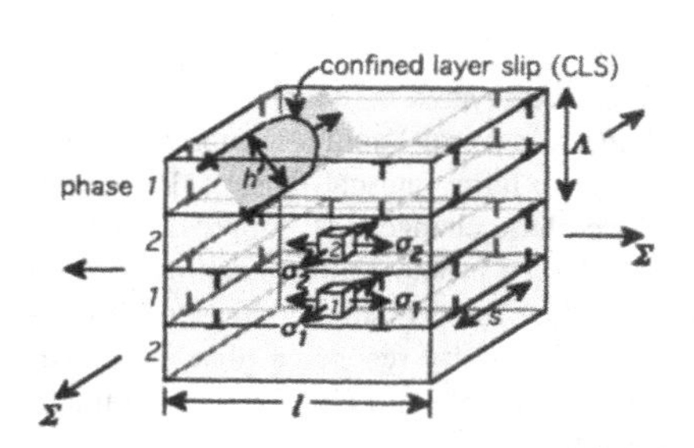

Fig. 1: Schematic of a multilayer composed of alternating bi-axial stress (σ_1, σ_2).

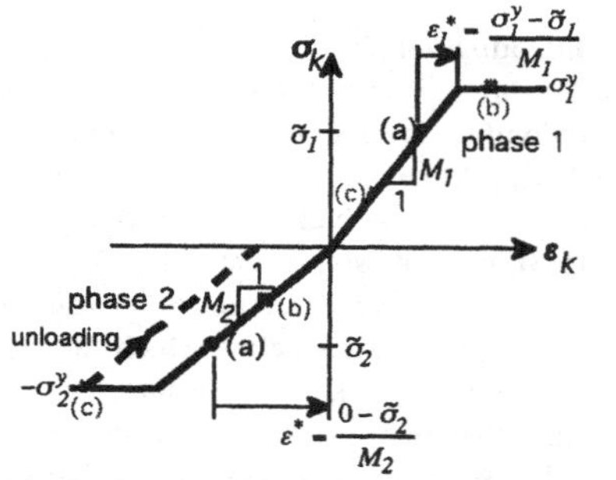

Fig. 2: Graphic of bi-axial responses of phase 1 (tensile) and phase 2 (compressive) and possible initial states (a), (b), and (c).

The elastic-perfectly plastic constitutive relation for the tensile portion of phase 1 and compressive portion of phase 2 is depicted in **Fig. 2**. Each phase has a bi-axial elastic modulus $M_k = E_k/(1 - \nu_k)$ (E_k, ν_k are the corresponding Young's modulus and Poisson's ratio) and a perfectly-plastic yield strength described by σ_k^y. During yield, dislocation loops propagate by confined layer slip (CLS) as shown in Fig. 1, so that arrays of dislocations with Burgers vector magnitude b_k and in-plane spacing s_k are deposited against interfaces that bound the yielding layer [5, 7, 16].

The corresponding elastic-plastic constitutive relations are

$$e_k = \frac{\sigma_k}{M_k} \text{ and } \varepsilon_k^{\mathrm{p}} = \ln\left(1 + \frac{b_k}{s_k}\right), \quad \frac{\sigma_k^{\mathrm{y}}}{M_k} = \frac{\ln(h_k / b_k)}{4\pi(1+\nu_k)h_k / b_k} \tag{2}$$

Here, σ_k^{y} is evaluated as the critical biaxial stress to introduce arrays of misfit dislocations as shown in Fig. 1. The misfit character reflects the effective content of dislocation arrays that serve to increase or decrease the l_k from their reference values. The value of σ_k^{y} reported here is for a elastically homogeneous system; image effects from elastic modulus mismatch would increase σ_k^{y} for the smaller elastic modulus layer and decrease it for the larger elastic modulus layer.

Numerous sources of dislocation loops from interfaces are envisioned [6, 7, 17] so that the yield strength of a confined layer is defined by σ_k^{y} rather than by source-limited plasticity. An analytic analysis of dislocation arrays in multilayered thin films suggests that σ_k^{y} does not increase with plastic strain provided $\varepsilon_k^{\mathrm{p}} < b_k / 10\, h_k$ and that pile-ups should not occur until $\varepsilon_k^{\mathrm{p}} \sim b_k / 2 h_k$ [18]. When an alternating stress state exists, the constitutive relation for an embedded layer is elastic-perfectly plastic, at least for $|\varepsilon_k^{\mathrm{p}}| < b_k / 10 h_k$. Unloading from $\sigma_k = \sigma_k^{\mathrm{y}}$ is assumed to be elastic until $\sigma_k = -\sigma_k^{\mathrm{y}}$, due to dislocation-dislocation entanglement. However, experimental evidence suggests that CLS may reverse direction when $|\sigma_k| < \sigma_k^{\mathrm{y}}$, in particular during thermal cycling involving elevated temperature [19].

During application of a macroscropic bi-axial tension Σ as shown in Fig. 1, the biaxial stress in phase 1 will evolve according to

$$\frac{\sigma_1}{M_1} = \frac{\Sigma - f_2 M_2 \left[\ln\frac{a_1^{\mathrm{o}}}{a_2^{\mathrm{o}}} + \ln\left(1 + \frac{b_1}{s_1}\right) - \ln\left(1 + \frac{b_2}{s_2}\right)\right]}{f_1 M_1 + f_2 M_2} \tag{3}$$

where $f_k = h_k/\Lambda$ denotes volume fraction and a_k^{o} the stress-free lattice parameter of phase k. Eq. (3) extends the result in [7] to large (logarithmic) strain theory and may be derived using Eq. (1), the first of eq. (2), and the macroscopic equilibrium relation, $\Sigma = f_1\sigma_1 + f_2\sigma_2$. The equality $l_1^{\mathrm{o}}/l_2^{\mathrm{o}} = a_1^{\mathrm{o}}/a_2^{\mathrm{o}}$ is also used, meaning that a coherent interface would be produced if the layers were stretched or compressed elastically to make $l_1 = l_2$.

All possible scenarios for multilayer yield are predicated on the assumption that dislocation loops will permeate throughout the multilayer once Σ is large enough to remove the alternating sign bi-axial stress state. Without loss of generality, $a_2^{\mathrm{o}}/a_1^{\mathrm{o}} > 1$ is assumed so that, prior to any macroscopic loading, the initial states $\tilde{\sigma}_1 > 0$ and $\tilde{\sigma}_2 < 0$ occur. Three possible scenarios for the initial state indicated in Fig. 2 are that the interfaces may be (*a*) coherent; or semi-coherent due to arrays of dislocations deposited on the interface by (*b*) yield of layer 1; or (*c*) yield of layer 2. Scenario (*a*) prevails if the coherent stress state $(\tilde{\sigma}_1^{\mathrm{c}}, \tilde{\sigma}_2^{\mathrm{c}})$ obtained from eq. (3) with $\Sigma = 0$ and $(b_1/s_1 = b_2/s_2 = 0)$ does not exceed yield (that is, $\tilde{\sigma}_1^{\mathrm{c}} < \sigma_1^{\mathrm{y}}$ and $|\tilde{\sigma}_2^{\mathrm{c}}| < \sigma_2^{\mathrm{y}}$). Alternately, scenario (*b*) occurs with $(\tilde{\sigma}_1, \tilde{\sigma}_2) = (\sigma_1^{\mathrm{y}}, -\sigma_1^{\mathrm{y}} f_1 / f_2)$ provided that $|\tilde{\sigma}_2| < \sigma_2^{\mathrm{y}}$. Finally, scenario (*c*) occurs with

$(\tilde{\sigma}_1, \tilde{\sigma}_2) = (\sigma_2^y f_2 / f_1, -\sigma_2^y)$ provided that $\tilde{\sigma}_1 < \sigma_1^y$. For scenario (*b*), the nonzero value of b_1/s_1 is determined by setting $\sigma_1 = \sigma_1^y$, $\Sigma = 0$, and $b_2/s_2 = 0$ in eq. (3). An upper limit, $(b_1/s_1)_{max} = (a_2^o/a_1^o) - 1$, is obtained using eq. (3) with $\Sigma = 0$ and $(\sigma_1, \sigma_2) = (0, 0)$, so that the interface is fully incoherent. Corresponding values of b_2/s_2 for scenario (*c*) are determined in a similar fashion, but with indices 1 and 2 reversed.

Two expressions for the bi-axial multilayer yield strength Σ_y are required since the compressive stress in layer 2 may be reduced to zero with or without any *pre-yield* of phase 1. This requires a comparison of the applied strain $\varepsilon^* = (0 - \tilde{\sigma}_2)/M_2$ to reduce σ_2 to zero and the corresponding value $\varepsilon_1^* = (\sigma_1^y - \tilde{\sigma}_1)/M_1$ to pre-yield phase 1. Provided $\varepsilon^* < \varepsilon_1^*$, no pre-yield occurs and the multilayer will extend with a full bi-axial modulus,

$$\Sigma_y = M_{full}\varepsilon^*, \quad M_{full} = f_1 M_1 + f_2 M_2 \quad \text{(valid for } \varepsilon^* \le \varepsilon_1^*) \tag{4}$$

If $\varepsilon^* > \varepsilon_1^*$, pre-yield of phase 1 occurs and the multilayer will extend with a full bi-axial modulus up to ε_1^* and a reduced bi-axial modulus for the remaining portion, $\varepsilon^* - \varepsilon_1^*$, when layer 1 is at yield,

$$\Sigma_y = M_{full}\varepsilon_1^* + M_{red}(\varepsilon^* - \varepsilon_1^*), \quad M_{red} = f_2 M_2 \quad \text{(valid for } \varepsilon^* \ge \varepsilon_1^*) \tag{5}$$

Results and Discussion

<u>Internal Stress Maps Prior to Loading</u>: The thin, shaded lines in **Figs. 3a,b** are contours of the initial stress $\tilde{\sigma}_k / M_k$ in layers 1 and 2, respectively, as a function of volume fraction f_1 and normalized bi-layer thickness Λ/b. Here, $M_2/M_1 = 0.91$ and $a_2^o/a_1^o = 1.01$ so that layer 2 has a smaller bi-axial elastic modulus and is in compression initially. For simplicity, the Burgers vectors are assumed to be $b_1 = b_2 = b$. The thick, darker lines define regions for which the initial interfacial structure is coherent, semi-coherent with initial phase 1 yield, and semi-coherent with initial phase 2 yield. The label $I = 0\%$ serves as the boundary between coherent and semi-coherent interfaces and the descriptions $I = 25\%, 50\%, ...$ define contours of increasing incoherency according to $(b_k/s_k) = I\,(b_k/s_k)_{max}$. The results imply that the magnitude of $\tilde{\sigma}_k / M_k$ increases with decreasing f_k, regardless of the phase or interfacial regime, and it is unaffected by Λ in the coherent regime. The shaded region indicates the regime for which layer thickness is too small (< $3b$ here) for σ_k^y in eq. (2) to be valid.

The positions and asymmetries of the contour lines in Fig. 3 depend on M_2/M_1 and a_2^o/a_1^o. According to eq. (3), an increase in a_2^o/a_1^o from the present value will shift the I contours to the left. Also, decreasing the modulus ratio from the present value of 0.91 will increase the size difference between the phase 1 and phase 2 semi-coherent regions, since $\sigma_k^y \sim M_k$ according to eq. (2). More detailed calculations that account for image effects are expected to reduce the magnitude of the asymmetry [20], since images effects tend to decrease line energy in the larger modulus phase and increase it in the smaller modulus phase, compared to the result for σ_k^y in eq. (2).

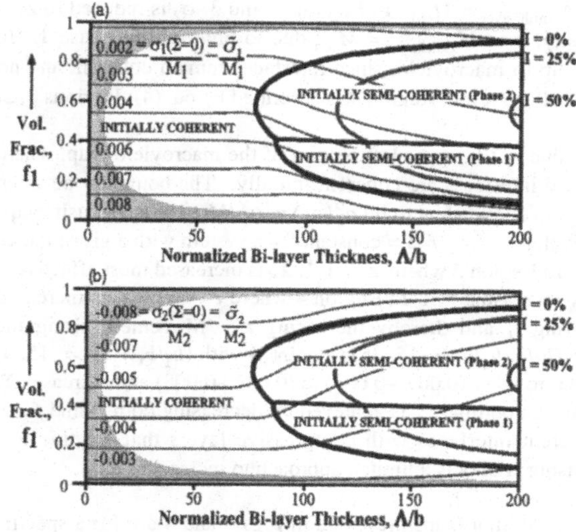

Fig. 3: Internal stress maps showing the bi-axial stress state $\tilde{\sigma}_k / M_k$ *at* $\Sigma = 0$ in *(a) phase 1 and (b) phase 2 , with a modulus mismatch* $M_2/M_1 = 0.91$, *lattice parameter mismatch* $a_2^o/a_1^o = 1.01$, *and* $b_1 = b_2 = b$. *Thin lines illustrate contours of constant strength and thicker lines are contours of constant interfacial structure, where* I *is the degree of incoherency. The model is not valid in the gray shaded region where layer thickness is < 3b.*

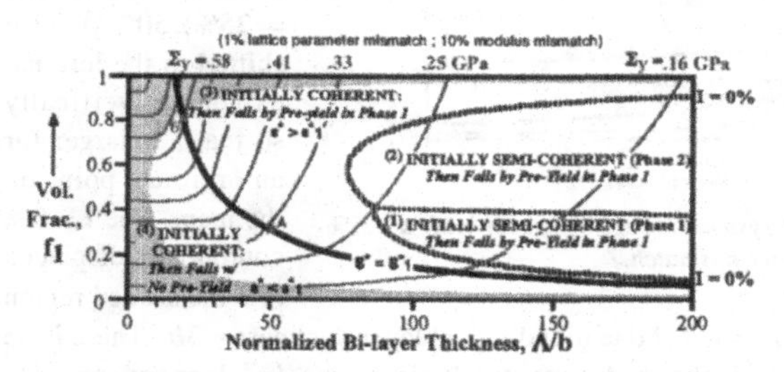

Fig. 4: Macroyield map displaying the applied bi-axial stress Σ_y *to induce co-deformation of phases 1 and 2, using the same parameters in Fig. 3 and also* $M_1 = 110$ GPa *and* $M_2 = 100$ GPa. *Thin lines illustrate contours of* Σ_y *and thicker lines define Regions 1 to 4, each with a different yield sequence. The black line* $\varepsilon^* = \varepsilon^*_1$ *indicates a path along which* Σ_y *is improved dramatically.*

Macroyield Maps: The lightly shaded contours in **Fig. 4** display the macroscopic biaxial tension Σ_y (in GPa) needed to fully yield the multilayer thin film, using $M_1 = 110$ GPa, $M_2 = 100$ GPa, and the parameter values used in Fig. 3. The largest yield strength arises when Λ and f_2 are decreased as to move into the upper left region of the map. A significant result is that decreasing Λ at constant f_k causes the yield strength to increase initially, and then reach a plateau at small Λ, similar to experimental results for several systems [12-14].

The yield process is understood by noting the four distinct regions defined inside the darker contour lines. Prior to loading, semi-coherent interfaces are expected at larger Λ (see Regions 1 and 2) and coherent interfaces are expected at smaller Λ (see Regions 3 and 4). Semi-coherent structure may exist initially due to tensile yield of phase 1 at smaller f_1 (Region 1) or compressive yield of phase 2 at smaller f_2 (Region 2). During loading, systems in Region 1 display plastic flow in phase 1 and the compressive stress in layer 2 is reduced to 0 with $M_{\text{multilayer}} = M_{\text{red}}$. In Regions 2 and 3, σ_2 is reduced to zero with $M_{\text{multilayer}} = M_{\text{full}}$ initially, then $M_{\text{multilayer}} = M_{\text{red}}$ due to pre-yield of phase 1. In Region 4, $M_{\text{multilayer}} = M_{\text{full}}$ up to macroyield, since interfaces remain coherent and no yield occurs before σ_2 reaches 0. Overall, Region 4 is described by eq. (4); Regions 1, 2, and 3 use eq. (5).

Σ_y reaches the largest value in the upper left section of the macroyield map, where Regions 3 and 4 dominate and interfaces are coherent initially. The boundary between these regions is given by $\varepsilon^* = \varepsilon_1{}^*$, equivalent to $\ln(a_2^o/a_1^o) = \sigma_1^y/M_1 \sim \ln(h_1)/h_1$ using eqs. (2-5). Thus, this line is equivalent to $h_1 \sim f_1\Lambda$ = constant for a system with a given lattice parameter mismatch a_2^o/a_1^o. In Region 3 where $\varepsilon^* > \varepsilon_1{}^*$, Σ_y is increased most effectively by decreasing Λ and thereby increasing $\varepsilon_1{}^*$. In Region 4 where $\varepsilon^* < \varepsilon_1{}^*$, Σ_y is increased most effectively by increasing f_1 and thereby increasing ε^*. Movement along the boundary from point A with $(h_1/b, h_2/b) = (15, 35)$ to point B with $(h_1/b, h_2/b) = (15, 4)$ changes $(\tilde{\sigma}_1/M_1, \tilde{\sigma}_2/M_2)$ from roughly (0.007,–0.003) to (0.002,–0.008) and increases Σ_y from ~0.3 to ~0.9 GPa. Thus, an optimal Σ_y is achieved by decreasing both Λ and f_2, so that the multilayer has coherent interfaces with compressive layers that are thin and highly stressed. Thus decreasing Λ only is a limited approach to increasing Σ_y.

Trends in Map Evolution: The design maps presented in Figs 3 and 4 are for a specific system described by fixed lattice parameter and bi-axial modulus mismatches (a_2^o/a_1^o and M_2/M_1). Fig. 5 shows the resulting macroyield map when the lattice parameter mismatch is increased to be $a_2^o/a_1^o = 1.025$. The contours of constant interface structure (I = 25%, 50%..) have shifted to the left and expanded vertically so that I is larger for an arbitrary point on the map. The $I = 0\%$ contour has displaced into the shaded region where the validity of σ_k^y is questioned due to individual layer thickness < $3b$. There is no coherent region predicted outside the shaded region. Increasing a_2^o/a_1^o does not appear to shift significantly the straight-line boundary separating Regions 1 and 2 nor is the magnitude of internal stress changed within the semicoherent regions on Fig. 3. Outside those semicoherent regions, the magnitude of internal stress is increased, with the largest increases and density of contours occurring at smaller Λ. The constant stress regions labeled "initially coherent" in Fig. 3 have collapsed into the small shaded region in which the validity of the model is questioned. Similar to the internal stress maps, the contours of Σ_y in Regions 1 and 2 of Fig. 4 are unchanged. Outside of these regions, the increase in a_2^o/a_1^o increases Σ_y, with the largest improvement occurring at smaller Λ. The $I = 0\%$

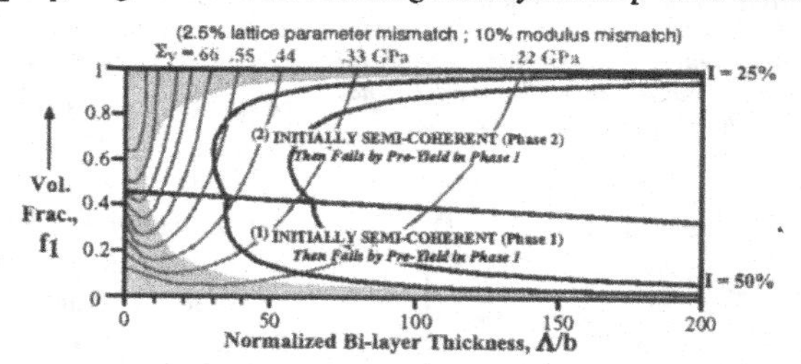

Fig. 5: Macroyield map of system in Fig. 4, except with a 2.5% instead of 1% lattice parameter mismatch.

and $\varepsilon^* = \varepsilon_1^*$ lines have displaced left into the shaded region, so that Regions 1 and 2 now cover the entire unshaded domain of the map.

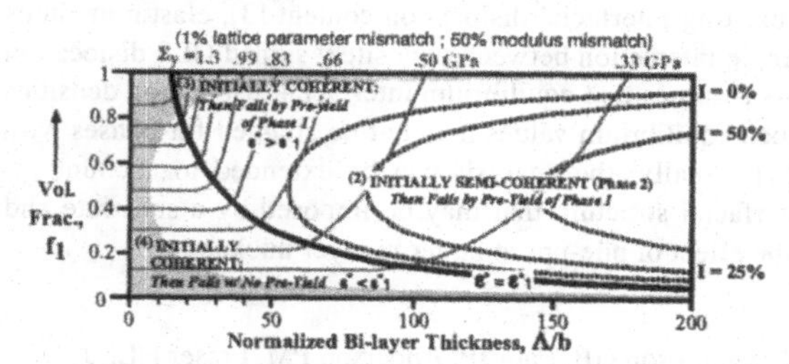

Fig. 6: Macroyield map of system in Fig. 4, except with M_1 = 220 GPa rather than 110 GPa.

Fig. 6 shows that increasing the bi-axial modulus mismatch via an increase in M_1 has a large impact on the macroyield strength. In particular, phase 1 becomes structurally stiffer, a larger amount of deformation is shifted to phase 2, and the stress magnitude increases in both phases for a given Λ, f_1, and I. Since, $\sigma_k^y \sim M_k$, the yield strength for phase 1 increases while that for phase 2 remains about the same. Thus, Region 2 grows in height and to the left so that in Fig. 6, it has displaced Region 1 completely. The boundary between Regions 3 and 4 does not move in this case. The initial stress is increased everywhere except within Region 2 of Fig. 3. The result is that Σ_y is increased for all regions of the map. The universal increase can be attributed to either a larger initial compression in phase 2 (Region 4 in Fig. 4), a larger stress to yield phase 1 (Region 2 in Fig. 4), or both (the remaining regions). The largest values of Σ_y still occur in the upper left corner of the map, but they are approximately double the values seen in Fig. 4.

Conclusions

Internal stress maps and macroscopic yield maps for two-phase multilayered thin films are developed assuming that (1) local yield is controlled by confined layer slip rather than source availability and (2) macroscopic yield occurs when the applied, in-plane, bi-axial stress eliminates any compressive stress state in alternating layers. The resulting maps predict that decreasing bi-layer thickness (Λ) is a limited approach to increasing the tensile yield strength of multilayers, valid only down to a critical Λ. Below that, decreasing the volume fraction of the compressively stressed phase is most effective. Other examples show that increasing the mismatch in stress-free lattice parameter or the modulus of the tensile (i.e., smaller lattice parameter phase) are viable strategies to increasing the bi-axial yield stress for multilayered thin films.

Several extensions to the analysis are warranted. First, applied states other than bi-axial tension may be considered, with the expectation that strategies to improve yield strength in one mode my likely degrade yield strength in another mode. For example, the strategy to improve bi-axial tension strength by creating thin layers with large compressive stress is likely to degrade bi-axial compressive strength. Second, the effect of image stresses from elastic moduli mismatch is expected to decrease σ_k^y in the larger modulus phase and increase it in the smaller modulus phase. Third, the criterion that macroscopic yield in bi-axial tension occurs when the compressive stress in phase 2 is eliminated is rather simple and does not incorporate structural interfacial barriers posed by interfacial slipping [23], existing interfacial dislocation content [3], elastic modulus mismatch [23], and elastic dipole interaction between oppositely signed CLS dislocation segments. Fourth, the analysis assumes that equilibrium interfacial dislocation densities prevail although in reality, such equilibrium values may not be attained for phases with low dislocation mobility [24]. Finally, the analysis can be extended to account for macroscopic loading and interfacial structure that may be imposed by a substrate and more accurately account for the effect of pile-ups at larger bi-layer thickness.

References

[1] Sperling EA, Banerjee R, Thompson GB, Fain JP, Anderson PM, Fraser HL. J Mater Res 2003;18(4):979.
[2] Huang H, Spaepen F. Acta Mater 2000;48(12):3261.
[3] Hoagland RG, Mitchell TE, Hirth JP, Kung H. Philos Mag A 2002;82(4):643.
[4] Thompson GB. Ph.D. thesis, The Ohio State University, Columbus Ohio, USA 2003.
[5] Nix WD. Metall Trans A 1989;20(11):2217.
[6] Was GS, Foecke T, Thin Solid Films 1996;286(1-2):1.
[7] Anderson PM, Foecke T, Hazzledine P. MRS Bulletin 1999;24(2):27.
[8] Misra A, Hirth HP, Kung H, Philos Mag A 2002; 82(16):2935.
[9] Yu D, Spaepen F. Work in progress, June 2003.
[10] Saif T. Work in progress, 2003.
[11] Anderson PM, Bingert JF, Misra A, Hirth JP. "Rolling textures in nanoscale Cu/Nb multilayers", accepted, Acta Mater: 2003.
[12] Clemens BM, Hung H, Barnett SA. MRS Bulletin 1999; 24(2):20.
[13] McKeown J, Misra A, Kung H, Hoagland RG, Nastasi M. Scripta Mater 2002;46(8):593.
[14] Tixier S, Boni P, Van Swygenhoven H. Thin Solid Films 1999;342(1-2):188.
[15] Banerjee R, Fain JP, Anderson PM, Fraser HL. Scripta mater 2001;44(11):2629.
[16] Embury JD, Hirth JP. Acta mater 1994;42(6):2051.
[17] Dehm G, Wagner T, Balk TJ, Artz E, Inkson BJ. J Mater Sci & Tech 2002;18(2):113.
[18] Kreidler ER, Anderson PM. Mat Res Soc Symp Proc: 1996;434:159.
[19] Baker SP, Kretschmann A, Artz E. Acta Mater 2001;49(12):2145.
[20] Kamat SV, Hirth JP. Script Metall 1987;21(11):1587.
[21] Sperling EA. M.S. thesis, The Ohio State University, Columbus Ohio, USA 2003.
[22] Lamm AV, Anderson PM. "Yield Maps for Nanoscale Metallic Multilayers", accepted, Scripta Mater: 2003.
[23] Anderson PM, Li ZY. Mater Sci & Eng A 2001;319-321:182.
[24] Freund LB. J App Mech 1987;54(3):553.

Acknowledgements
The authors gratefully acknowledge the support of the Air Force Office of Scientific Research Metallic Materials Program (F49620-01-1-0092) and the National Science Foundation Mechanics & Materials and Metallic Materials Programs (0072010).

Processing and Properties of Structural Nanomaterials
Edited by Leon L. Shaw, C. Suryanarayana and Rajiv S. Mishra
TMS (The Minerals, Metals & Materials Society), 2003

Molecular and Atomic Polarizabilities of Model Carbon Nanotubes

Francisco Torrens

Institut Universitari de Ciència Molecular, Universitat de València, Dr. Moliner-50,
E-46100 Burjassot (València), Spain

Keywords: Polarizability, Induced dipole polarization, Elliptical deformation, Carbon nanotube

Abstract

The interacting induced dipole polarization model implemented in program POLAR is used for the calculation of the polarizability α. The method is tested with single-wall carbon nanotube (SWNT) models as a function of SWNT radius and elliptical deformation. The α follows the same trend as reference calculations performed with program PAPID. For the zigzag tubes, α follows a simple law. PAPID differentiates more effectively than POLAR among SWNT models with increasing radial deformation. The α can be modified reversibly by external radial deformation. Different effective α^{eff} are calculated for the atoms at the highest and lowest curvature sites. MOPAC-AM1 standard heat of formation per C atom $\Delta H_f^{\circ}/v$ shows that SWNT models are less stable than acenes. While SWNT models are stabilized with increasing number of vertices, acenes are destabilized. For SWNT models, the ratio number of trivalent vertices/number of divalent vertices v_3/v_2 is one, which is greater than for the corresponding acenes.

Introduction

Single-wall carbon nanotubes (SWNT) provide a system where the electronic properties can be controlled by structure [1]. It is desirable to have a good understanding of their electronic and mechanical properties [2]. The electronic properties of SWNTs are modified by elliptical deformation [3]. The energy gap of an insulating SWNT is decreased and eventually vanishes at an insulator–metal transition, with increasing applied radial strain [4]. The elliptical deformation necessary to induce metallicity was found in the elastic range [5]. This allows the *fine-tuning* of the properties of SWNTs *via* reversible deformation, and leads to variable and reversible quantum devices such as metal–insulator and rectifying junctions [6]. Gülseren *et al.* found that the elliptical distortion disturbs the uniformity of charge distribution of SWNTs [7].

A SWNT consists of a graphene sheet rolled up in one dimension to form a cylinder. When the graphene layer is rolled on itself, the atom at the origin is superposed to some of the other atoms on the plane. The lattice vector that joins these atoms is designated *chiral vector*, and is defined as: $\mathbf{C}_h = n\ \mathbf{a}_1 + m\ \mathbf{a}_2$, where $\mathbf{a}_1$ and $\mathbf{a}_2$ are the unit vectors of two-dimensional hexagonal network, and n and m are integers. The angle that $\mathbf{C}_h$ forms with $\mathbf{a}_1$ is designated *chiral angle*. The diameter of SWNT is related to the chiral vector; it grows with n and m. The structure of a SWNT is given once specified n and m; these two numbers are used to designate each SWNT, in the form (n,m). SWNTs can be classified in three types: (1) *armchair* (n,n), since the edge presents the form of an armchair with arms, (2) *zigzag* $(n,0)$, by the form of their edge (both sorts of tubes present a symmetry such that both turning senses about the axis of the tube are equivalent), and (3) *chiral* (n,m), because such symmetry is broken.

Computational Method

The molecular polarizability, α_{ab}^{mol}, is defined as the linear response to an external electric field,

$$\mu_\alpha^{\text{ind}} = \alpha_{\alpha\beta}^{\text{mol}} E_\beta^{\text{ext}} \tag{1}$$

where $\mu_{\alpha}^{\text{ind}}$ is the induced molecular dipole moment. Considering a set of N interacting atomic polarizabilities, the atomic induced dipole moment has a contribution also from the other atoms,

$$\mu_{p,\alpha}^{\text{ind}} = \alpha_{p,\alpha\beta}\left(E_{\beta}^{\text{ext}} + \sum_{q \neq p}^{N} T_{pq,\beta\gamma}^{(2)} \mu_{q,\gamma}^{\text{ind}}\right) \tag{2}$$

where $T_{pq,\beta\gamma}^{(2)}$ is the interaction tensor as modified by Thole [8]

$$T_{pq,\alpha\beta}^{(2)} = \frac{3v_{pq}^{4} r_{pq,\alpha} r_{pq,\beta}}{r_{pq}^{5}} - \frac{\left(4v_{pq}^{3} - 3v_{pq}^{3}\right)\delta_{\alpha\beta}}{r_{pq}^{3}} \tag{3}$$

where $v_{pq} = r_{pq}/s_{pq}$ if $r_{pq} < s_{pq}$; otherwise $v_{pq} = 1$. The term s is defined as $(\Phi_p \ \Phi_q)^{1/4}$ where Φ_p is a fitting parameter proportional to the atomic second-order moment. The molecular polarizability can then be written as

$$\alpha_{\alpha\beta}^{\text{mol}} = \sum_{q,p}^{N} B_{pq,\alpha\beta} \tag{4}$$

where **B** is the relay matrix defined as (in a supermatrix notation)

$$\mathbf{B} = \left(\alpha^{-1} - T^{(2)}\right)^{-1} \tag{5}$$

The following improvements have been implemented in the model. (1) In order to build the relay matrix **B**, the atomic polarizability tensor $\alpha^{i} = \alpha_{s}^{i} + \alpha_{p}^{i}$ has been used instead of the scalar polarizability α^{i}. (2) The interaction between bonded atoms and between atoms within a distance lying in an interval defined by $[r^{\text{inf}}, r^{\text{sup}}]$ has been neglected. (3) A damping function has been used in the calculation of the interaction tensor to prevent the polarizability from going to infinity.

An optimized version of our program POLAR [9] has been implemented in the program molecular mechanics (MM2) [10], in its extension to transition metals (MMX) [11], and in the empirical conformational energy program for peptides (ECEPP2) [12]. The new versions are called MMID2 [13], MMXID [14] and ECEPPID2 [15].

Calculation Results and Discussion

Undeformed $(n,0)$ zigzag SWNT models are studied with the number of C atoms, v varying from 28 to 48 for $n = 7$ to 12, respectively. Molecular geometries have been optimized with MMID2. Figure 1 illustrates the variation of the elementary polarizability per atom $<\alpha>$ of the undeformed $(n,0)$ zigzag SWNT models. The radius of the tube model R increases with the number of rings in the tube. Contrary to POLAR *(α)*, PAPID [16] *(α* Ref.) increases monotonically with R. POLAR results are doubled with respect to PAPID reference. However, the general shape is similar for both curves. In particular, for the (9,0) zigzag SWNT model, PAPID $<\alpha> = 1.361\text{Å}^3$ is in agreement with the experiment (1.304Å^3) by Jensen *et al.* [17]. The $<\alpha>$ increases with increasing radius of the tube model, and eventually saturates at a value corresponding to that on graphene sheet. However, the appearance of Figure 1 should be taken with care; the effect of the curvature appears artificially smoothed due to the long length of the y-axis, in order that both lines fit in Figure 1. The same effect occurs in Figures 2 and 3. The variation of $<\alpha>$ with the radius of the zigzag tube model (Figure 1) fits to

$$\frac{1}{\alpha} = a + \frac{b}{R} \quad (6)$$

where α is the inverse $<\alpha>$ of the graphene. With POLAR (five points with greater R) it results

$$\frac{1}{\alpha} = 0.264 + \frac{0.147}{R} \qquad N = 5 \qquad r = 0.538 \quad (7)$$

and $<\alpha>$ for the graphene layer is extrapolated as 3.782Å^3. For PAPID references,

$$\frac{1}{\alpha} = 0.580 + \frac{0.527}{R} \qquad N = 6 \qquad r = 0.989 \quad (8)$$

and $<\alpha>$ for the graphene plane extrapolates as 1.723Å^3. The $<\alpha>$ calculated for undeformed $(n,0)$ zigzag SWNT models with $n \leq 8$ deviates from the above simple scaling, perhaps due to the fact that the singlet π^* band, which is normally in the conduction band, falls into the band gap as a result of increased σ^*–π^* mixing at high curvature [18]. While the band gap shows significant change with n, $<\alpha>$ varies smoothly with R^{-1}. Increasing $<\alpha>$ with decreasing R shows that, for small R, the character of the surface deviates from that of the graphene sheet. This finding suggests that, creating regions of different curvature on a single SWNT model by radial deformation, one can attain different values of $<\alpha>$. The radial deformation decreases the radius in the y direction while elongates it along the x direction. As a result, the circular cross section is distorted to an elliptical one with major and minor axes, a and b, respectively. The elliptical radial deformation can be defined as $\varepsilon_{yy} = (R - b)/R$, where R is the radius of the undeformed nanotube. For different values of ε_{yy}, full structural optimization was carried out under the constraint that the major axis is kept fixed at a preset value, by freezing the carbon atoms at both ends of the major axis.

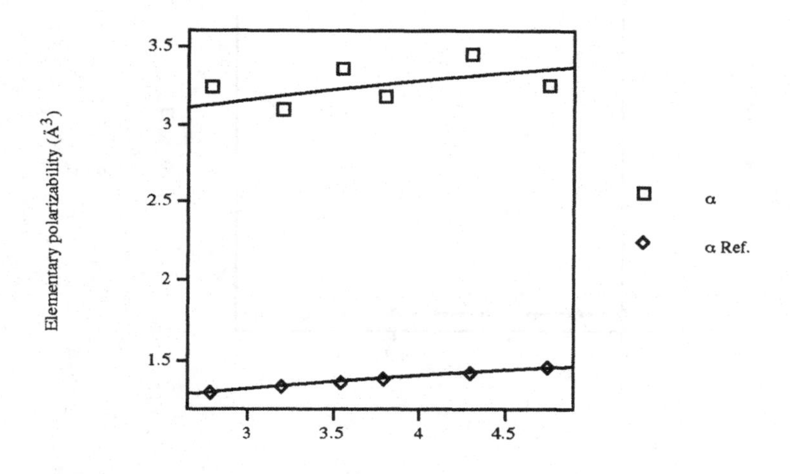

Figure 1: Elementary polarizability of zigzag $(n,0)$ SWNT models $vs.$ the radius of tubes.

The molecular polarizability α of the (8,0) zigzag surface with the applied radial strain is shown in Figure 2. The values computed with POLAR and PAPID slightly increase from $\varepsilon_{yy} = 0$ to 0.32. POLAR α is increased by only 0.04Å^3 (0.04%) for $\varepsilon_{yy} = 0.28$, and PAPID α is augmented by 0.09Å^3 (0.2%) for $\varepsilon_{yy} = 0.24$. Cubic correlation models have been fitted for POLAR and PAPID. For POLAR, the best regression results

$$\alpha = 98.9 - 0.410\varepsilon_{yy} + 4.48\varepsilon_{yy}^2 - 9.38\varepsilon_{yy}^3 \qquad N = 9 \qquad r = 0.979 \qquad (9)$$

With PAPID reference, for the seven points with lower ε_{yy}, the best fit gives

$$\alpha = 42.7 - 0.0829\varepsilon_{yy} + 1.83\varepsilon_{yy}^2 + 0.177\varepsilon_{yy}^3 \qquad N = 7 \qquad r = 0.999 \qquad (10)$$

Figure 3 displays the effective polarizability α^{eff} of atoms in the (8,0) zigzag SWNT models as a function of the elliptic radial deformation for two cases. (A) The α^{eff} of the C atom in the highest curvature side of the surface (the *sharp site* at one of the ends of the major axis, $x = a$, $y = 0$). (B) The α^{eff} of the C atom in the lowest curvature side at one of the ends of the minor axis ($x = 0$, $y = b$, *flat site*). POLAR α^{eff} at the sharp site is greater than at the flat site. This is in qualitative agreement with PAPID. In particular, POLAR α^{eff} at the sharp site is 9% greater than that at the flat site for $\varepsilon_{yy} = 0.32$. The same happens for PAPID (5% for $\varepsilon_{yy} = 0.28$). The difference between both POLAR α^{eff} curves increases with ε_{yy}. However, the difference between both PAPID α^{eff} reference curves passes through a smooth maximum at $\varepsilon_{yy} = 0.28$.

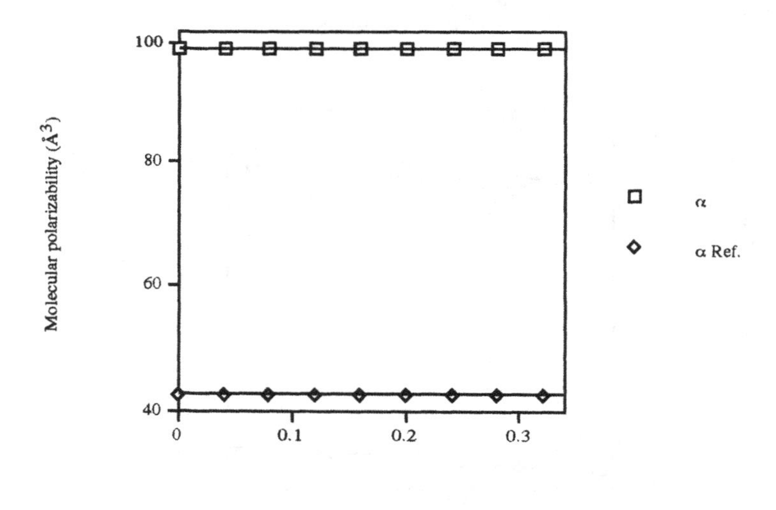

Figure 2: Molecular polarizability of zigzag (8,0) SWNT model *vs*. deformation ε_{yy}.

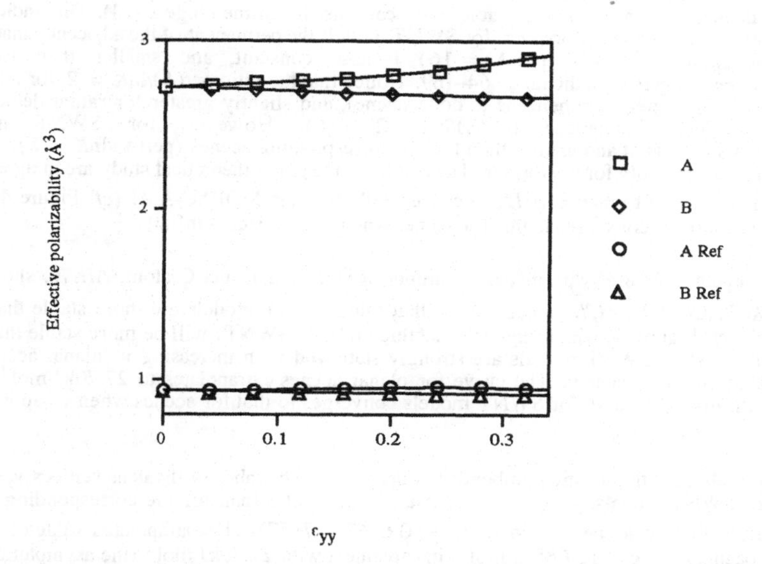

Figure 3: Effective atomic polarizability at the end of the axes *vs.* deformation ε_{yy}.

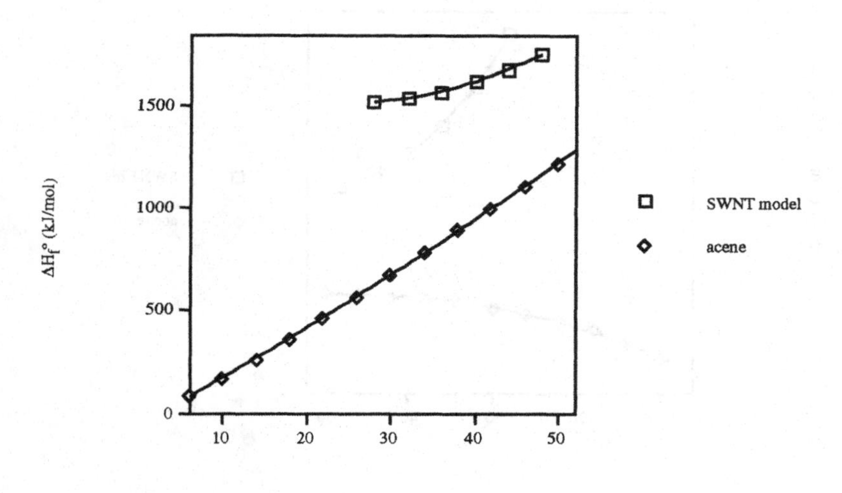

Figure 4: Standard heats of formation calculated with MOPAC-AM1 *vs.* v.

A graph-theoretical study of the zigzag SWNT models has been performed. For SWNT models, the Kekulé structure count *($K = 2$)* is constant, and smaller than for the corresponding planar acenes (heptacene to dodecacene linear polybenzenoid hydrocarbons, K in the range 8–13). This indicates a smaller stability for the former. Moreover, for SWNT models the permanent of the adjacency matrix of the hydrogen-suppressed graph (per(**A**) = 16) is also constant, and smaller than for the corresponding acenes, per(**A**) in the range 64–169. Notice that the ratio per(**A**)/ln$K$ = 2 for alternant hydrocarbons (*e.g.*, acenes from benzene to dodecacene), and slightly greater for rather delocalized hydrocarbons (*e.g.*, corannulene per(**A**)/ln$K$ = 2.0136). However, for SWNT models per(**A**)/ln$K$ = 4 is constant and greater than for the corresponding acenes (per(**A**)/ln$K$ = 2), which confirms the smaller stability for the former. The results of the graph-theoretical study are in agreement with the standard heats of formation ΔH_f°, calculated with program MOPAC-AM1 (*cf.* Figure 4). The curve for SWNT models converges to that for acenes when v approaches infinity.

In order to compare structures with different number of atoms, ΔH_f° per C atom, $\Delta H_f^\circ/v$, should be contrasted. MOPAC-AM1 $\Delta H_f^\circ/v$ results show that thick SWNT models are more stable than thin SWNT models (*cf.* Figure 5), which suggests that thick infinite SWNTs will be more stable than thin infinite SWNTs. While SWNT models are strongly stabilized with increasing v, planar acenes are slightly destabilized for greater v. The curve for planar acenes extrapolates to 27.70kJ·mol^{-1} for v approaching infinity. The curve for SWNT models converges to that for acenes when v approaches infinity.

$\Delta H_f^\circ/v$ also correlates with the ratio number of trivalent vertices/number of divalent vertices v_3/v_2 (*cf.* Figure 6). For SWNT models, $v_3/v_2 = 1$ is constant and greater than for the corresponding planar acenes (heptacene to dodecacene v_3/v_2 in the range 0.6667–0.7857). The extrapolated value of $\Delta H_f^\circ/v$ for v_3/v_2 approaching one gives 27.85kJ·mol^{-1}, in agreement with 27.70kJ·mol^{-1} (the asymptotic value in Figure 5).

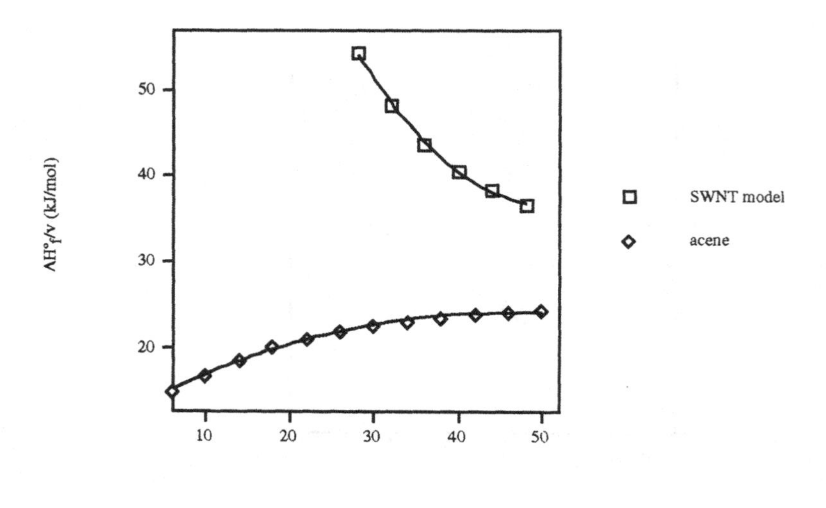

Figure 5: Standard heats of formation per C atom calculated with MOPAC-AM1 *vs.* v.

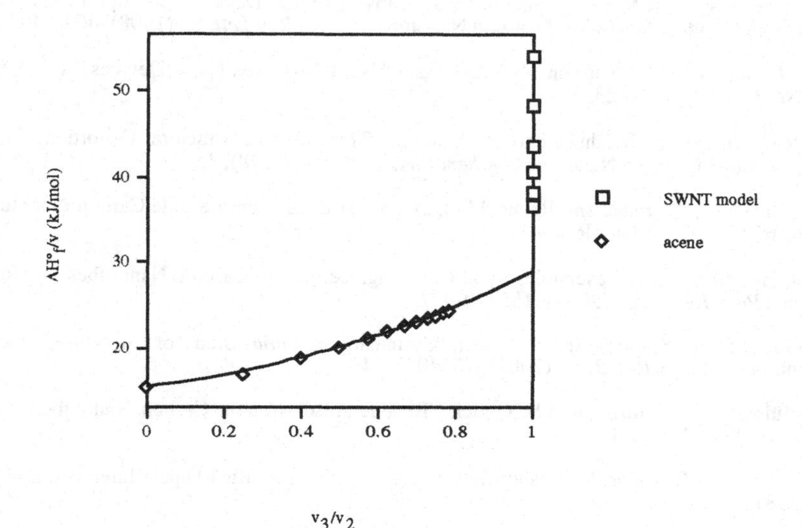

Figure 6: Standard heats of formation per C atom calculated with MOPAC-AM1 *vs*. v_3/v_2.

Conclusions

From the previous results the following conclusions can be drawn.

1. Due to the differences between POLAR and PAPID results, it may become necessary to recalibrate POLAR. Results of good quality *ab initio* calculations might be primary standards for such a calibration.

2. The α of a zigzag SWNT model can be modified reversibly by radial deformation. PAPID differentiates more effectively than POLAR among SWNT models with increasing radial deformation.

3. Different effective polarizabilities are calculated for the atoms at the sharp and flat sites. POLAR discriminates more efficiently than PAPID between α^{eff} of the sharp and flat sites. The physics and chemistry of SWNT models permit a difference for undeformed and elliptically deformed SWNTs. While all positions in the undeformed SWNT are identical, a variation of C atoms is found in deformed SWNTs. The different atoms display different chemical reactivities. These different reactivities were observed in the chemistry of fullerenes C_{60} and C_{70}.

4. A graph-theoretical study and MOPAC-AM1 ΔH_f° show that SWNT models are less stable than planar acenes. However, both curves converge for infinite v. While SWNT models are stabilized with increasing v, planar acenes are destabilized. For SWNT models, $v_3/v_2 = 1$ is constant and greater than for the corresponding acenes, for which $\Delta H_f^\circ/v$ correlates with v_3/v_2.

Acknowledgements

The author acknowledges financial support of the Spanish MCT (Plan Nacional I+D+I, Project No. BQU2001-2935-C02-01).

References

1. A. Bezryadin, A.R.M. Verschueren, S.J. Tans, and C. Dekker, "Multiprobe Transport Experiments on Individual Single-Wall Carbon Nanotubes," *Phys Rev Lett*, 80 (1998), 4036-4039.

2. J.P. Lu and J. Han, "Carbon Nanotubes and Nanotube-Based Nano Devices," *Int J High Electron Syst*, 9 (1998), 101-123.

3. A. Rochefort, D.R. Salahub, and P. Avouris, "The Effect of Structural Distortions on the Electronic Structure of Carbon Nanotubes," *Chem Phys Lett*, 297 (1998), 45-50.

4. Ç. Kiliç et al., "Variable and Reversible Quantum Structures on a Single Carbon Nanotube," *Phys Rev B*, 62 (2000), R16345-R16348.

5. O. Gülseren et al., "Reversible Band-Gap Engineering in Carbon Nanotubes by Radial Deformation," *Phys Rev B*, 65 (2002), 155410–1-7.

6. O. Gülseren, T. Yildirim, and S. Ciraci, "Systematic *ab Initio* Study of Curvature Effects in Carbon Nanotubes," *Phys Rev B*, 65 (2002), 153405–1-4.

7. O. Gülseren, T. Yildirim, and S. Ciraci, "Tunable Adsorption on Carbon Nanotubes," *Phys Rev Lett*, 87 (2001), 116802–1-4.

8. B.T. Thole, "Molecular Polarizabilities Calculated with a Modified Dipole Interaction," *Chem Phys*, 59 (1981), 341-350.

9. F. Torrens, J. Sánchez-Marín, and I. Nebot-Gil, "Interacting Induced Dipoles Polarization Model for Molecular Polarizabilities. Reference Molecules, Amino Acids and Model Peptides," *J Mol Struct (Theochem)*, 463 (1999), 27-39.

10. N.L. Allinger, "Conformational Analysis. 130. MM2. A Hydrocarbon Force Field Utilizing V_1 and V_2 Torsional Terms," *J Am Chem Soc*, 99 (1977), 8127-8134.

11. J.A. Deiters et al., "Computer Simulation of Phosphorane Structures," *J Am Chem Soc*, 99 (1977), 5461-5471.

12. G. Némethy, M.S. Pottle, H.A. Scheraga, "Energy Parameters in Polypeptides. 9. Updating of Geometric Parameters, Nonbonded Interactions, and Hydrogen Bond Interactions for the Naturally Occurring Amino Acids," *J Phys Chem*, 87 (1983), 1883-1887.

13. F. Torrens et al., "Conformational Aspects of Some Asymmetric Diels-Alder Reactions. A Molecular Mechanics + Polarization Study," *Tetrahedron*, 48 (1992), 5209-5218.

14. F. Torrens, "Nature of Fe^{III}–O_2, Fe^{II}–CO and Fe^{III}–CN Complexes of Hemoprotein Models," *Polyhedron*, 22 (2003), 1091-1098.

15. F. Torrens, "Polarization Force Fields for Peptides Implemented in ECEPP2 and MM2," *Mol Simul*, 24 (2000), 391-410.

16. C. Voisin, A. Cartier, and J.-L. Rivail, "Computation of Accurate Electronic Molecular Polarizabilities," *J Phys Chem*, 96 (1992), 7966-7971.

17. L. Jensen et al., "Static and Frequency-Dependent Polarizability Tensors for Carbon Nanotubes," *J Phys Chem B*, 104 (2000), 10462-10466.

18. X. Blasé et al., "Hybridization Effects and Metallicity in Small Radius Carbon Nanotubes," *Phys Rev Lett*, 72 (1994), 1878-1881.

STRUCTURE AND PROPERTY RELATIONSHIPS

Processing and Properties of Structural Nanomaterials
Edited by Leon L. Shaw, C. Suryanarayana and Rajiv S. Mishra
TMS (The Minerals, Metals & Materials Society), 2003

Grain Size Distribution and Mechanical Properties of Nanostructure Materials

Carl C. Koch[1] and Ronald O. Scattergood[1]

[1]Materials Science and Engineering Department, North Carolina State University; Raleigh NC, 27695, USA.

Keywords: Grain size distribution; ductility, nanostructured metals

Abstract

The importance of optimizing strength and ductility in nanostructured materials is emphasized in this paper. Examples from the literature and the authors' laboratory are given which illustrate the desired combinations of strength and ductility in metals and alloys which have at least part of their microstructures at the nanoscale. It is concluded that a grain size distribution that allows for strain hardening is in general a common factor that induces both some ductility along with increased strength.

Introduction

The possibility of achieving a combination of greatly enhanced strength (hardness) along with good ductility has been the main driving force behind the significant research activity on the mechanical behavior of nanostructured metals. Large increases in hardness and strength have been documented for nanostructured materials—most clearly for elemental metals [1], less so for alloys or compounds [2]. In conventional grain size (cgs) metals and alloys, the extrapolation of ductility (e.g. percent elongation to failure in tension) to smaller grain sizes suggests that ductility also might increase as the grain size is reduced to the nanoscale. However, it has been generally observed that the ductilities of nanostructured metals are disappointingly low—typically less than 2% elongation to failure for most metals with grain sizes less than about 25 nm [3], where for the same cgs metals the ductilities are usually large—40 to 70 % elongation.

Ductility can be discussed in terms of the ability of a material to change shape by plastic deformation without fracture. Two major limitations to ductility in materials that can be applicable to nanostructured materials are: 1. a crack nucleation or propagation instability, and 2. a force instability in tension. Crack nucleation or propagation instability involves the imposed stress concentration at an existing flaw exceeding the critical toughness value of the material. This approach considers the stress intensity (K) at a pre-existing flaw (or J, the energy or work input required to reach that state) as the sample is increasingly loaded [4]. Work must be supplied from the external source to i. produce the elastic concentration at the notch tip, ii. produce the local plastic strain as the notch starts to deform, and iii. to achieve global plasticity in the case where the material shows some overall ductility. These three terms all vary with the initial flaw size, which may be considered to be the same as the grain size, in the absence of other flaws, artifacts from processing. If it is assumed that the flow stress and fracture toughness are constants, it should be possible at sufficiently fine grain sizes to achieve global plasticity before brittle, local crack propagation sets in. However, given the strong increase in flow stress with decreasing grain size at the nanoscale, the competition between plastic flow and fracture is difficult to predict. Therefore, if the flow stress increases to above the fracture stress, ductility may be limited. However, for many of the processing methods that provide the finest nanoscale grain sizes, it is difficult to eliminate artifacts, which can mask the inherent mechanical behavior. Many of these processing methods involve synthesis of nanoscale particulates [5], or powders with a nanocrystalline internal grain structure [6], followed by consolidation to bulk samples [7]. These have been referred to as "two-step" methods. The goal of particulate/powder compaction is to obtain theoretical density and complete particle bonding without significant coarsening of the nanoscale microstructure. Porosity was a major artifact in the earlier studies of metals made by the inert gas condensation technique. However, even when theoretical density is obtained in these "two-step" processes, complete particle bonding may be lacking and tensile tests reveal fracture along the prior powder boundary interfaces. "One-step" processes such as electrodeposition and the variety of severe plastic deformation methods remove the need for a particulate consolidation step. However, these methods are not necessarily artifact free either. Electrodeposition can introduce impurity atoms (such as sulfur or carbon) that can influence mechanical behavior. The high dislocation densities introduced by the severe plastic deformation methods can result in suppression of subsequent strain hardening and lead to the force instability limit to ductility.

If artifacts are absent, and the flow stress is not exceedingly high compared to the fracture stress, force instability in tension can limit ductility. Necking generally begins at maximum load during tensile testing; the amount of uniform elongation depends upon strain hardening such that true uniform strain $\varepsilon_u = n$ in a cylindrical specimen (or $\varepsilon_u = 2n$ for a sheet) where n is the strain hardening coefficient. For an ideally plastic material (such as metallic glasses) where $n = 0$, the necking instability would begin just as soon as yielding occurred. This criterion implies simply that the sample is mechanically stable until the rate of work hardening falls to a level determined by the flow stress (and prior strain) at that time. Materials with a high capacity for strain hardening are therefore stable, while those with little capacity for strain hardening are potentially unstable. For the extreme case of an amorphous alloy, in tension or unconstrained compression, the first plastic deformation takes place locally and is confined to a shear band which then can propagate across the sample to failure, giving essentially no macroscopic ductility.

This paper will present examples of nanostructured metals which exhibit both increased strength and hardness compared to their cgs counterparts as well as good ductility. The importance of strain hardening in these examples, which in some cases apparently result from a grain size distribution, will also be discussed.

Examples of Materials with Combined Strength and Ductility

Electrodeposited Cobalt

Electrodeposited cobalt metal has been prepared with a small average grain size of about 12 nm and with a fairly narrow grain size distribution of ± 7 nm [8]. This material had the hcp structure with no trace of the fcc phase, that is it had the equilibrium structure for room temperature. The hardness, yield strength, and ultimate tensile strength for this nanostructured Co were 2-3 times higher than for cgs Co. On decreasing the strain rate in the tensile tests the ultimate tensile strength and work hardening exponent of the nanostructured material clearly increased, while the yield strength did not noticeably change. Of most interest is the fact that the nanocrystalline Co exhibited elongation to fracture values of about 6 to 9%. This is the only nanocrystalline material with a narrow grain size distribution and average grain size below 20 nm that exhibits both significantly higher strength than the cgs metal combined with good ductility. The differences in the nanostructured Co from the cgs material included the fact that it was 100% hcp structure, which is not easily obtained by conventional thermal/mechanical processing. In addition it exhibited a marked crystallographic texture with a strong x-ray diffraction 0002 peak such that the basal planes were preferentially aligned parallel to the surface of the sample. The authors suggest that the deformation behavior of this nanostructured Co may be controlled by time dependent mechanisms, such as creep. More studies are needed to clarify the very interesting, and until now, unique behavior of this nanostructured material.

Nanostructured/Ultrafine Grained Copper

Copper has been one of the most studied nanostructured metals and recent results of Ma and co-workers [9] have provided a dramatic example of combinations of high strength and high tensile ductility. The promising aspects of this work include not only the favorable mechanical properties observed, but the manner in which the nanostructured/ultrafine grained/micron samples which gave the optimum properties were prepared. Copper was rolled at liquid nitrogen temperature to high strains of 93% to create a high dislocation density that does not dynamically recover. The as-rolled microstructure showed the high dislocation density along with some resolvable nanoscale grains with dimensions less than 200 nm. Annealing for short times at temperatures up to 200°C provided for the development of grains with high angle boundaries that were in the nanoscale or submicron size range. Some abnormal recrystallization was observed such that for annealing at 200°C for 3 minutes about 25% volume fraction of the samples consisted of grains 1-3 μm in diameter. Rolling the Cu at room temperature did not provide sufficient dislocation density to accomplish the subsequent nanoscale/submicron grain sizes on annealing. This work on Cu gave stress – strain curves for annealed coarse-grained Cu, Cu rolled to 95% at room temperature, Cu rolled to 93% at liquid nitrogen temperature and these samples annealed for 3 minutes at either 180 or 200°C. The optimum properties were obtained for the mixed grain size material with the 1-3 μm grains imbedded in the matrix of nanoscale and submicron size grains. This material had a high yield stress of about 340 MPa, a total elongation to failure of 65% and uniform elongation of about 30%. The ductility was thus comparable to that of annealed cgs Cu, but the yield strength was almost 7 times higher. In figure 1 we have plotted the yield strength for selected Cu samples against the percent elongation to failure in tension. Elongation to failure was selected in order to make comparisons with other samples where the uniform elongation was not reported. If uniform elongation were used then the data of [9] would stand out even more from the others. A curve separating the data that exhibit the expected trend of high yield strength with low ductility is drawn from the very high strength nanoscale Cu [10] through the value for Cu cold-rolled at room temperature [9] and annealed cgs Cu [9]. Several data sets stand out for discussion. The datum point of reference [11] is from the work of Lu et al on electrodeposited high purity Cu. This material, had nanoscale grains with low angle boundaries, really subgrains, within micron size grains. This material had good ductility (30% elongation to failure) but yield strength only comparable to that of cgs Cu. Most of the other data reported for nanostructured Cu [10,12-14] are clustered at the low ductility end of the graph, with most having % elongation to failure <2%. The other datum point that exhibits both high strength and ductility is from the work of Valiev et al [15]. This is for Cu processed by severe plastic deformation by the method of equal channel angular pressing (ECAP) to the high strain level (16 ECAP passes in a 90° die). This material is marginally nanostructured with an average grain size of 100 nm, but with some grains less than 100 nm. Details of the grain size distribution were not given.

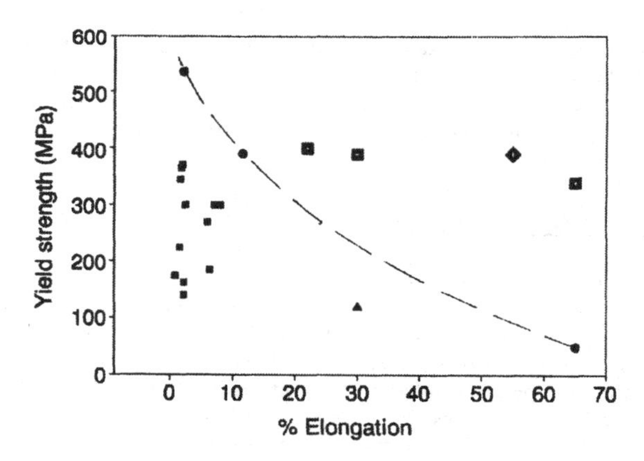

Figure 1. Yield strength (MPa) vs. % elongation to failure for nanostructured Cu.
Data from references: ◘ [9], ● [9,10], ▲ [11], ■ [12-14], ◆ [15]

The data on "nanostructured" Cu that exhibits both high yield strength and high values of elongation to failure all are from samples that have a range of grain sizes encompassing both the nanoscale and the submicron, and in the case of Ma and co-workers datum point [9], some micron size grains. The tentative conclusion regarding the optimum properties of these materials is that the nanoscale grain size matrix provides the high strength while the larger, submicron or micron, grains allow for strain hardening which mitigates the plastic instability that can induce premature failure in otherwise ductile metals. The details of the precise mechanisms giving rise to this behavior are not yet understood. There is additional data in the literature on ultrafine grained metals that exhibit good combinations of strength and ductility; however, we restrict the discussions in this paper to materials with at least some nanoscale grains in the microstructure.

Nanostructured Aluminum-Magnesium Alloys

Lavernia and co-workers have prepared a commercial aluminum alloy, 5083, [16] and an Al-7.5 Mg alloy [17] by cryomilling followed by powder compaction by hot isostatic pressing and extrusion. The cryomilling of Al alloy 5083[16] resulted in a nanoscale microstructure with average grain size about 30 nm. After hipping and extrusion the grain size remained mostly nanoscale at about 35 nm. However, some larger grains were also observed in the TEM analysis. The stability of the nanoscale grain size during the elevated temperature compaction steps was attributed to the large number of various precipitates including several intermetallic compounds such as Mg_2Si and Al_3Mg_2 and AlN and Al_2O_3 which presumably retard grain growth by Zener pinning. A few larger, micron size grains formed by secondary recrystallization. These large grains were believed responsible for the good ductility observed in these materials along with large increases in strength. The yield strength vs. % elongation to failure for this sample is plotted in figure 2. In figure 2 we also plot data on cgs 5083 with various strain-hardening and annealing treatments [18], data on "ultrafine grained" 5083 [19] wherein the grain sizes are about 500 nm and slightly higher, and data from [17] on nanostructured Al-7.5 Mg alloys. It is apparent that the data for the ultrafine grained and nanostructured alloys exhibit significantly superior behavior compared to the yield strength vs. % elongation for the cgs 5083. Guided by the results reported in [16], the Al-7.5 Al alloy was cryomilled to nanostructured grain sizes. The cryomilled powder was then combined with either 15 or 30% by volume of unmilled alloy powder, which was made by powder atomization and presumably with micron scale grain sizes. The powders were then consolidated by hipping and extrusion to bulk samples for tensile testing.

The grain size distributions for these samples were, after consolidation, about 100 to 300 nm for the cryomilled powders, with some larger grains up to one micron. The samples containing the unmilled powders had similar grain structures but with more micron size grains, some in bands.

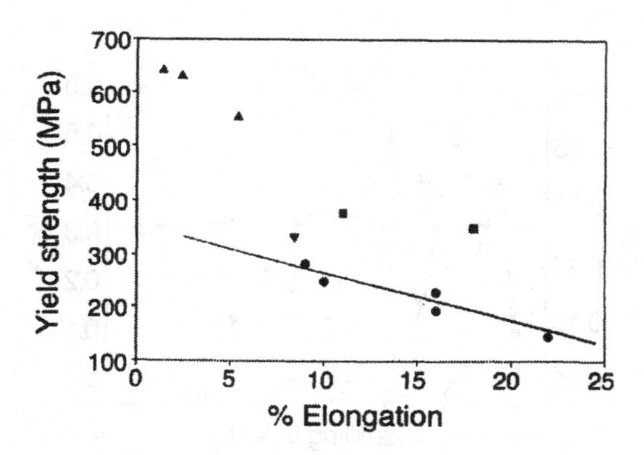

Figure 2. Yield strength vs. % elongation to failure for Al-Mg alloys. ● cgs alloy 5083 [18], ▼ nc/ufg alloy 5083 [16], ■ ufg alloy 5083 [19], ▲ nc/ufg Al-7.5 Mg [17]

It is apparent that for these Al–Mg alloys, a mixture of nanoscale with submicron and micron size grains provides optimum values of strength and ductility.

Cryomilled Zinc with Distributions of Nanoscale Grain Sizes.

Zinc metal that was ball milled at liquid nitrogen temperature—cryomilled—was found to exhibit an oscillatory variation of the hardness with milling time [20]. Transmission electron microscopy showed that large variations in the dislocation density and grain-size distribution occurred during cryomilling. The observations suggest that recrystallization takes place in larger grains when the dislocation density due to strain-hardening reaches a critical level. A reaction-rate model was developed which accounts for the dynamic recrystallization effect and the observed oscillations in hardness. In addition to hardness data, the miniaturized disk bend test (MDBT) was used to obtain additional mechanical property data on these cryomilled Zn samples [21]. Yield stress data extracted from the MDBT tests showed the same modulated oscillatory variation with cryomilling time as did hardness. Comparison between the hardness data and 3x the yield strength data gave good agreement, indicating that the MDBT test which represents biaxial tension can provide consistent yield strength data. The normalized displacement in the MDBT test is proportional to the equibiaxial strain. It was found that maxima in yield strength with milling time corresponded to minima in "% strain" (normalized displacement). These data, yield strength vs. "% strain" are plotted against cryomilling time in Figure 3. This clearly illustrates the inverse nature of the yield strength and ductility data. At the cryomilling time of 4h it is noted that the strength value has reached a level comparable to those for longer milling times, that is about 180 MPa, while the "% strain" value is at a maximum. This can be contrasted to the data for 2h of cryomilling where the strength is much lower, about 130 MPa, and the ductility is comparable. In addition the 12h milled sample has about the same yield strength value but much lower ductility. Thus, the sample cryomilled 4h represents another example of an optimum combination of strength and ductility. The microstructure of this sample with the optimum mechanical properties was nanoscale with a grain size distribution such that about 30% of the volume fraction of grains had sizes >50 nm. The largest grain sizes were about 150 nm. This combination of small, <50 nm, and larger grains provided the best combination of properties.

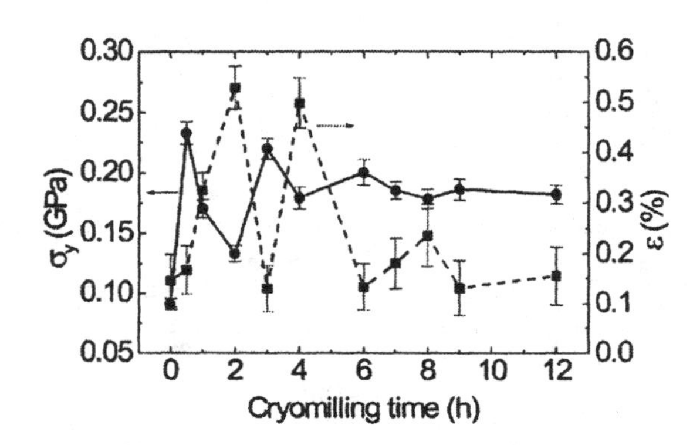

Figure 3. Yield strength and "% strain" vs. cryomilling time for Zn [21].

Strain Hardening and Ductility

It is clear that in order to attain significant uniform elongations in otherwise ductile metals, substantial strain hardening is required. Some examples of the importance of strain hardening on the ductility of nanostructured or ultrafine grained metals will be given below.

The nanocrystalline electrodeposited Co described earlier [8] exhibits strain hardening behavior comparable to that of cgs Co. At a strain rate of 5 x 10^{-4} the strain hardening exponents were essentially the same at n = 0.2.

Lu et al [22] reported an abnormal strain rate effect for electrodeposited nanocrystalline Cu in that, unlike cgs Cu, the fracture strain increased significantly with increasing strain rates. There was substantial strain hardening observed which may have increased slightly with increasing strain rate, but was fairly constant. It should be pointed out again that these samples had nanocrystallites about 20 nm in size but with low angle grain boundaries.

Zhang et al [23, 24] prepared nanocrystalline, submicron grain size Zn by ball milling at room temperature. In this case due to the good ductility of Zn at room temperature the ball milled powder consolidated into spherical balls during milling and showed homogeneous deformation throughout the spheres. Disks pressed from these spheres had theoretical density and were suitable for tensile specimen preparation. The grain sizes and grain size distributions were larger than in cryomilled Zn, ranging from average grain sizes of 240 nm for 3h milling time to 23 nm for 25h. The samples milled for 3h with 240 nm average grain size exhibited higher strength, higher strain hardening, and larger strain to failure with increasing strain rate [24]. A sample with 60 nm average grain size exhibited strain hardening and percent elongation to failure of about 50% while the sample with 23 nm grain size, with similar strength levels, had essentially no strain hardening, no uniform elongation, but had percent elongation to failure of about 20%. The 23 nm average grain size sample had a grain size distribution with no grains larger than 100 nm while the 60 nm average grain size sample had a grain size distribution with some grains up to 200 nm in size. Both samples also had grains at the very smallest nanoscale, that is less than 10 nm in size. In this case, a larger grain size distribution promoted strain hardening and with it a better combination of strength and ductility.

Concluding Remarks

While the optimization of strength and ductility was the subject of this paper, in many applications it may be toughness, not ductility per se, that is important. Toughness is defined several ways, including the area under the stress-strain curve. In the extreme case of metallic glasses, which exhibit essentially no macroscopic plastic strain in tension, by this definition their toughness would be negligible. However, using standard crack extension toughness tests in bulk metallic glasses large toughness values of 55 $MPa\sqrt{m}$ have been measured [25]. There is little fracture toughness data available on nanocrystalline materials and this is clearly an area for more study.

Acknowledgements

The authors' research on this topic is supported by the Department of Energy under grant number DE-FG02-02ER46003.

References

1. J. R. Weertman, "Mechanical Behavior of Nanocrystalline Metals," *Nanostructured Materials: Processing, Properties and Applications*, ed. C. C. Koch (Norwich, New York: William Andrew Publishing, 2002), 397.

2. T.D. Shen and C.C. Koch, Acta Mater., 44 (1996) 753.

3. C.C. Koch, D.G. Morris, K. Lu, and A. Inoue, MRS Bulletin, 24 (1999) 54.

4. D.G. Morris, "Mechanical Behavior of Nanostructured Materials" (Uetikon-Zurich, Switzerland, Trans Tech Publications, 1998).

5. R. Birringer, H. Gleiter, H.P. Klein, and P. Marquart, Phys. Lett., 102A (1984) 365.

6. C.C. Koch, NanoStructured Mater, 2 (1993) 109.

7. J.R. Groza, "Nanocrystalline Powder Consolidation Methods," *Nanostructured Materials: Processing, Properties, and Applications*, ed. C. C. Koch (Norwich, NY: William Andrew Publishing, 2002), 115.

8. A. A Karimpoor, U. Erb, , K. T. Aust Z. Wang, and G. Palumbo, Mater. Sci. Forum 386-388 (2002) 415.

9. Y. Wang, M. Chen, F. Zhu, and E. Ma, Nature, 419 (2002) 912.

10. M. Legros, B. R. Elliott, M. N. Rittner, J. R. Weertman, and K. J. Hemker, Phil Mag A, 80 (2000) 1017.

11. L. Lu, L. B. Wang, B. Z. Ding, and K. Lu, J. Mater. Res., 15 (2000) 270.

12. G.W. Nieman, J.R. Weertman, and R.W. Siegel, NanoStructured Mater., 1 (1992) 185.

13. B. Gunther, A. Baalman and H. Weiss, MRS Symp. Proc. 195 (1990) 611.

14. P.G. Sanders, J.A. Eastman, and J.R. Weertman, Acta Mater., 45 (1997) 4019.

15. R. Z. Valiev, I. V. Alexandrov, Y. T. Zhu, and T. C. Lowe, J. Mater. Res., 17 (2002) 5.

16. V.L. Tellkamp, A. Melmed, and E.J. Lavernia, Metall. Mater. Trans. A, 32A (2001) 2335.

17. D. Witkin, Z. Lee, R. Rodriguez, S. Nutt, and E. J. Lavernia, Scripta Mater., 49 (2003) 297.

18. J. R. Davis, P. Allen, S. R. Lampman, T. B. Zorc, eds. Metals Handbook, 10th edition, Vol. 2, Properties and Selection: Non-Ferrous Alloys and Special-Purpose Materials (Metals Park, OH; ASM International, 1990) 93.

19. M. V. Markushev, C. C. Bampton, M. Yu. Murashkin, D. A. Hardwick, Mater. Sci. Engr. A, 234-236 (1997) 927.

20. X. Zhang, H. Wang, R. O. Scattergood, J. Narayan, and C. C. Koch, Acta Mater., 50 (2002) 3995.

21. X. Zhang, H. Wang, R. O. Scattergood, J. Narayan, and C. C. Koch, Acta Mater., 50 (2002) 3527.

22. L. Lu, S.X. Li, and K. Lu, Scripta Mater., 45 (2001) 1163.

23. X. Zhang, H. Wang, R. O. Scattergood, J. Narayan, C. C. Koch, A. V. Sergueeva, and A. K. Mukherjee, Appl. Phys. Lett., 81 (2002) 823.

24. X. Zhang, H. Wang, R. O. Scattergood, J. Narayan, C. C. Koch, A. V. Sergueeva, and A. K. Mukherjee, Acta Mater., 50 (2002) 4823.

25. C.J. Gilbert, R.O. Ritchie, and W.L. Johnson, Appl. Phys. Lett. 71 (1997) 476.

Processing and Properties of Structural Nanomaterials
Edited by Leon L. Shaw, C. Suryanarayana and Rajiv S. Mishra
TMS (The Minerals, Metals & Materials Society), 2003

Dependence of Microhardness on Internal Strains of a Nanocrystalline Al Alloy

L. Shaw[1], J. Villegas[1], H. Luo[1] and D. Miracle[2]

[1] Department of Metallurgy and Materials Engineering
University of Connecticut, Storrs, CT, USA
[2] Air Force Research Laboratory,
Materials and Manufacturing Directorate, Wright-Patterson AFB, OH, USA

Keywords: Nanostructured materials, Aluminum alloy, Internal strains, Microhardness

Abstract

Effects of internal strains on the microhardness of nanostructured materials were investigated using an $Al_{93}Fe_3Ti_2Cr_2$ alloy prepared via mechanical alloying starting from elemental powders. It was found that the hardness of the nanostructured Al alloy decreased with decreasing internal strain at a given grain size, even though dislocation activity within the grain interior is absent in all cases.

Introduction

Nanocrystalline (nc) materials (with grain sizes < 100 nm) possess many interesting properties including superior strength, hardness, wear resistance and superplasticity [1-6]. Among all the properties investigated, the hardness of nc materials has drawn the most attention because of its relatively simple requirement on the sample preparation and its applicability to small samples. The hardness of nc materials at the higher end of the nanograin size is found to follow the Hall-Petch (HP) relation [7-9]:

$$H = H_0 + k_H d^{-1/2} \tag{1}$$

where H is the hardness, H_0 is a constant related to the frictional stress resisting the motion of gliding dislocations or the internal back stress, k_H is the Hall-Petch slope, and d is the average grain size. This semi-empirical equation breaks down at a critical grain size below which an inverse HP relation is observed [6,9,10]. The inverse HP relation has been attributed to a variety of possible softening mechanisms that could occur at very small grain sizes [9-15]. Although the grain size dependence of the hardness of nc materials has been established, little attention is paid to the dependence of the hardness on other materials parameters. The purpose of this paper is to elucidate the effect of internal strains on the hardness of nc materials prepared via mechanical alloying (MA). As will be shown below, the hardness of MA-processed nc materials is not just determined by its grain size, but also influenced by the internal strain.

Experimental Procedure

Crystalline elemental powders were used to prepare an Al alloy with a nominal composition of $Al_{93}Fe_3Ti_2Cr_2$. The aluminum powder used had a purity of 99.5% with a mean particle size of 70 μm, while the corresponding values for iron, chromium and titanium powders used were 99.0%, 98.5% and

99.5% as well as 50, 30 and 30 μm, respectively. Mechanical alloying was performed in a Spex 8000 laboratory mill using stainless steel balls and vial with a ball-to-powder weight ratio of 5:1. The vial loaded with powders of the desired average composition was sealed in a glove box under an argon atmosphere, and milling was conducted in a stationary argon atmosphere. The duration of the continuous milling process was 16 hours.

No process control agents were used in milling. Other studies [16-18] have shown that excessive cold welding of Al particles occurs if ball milling of Al alloys is conducted without a process control agent. This excessive cold welding behavior was utilized here to form large $Al_{93}Fe_3Ti_2Cr_2$ alloy particles so that the microhardness of these powder particles in the as-milled condition could be determined using Vickers indentation. It was found in this study that with no process control agents, both large (1 – 5 mm) and small (0.2 – 0.5 mm) particles were produced. The large particles so-obtained were used for the measurement of their microhardness in the as-milled condition once these particles were mounted in an epoxy and polished to provide a flat surface. Some of the large particles obtained from the milling process were further subjected to a heat treatment before the microhardness measurement. The heat treatment comprised heating the milled alloy particles to a desired temperature at a heating rate of 10^0C/min and holding at that temperature for 1 hour before cooling to room temperature with a cooling rate of 50^0C/min. These heat-treated powder particles were then mounted and polished so that their microhardness could be measured. Each hardness value reported was the average of fifteen measurements.

Extensive characterization of the microstructure and phase transformation behavior of the MA-processed $Al_{93}Fe_3Ti_2Cr_2$ alloy was performed previously using transmission electron microscopy (TEM), scanning electron microscopy (SEM), energy dispersive spectrometry (EDS), X-ray diffraction (XRD), differential scanning calorimetry (DSC), and thermogravimetric analysis (TGA) [19]. Interested readers are referred to the previous study for experimental details and results. Here, only the two most critical analytical techniques, TEM and XRD, closely related to this study are described briefly. TEM analyses were performed using a TEM (Philips CM200 FEG) operating at 200kV. Both bright and dark field image techniques coupled with selected area diffraction (SAD), convergent beam electron diffraction (CBED) and EDS with an electron beam of 3.5 nm were used to characterize particles.

For XRD, a Rigaku RU-200B series diffractor with $CuK\alpha$ radiation was utilized. The XRD peak broadening was attributed to the refinement of crystals, introduction of internal strains and instrumental effects. The correction for instrumental broadening was done using an annealed micron-sized Al powder standard and assuming a Gaussian peak shape for both the instrumental and the sample peak functions. Thus, the breadth at half maximum intensity of the sample peak, $\beta_g(2\theta)$, which excluded instrumental broadening, was calculated with the aid of [20]

$$\beta_g^2(2\theta) = \beta_h^2(2\theta) - \beta_f^2(2\theta) \qquad (2)$$

where $\beta_h(2\theta)$ is the half-maximum breadth measured from the sample and $\beta_f(2\theta)$ is the one from the standard. The (111) and (200) reflections of fcc-Al were used to estimate the average grain size, D, of fcc-Al through the Scherrer formula [21]

$$\beta_g(2\theta) = \frac{0.9\lambda}{D\cos(\theta)} \qquad (3)$$

where λ is the wave length of the X-ray radiation and θ is the Bragg angle. The effective internal strain, 2ξ, within fcc-Al was estimated using the (220) and (311) reflections of fcc-Al with the aid of the Stokes and Wilson formula [22]

$$\beta_g(2\theta) = 2\xi \tan\theta \quad (4)$$

In this study, the Hall-Williamson plots [i.e. $\beta_g(2\theta)\cdot\cos\theta$ versus $\sin\theta$] of fcc-Al in the as-milled and annealed conditions were upward-concave curves, similar to those obtained by Williamson and Hall working on filed aluminum powder [23]. However, the determination of the grain size and internal strains by their method [23,24] produced inconsistent results as the annealing temperature varied, and therefore was not used here. The inconsistence was likely due to the overlap of Al(200), Al(220) and Al(222) reflections with peaks of the remaining crystalline Fe and Cr and Al_3Ti that formed during annealing. Since the XRD broadening at low angles is dominated by small grain sizes and at high angles by internal strains [23], the low angle reflections (111 and 200 of fcc-Al) were used to estimate the average grain size using the Scherrer formula, and the high angle reflections (220 and 311) to estimate the internal strain via the Stokes and Wilson formula, as described above. The Al(222) reflection was substantially distorted by Fe(211) and Cr(211) reflections and thus not used in this study. The estimations so obtained are bound to result in finer grain sizes and higher internal strains than what they are in fcc-Al. Nevertheless, as will be shown in this study, the TEM analysis essentially substantiates the grain sizes obtained from the Scherrer formula, suggesting that the error introduced by the separate estimation of grain sizes and internal strains via different X-ray reflections is small in the present study.

Results and Discussion

Figure 1 shows several X-ray diffraction patterns of the MA-processed $Al_{93}Fe_3Ti_2Cr_2$ alloy in the as-milled condition and after exposing for one hour to various temperatures as indicated. The as-milled alloy consists of a fcc-Al solid solution, as indicated by the peak shifting of the MA-processed Al alloy compared with the XRD pattern of pure aluminum. A previous study [19] using the EDS analysis of TEM foils has established that this solid solution is supersaturated with Fe, Cr and Ti solutes. The average grain size of fcc-Al in the as-milled condition estimated using eq. (3) is 14 nm, whereas the effective internal strain, 2ζ, estimated using eq. (4) is 1.8%. Upon heating, no new phases are formed until 330^{0}C is reached. At 330^{0}C orthorhombic-Al_6Fe is identified and at 450^{0}C an additional new phase, tetragonal-Al_3Ti, is present.

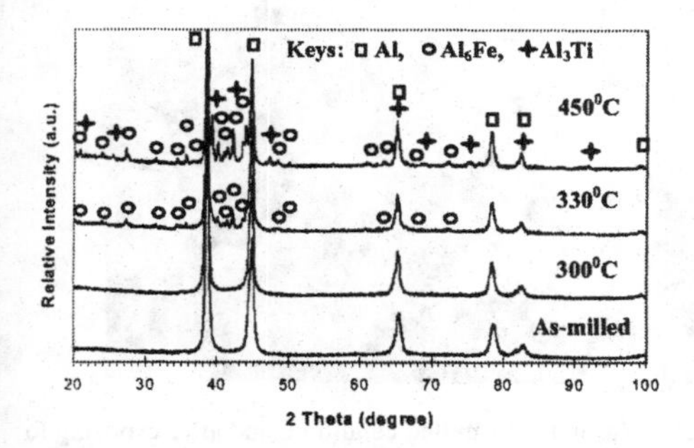

Fig. 1 XRD patterns of the MA-processed Al alloy as a function of the annealing temperature.

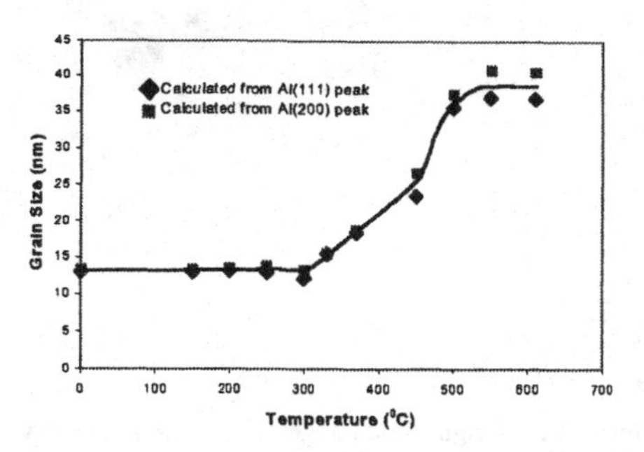

Fig. 2 Grain size of fcc-Al in the MA-processed alloy as a function of the annealing temperature, estimated using eq. (3).

The grain size of fcc-Al as a function of temperature for one-hour anneal estimated using eq. (3) is shown in Figure 2. The grain size of fcc-Al changes little below 300^0C, but starts to increase gradually above 300^0C, exhibits a relative large jump at about 500^0C, and finally stabilizes at about 40 nm at temperatures above 500^0C. The average grain size and its variation with the annealing temperature derived from XRD are confirmed by the TEM analysis. Shown in Figure 3 are TEM images of the MA-processed Al alloy at conditions critical to this study. Note that the grain size of the as-milled alloy ranges from 6 to 45 nm, substantiating the finding of the grain size of 14 nm estimated from the XRD analysis. Furthermore, there is no grain size change up to 300^0C, a result consistent with the XRD analysis. At 330^0C a discernable grain size increase is observed and nano-grains continue to grow above this temperature, again in good agreement with the XRD analysis. Finally, it is noted that dislocation activity is absent in the MA-processed Al alloy even though its nano-grain microstructure is achieved by severe plastic deformation. This phenomenon is consistent with many other studies [7,8,25,26] that have shown the absence of dislocation activity when the grain size is smaller than 100 nm.

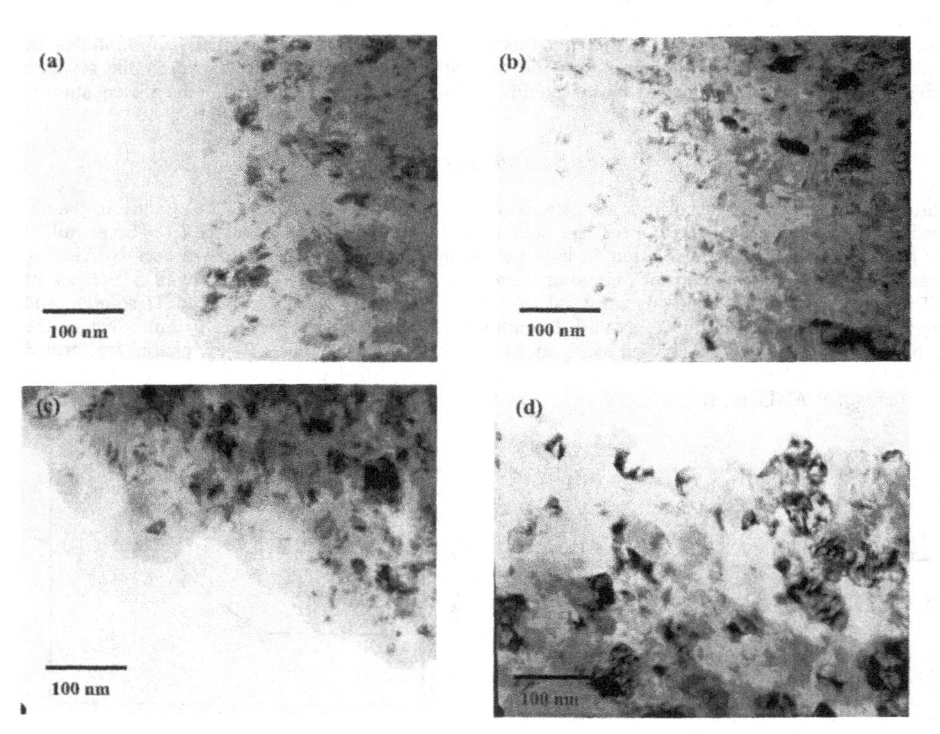

Fig 3. TEM bright-field images of the MA-processed Al alloy (a) in the as-milled condition, and after exposing to (b) 300^0C, (c) 330^0C and (d) 450^0C for one hour.

Shown in Figure 4 is the effective internal strain within fcc-Al as a function of the annealing temperature estimated using eq. (4). The internal strain decreases little below 150^0C, but decrease continuously above this temperature, and display a relatively large reduction between 450 and 500°C before it stabilizes above 500°C. Based on Figures 2 to 4, it can be concluded that the onset temperature of grain growth for the MA-processed $Al_{93}Fe_3Ti_2Cr_2$ alloy is 330°C, whereas the release of internal strains starts about 150°C. As such, between 150 and 300°C the MA-processed $Al_{93}Fe_3Ti_2Cr_2$ alloy has the same grain size but different internal strains. In addition, a previous study [19] using modulated DSC indicates that the heat released due to the internal strain reduction between 150 and 300°C is about 2.6% of the enthalpy of fusion of pure Al.

Figure 5 shows the Vickers hardness of the MA-processed alloy in the as-milled condition and after exposing for one hour to different temperatures. The hardness decreases with increasing the annealing temperature even before grain growth. Further, a relatively large drop in hardness occurs at 330°C, coinciding with the onset of grain growth. The decrease in hardness at and below 300°C is attributed to the release of internal strains introduced in the MA process since there are no grain growth and intermetallic precipitation at and below this temperature. The dependence of yield strength on internal strains was also observed previously in fine-grained Cu (d = 170 to 210 nm) prepared via torsion straining and equal-channel angular pressing [27]. In this previous study, fine-grained Cu was annealed at 473K for 3 minutes, which decreased internal stresses but induced no grain growth; however, the yield strength dropped substantially (~ 75%).

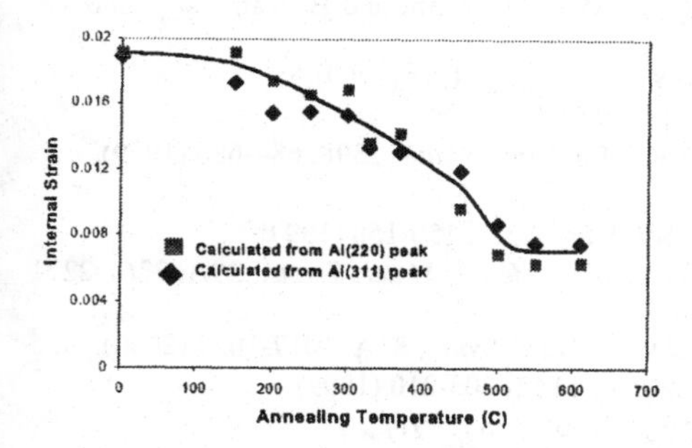

Fig. 4 Internal strain of fcc-Al in the MA-processed Al alloy as a function of the annealing temperature.

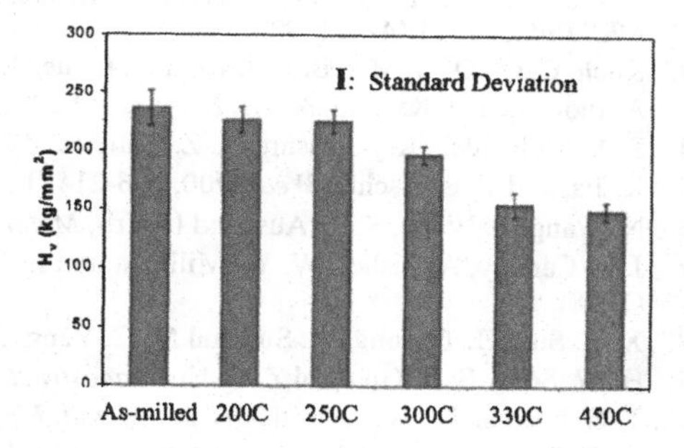

Fig. 5 Room-temperature Vickers hardness of the MA-processed Al alloy as a function of the annealing temperature as indicated.

The release of internal strains in the present alloy is most likely due to the relaxation of lattice bending since no changes in dislocations and solute concentration within the grain interior have been identified [19] and excess vacancies in fcc-Al normally migrate out of the lattice at room temperature [28]. Lattice bending has been found in the interior of fine grains formed via severe plastic deformation [29-31]. Many of the fine grains with sizes of ~ 0.5 μm are free of dislocations in the interior of the grains, yet lattice bending is present which leads to high internal stresses [30,31]. The presence of lattice bending is due to the non-equilibrium state of the strain-induced grain boundaries where excess dislocations are inhomogeneously distributed and cannot be fully relaxed [29-31]. The relaxation of lattice bending leads to less internal stresses which in turn results in a lower resistance to plastic deformation and therefore a lower Vickers hardness.

Concluding Remarks

The present study indicates that the hardness of the nanostructured $Al_{93}Fe_3Ti_2Cr_2$ alloy prepared via mechanical alloying is not just determined by its grain size, but also influenced by its internal strain. Hardness decreases with decreasing the internal strain at a given grain size. As such, caution should be exercised when comparing hardnesses and strengths of nc materials with the same grain size. It is expected that the effect of internal strains on hardness and strength is especially potent for nc materials prepared via techniques associated with severe plastic deformation because these techniques introduce large internal strains to nc materials.

Acknowledgements – LS is grateful to the University of Connecticut for granting a sabbatical leave to conduct research at the Air Force Research Laboratory. The partial support from the Materials and Manufacturing Directorate of the Laboratory Director's Fund of the Air Force Research Laboratory is greatly appreciated. The authors also appreciate the help provided by Dr. Robert Wheeler, UES, Inc. in initiating the TEM experiment.

References

1. J. R. Weertman, D. Farkas, K. Hemker, H. Kung, M. Mayo, R. Mitra and H. Van Swygenhoven, *MRS Bull.*, 24 [2] 44-50 (1999).
2. Koch, C. C., D. G. Morris, K. Lu and A. Inoue, *MRS Bull.*, 24 [2] 54-58 (1999).
3. A. Inoue and H. Kimura, *Mater. Sci. Eng.*, A286 [1] 1-10 (2000).
4. S. X. McFadden, R. S. Mishra, R. Z. Valiev and A. K. Mukherjee, *Nature*, 398, 684-686 (1999).
5. K. Jia, and T. E. Fischer, *Wear*, 200, 206-214 (1996).
6. N. Wang, Z. Wang, K. T. Aust and U. Erb, *Mater. Sci. Eng.*, A237, 150-158 (1997).
7. J. E. Carsley, A. Fisher, W. W. Milligan and E. C. Aifantis, *Metall. Mater. Trans.*, 29A, 2261-2271 (1998).
8. X. K. Sun, H. T. Cong, M. Sun and M. C. Yang, *Metall. Mater. Trans.*, 31A, 1017-1024 (2000).
9. H. W. Song, S. R. Guo and Z. Q. Hu, *Nanostruct. Mater.*, 11 [2] 203-210 (1999).
10. T. G. Nieh and J. Wadsworth, *Scripta Metall. Mater.*, 25, 955-958 (1991).
11. R. O. Scattergood and C. C. Koch, *Scripta Metall. Mater.*, 27, 1195-1200 (1992).
12. N. Wang, Z. Wang, K. T. Aust and U. Erb, *Acta Metall. Mater.*, 43 [2] 519-528 (1995).
13. S. Yip, *Nature,* 391 [5] 532-533 (1999).
14. J. Schiotz, F. D. Di Tolla and K. W. Jacobsen, *Nature,* 391 [5] 561-563 (1999).
15. J. Lian, B. Baudelet and A. A. Nazarov, *Mater. Sci. Eng.*, A172, 23-29 (1993).
16. J. S. Benjamin and M. J. Bomford, *Metall. Trans.*, 8A, 1301-1305 (1977).
17. R. F. Singer, W. C. Oliver and W. D. Nix, *Metall. Trans.*, 11A, 1895-1901 (1980).
18. P. S. Gilman and W. D. Nix, *Metall. Trans.*, 12A, 813-824 (1981).
19. L. Shaw, J. Villegas, H. Luo and D. Miracle, Acta Mater., 51 [9] 2647-2663 (2003).
20. B. E. Warren, X-Ray Diffraction, Dover Publications, Inc., New York, 1990.
21. H. P. Klug and L. E. Alexander, X-Ray Diffraction Procedures for Polycrystalline and Amorphous Materials, John Wiley & Sons, Inc., London, 1954, pp. 491-494.
22. A. R. Stokes and A. J. C. Wilson, *Proc. Phys. Soc. London*, 56, 174-181 (1944).
23. G. K. Williamson and W. H. Hall, *Acta Metall.*, 1, 22-31 (1953).
24. W. H. Hall, *Proc. Phys. Soc. London,* A62, 741-743 (1949).
25. D. G. Morris and M. A. Morris, *Acta Metall. Mater.*, 39 [8] 1763-1770 (1991).

26. R. W. Hayes, R. Rodriguez and E. J. Lavernia, *Acta Mater.*, 49, 4055-4068 (2001).
27. V. Y. Gertsman, R. Z. Valiev, N. A. Akhmadeev and O. V. Mishin, Materials Science Forun, 225-227, 739-744 (1996).
28. R. W. Cahn, in Physical Metallurgy, Eds. R. W. Cahn, North-Holland Publishing Company, Amsterdam, Netherlands, 1965, pp. 925-987.
29. A. Belyakov, T. Sakai, H. Miura and R. Kaibyshev, *Phil. Mag. Lett.*, 80 [11] 711-718 (2000).
30. A. Belyakov, T. Sakai, H. Miura and R. Kaibyshev, *Scripta Mater.*, 42, 319-325 (2000).
31. A. Belyakov, T. Sakai, H. Miura and K. Tsuzaki, *Phil. Mag. A*, 81 [11] 2629-2643 (2001).

Processing and Properties of Structural Nanomaterials
Edited by Leon L. Shaw, C. Suryanarayana and Rajiv S. Mishra
TMS (The Minerals, Metals & Materials Society), 2003

Experiments and Modeling of the Surface Nanocrystallization and Hardening (SNH) Process

J. Villegas[1], K. Dai[1], L. Shaw[1], P. Liaw[2]

[1]Department of Metallurgy and Materials Engineering, Institute of Materials Science, University of Connecticut, Storrs, CT 06269
[2]Department of Materials Science and Engineering, University of Tennessee, Knoxville, TN 37996

Keywords: Surface nanocrystallization and hardening, Nickel alloys, Severe plastic deformation, Nanostructured materials.

Abstract

A surface nanocrystallization and hardening (SNH) process is developed to process bulk metallic components with a nanocrystalline surface and coarse-grained interior. The SNH process entails impacting metallic components repeatedly with high-energy balls under a controlled atmosphere. The efficacy of the SNH process is demonstrated using a nickel-based Hastelloy C-2000® alloy. The hardening layer and grain size refinement are evaluated as a function of the processing time. The result suggests that surface strengthening of C-2000® alloy obtained via the SNH treatment (about 150% increase) is mainly due to work hardening, whereas the contribution of grain refinement to strengthening, if any, is small. Such a conclusion is also consistent with the result obtained from a three-dimensional finite element model developed to simulate the SNH process.

Introduction

Severe plastic deformation (SPD) has been known to be one of the effective methods for making nanostructured materials. Examples of this kind include mechanical alloying and mechanical attrition to prepare nanostructured powders [1-3]. Recently, such material processing strategy has been extended to make bulk materials with a nanocrystalline surface and coarse-grained interior. There are several variants of such surface SPD-based technique, including ultrasonic shot peening (USSP) [4,5], high-energy shot peening (HESP) [6], surface mechanical attrition (SMA) [7,8] and surface nanocrystallization and hardening (SNH) [9]. The detail of nanocrystallization mechanisms in these surface SPD-based processes has been found to depend on the stacking fault energy (SFE) of the material [8]. For materials with high SFEs, grain refinement is dominated by dislocation activities, entailing generation of high dislocation densities, formation of subgrains, and evolution of subgrain boundaries to highly misoriented grain boundaries [5,7]. This surface nanocrystallization mechanism is very similar to that identified in the grain refinement process during mechanical alloying and attrition [1,10,11].

Surface strengthening has been achieved in the aforementioned surface SPD-based techniques [4,6]. Furthermore, the surface strengthening achieved can lead to improvements in the tensile yield strength of bulk plates with minimum degradation of ductility and toughness [6]. It has been suggested that surface strengthening may be due to work-hardening and surface nanostructure, of which the contributions cannot be separated [6]. In this study, a nickel-based Hastelloy C-2000® alloy has been used as a model material to investigate the strengthening behavior and mechanism in surface SPD-based processes. In order to assist the analysis of experimental results, a finite element model is developed to evaluate work hardening contribution. As will be shown below, both experiments and modeling suggest that the surface strengthening achieved in surface SPD-based processes is mainly due to work hardening, whereas the contribution from surface nanostructure, if any, is small.

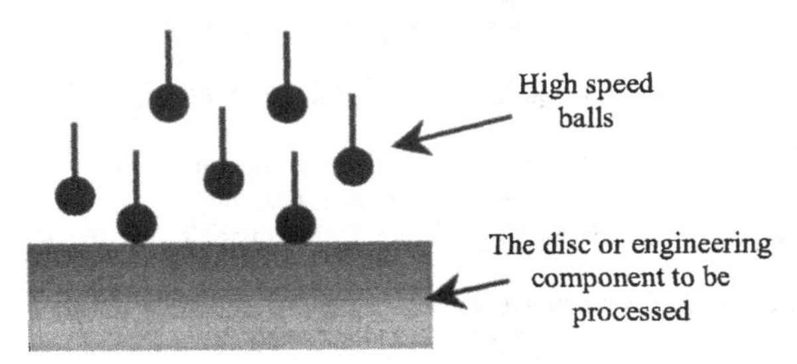

Figure 1. Schematic of the Surface Nanocrystallization and Hardening (SNH) process.

Experimental Procedure

The SNH process, which entails impacting metallic components with high-energy balls repeatedly under a controlled atmosphere is shown schematically in Fig. 1. The discs to be processed were cut out from a nickel-based Hastelloy C-2000® alloy plate with a nominal thickness of 3.25 mm. These discs 49 mm in diameter were first cleaned with acetone, then with ethanol and finally processed with the SNH treatment. In each SNH treatment, a C-2000® alloy disc was held at one end of a cylindrical container with internal dimensions of 38.0 mm in diameter and 57.0 mm in depth made of a hardened steel. Five WC/Co balls (with a composition of 94%WC – 6%Co) 7.9 mm in diameter were loaded into the container to provide the desired impact on the surface of the C-2000® alloy disc. The container was filled with argon before the SNH treatment. The high velocity of the balls in the SNH process (~ 5 m/s) was achieved by shaking the container three dimensionally using a Spex 8000 Mill. Such 3D shaking provided kinetic energy to the balls and generated a complex pattern of motion of the balls inside the container. Various processing times, ranging from 5 minutes to 180 minutes, were applied and no temperature control was employed during the SNH process.

The cross section of the SNH-processed C-2000® alloy was examined using both optical and scanning electron microscopy (SEM) in order to follow the microstructural evolution of the processed samples. X-ray diffraction (XRD) patterns of the processed surfaces were also collected on a Bruker AXS D5005 X-Ray Diffractometer with Cu$K_{\alpha 1}$ radiation to monitor phase changes and evolution of grain size during the SNH treatment. The estimation of grain size was carried out based on broadening of the Ni (100) reflection using the Scherrer formula [12]

$$\beta_g(2\theta)=\frac{0.9\lambda}{D\cos(\theta)} \quad (1)$$

where D was the average crystallite size, λ the wave length of the X-ray radiation, θ the Bragg angle, and $\beta_g(2\theta)$ the breadth at the half-maximum intensity of the sample peak after excluding instrumental broadening. The contribution of internal strains to the broadening was neglected in the estimation. Vickers microhardness tests were performed on the processed samples using a LECO DM-400FT hardness tester with a 200-gf load and a dwell time of 15 seconds in order to establish the microhardness dependence on the impacting time and the location measured from the impacted surface.

Description of Finite Element Modeling

A three-dimensional (3D) dynamic finite element model (FEM) based on an explicit integration method was built to develop a fundamental understanding of evolution of the plastic deformation zone under multiple impact conditions. The model entailed a solid WC/Co ball with a diameter of 7.9 mm impacting the center of a C-2000® plate with dimensions of 20 × 20 × 3.2 mm^3 at a speed of 5 m/s (Fig. 2). The plate was simulated as an elasto-plastic material with piecewise linear isotropic hardening behavior, whereas the ball was modeled as a rigid material because it was much harder than the plate. The Young's modulus, density and Poisson's ratio were 600 GPa, 14.5 g/cm^3 and 0.3 for the WC/Co ball, and 200 GPa, 8.5 g/cm^3 and 0.3 for the Hastelloy C-2000® plate, respectively. Furthermore, based on the tensile tests conducted with a strain rate of 10^{-4} s^{-1}, the initial yield strength of Hastelloy C-2000® was chosen to be 381 MPa and its plastic true stress – strain curve was described by a power law hardening behavior:

$$\sigma = 1990\varepsilon^{0.7}\ \text{MPa} \tag{2}$$

All the nodes are free to move in X, Y and Z directions except that the nodes at the bottom surface of the specimen are fixed in its thickness direction, i.e., the Z direction in Fig. 2. To ensure the calculation accuracy, the mesh in the region being impacted by the ball is much finer than that in the other regions. The dimension of the element in the impacted region is 0.333 × 0.333 × 0.15 mm^3. The numerical simulation is carried out using the LS-DYNA code.

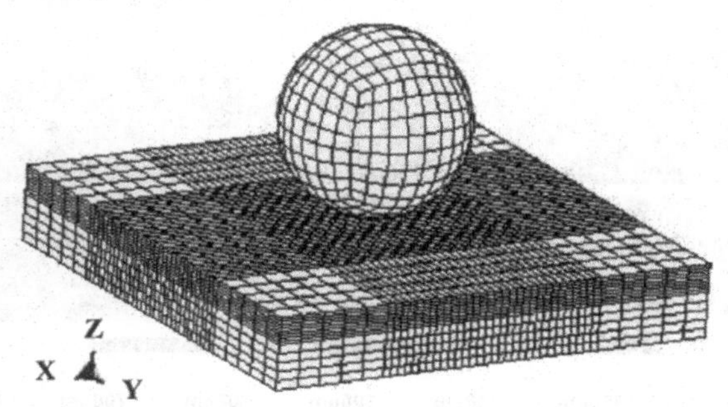

Figure 2. Finite element model describing multiple impacts at the same location on a Hastelloy C-2000® plate by a WC/Co ball.

Experimental Results and Discussion

Shown in Fig. 3 are Vickers microhardness data of C-2000® alloy specimens processed with five WC/Co balls 7.9mm in diameter. It can be seen that microhardness changes little with 5-minute processing. However, as the processing time exceeds 5 minutes, surface hardening becomes discernible. With 30-minute processing the microhardness at the very surface has been increased by 150% in comparison with the untreated surface. Furthermore, impacting beyond 30-minute appears to give no significant additional hardening because the overall hardness profile of the sample processed for 180 minutes overlaps with that processed for 30 minutes. The thickness of the surface hardening layer is about 700 μm for the SNH processing condition employed.

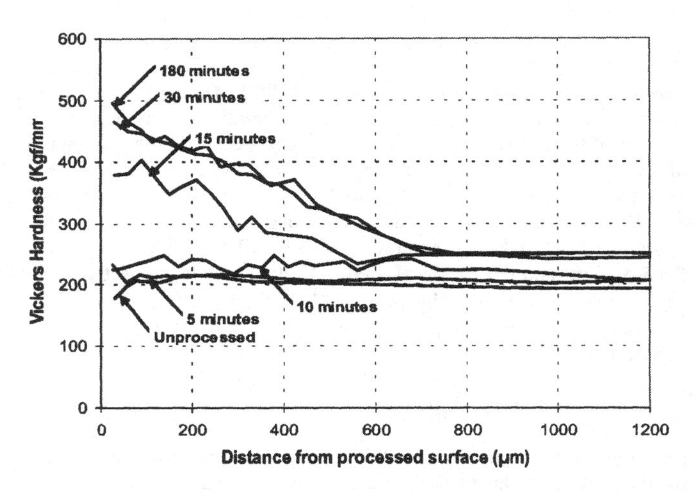

Figure 3: Hardness evolution as a function of processing time for a Hastelloy C-2000® alloy discs SNH-treated with five 7.9-mm balls.

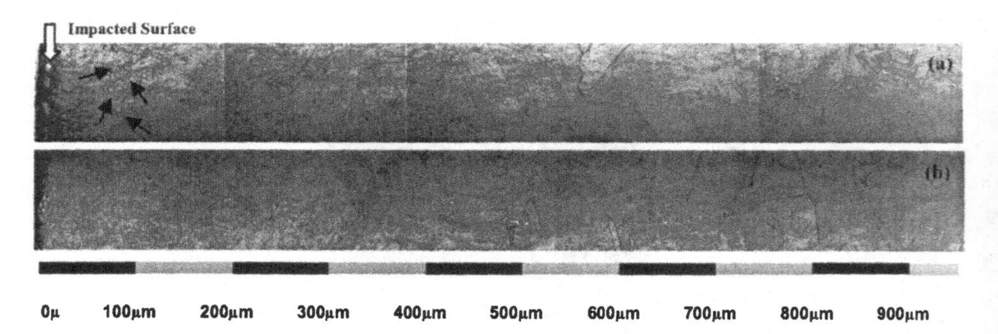

Figure 4: Optical micrographs of cross-sectioned, SNH-treated, Hastelloy C-2000® alloy discs showing the extent and type of microstructural changes from the treated surface to the bulk: (a) impacted for 70 minutes and (b) impacted for 180 minutes.

Figure 4 presents optical micrographs of the samples SNH-processed for 70 and 180 minutes. Note that proper electrochemical etching after polishing the cross section of the samples can reveal slip lines, a high magnification of which is shown in Fig. 5. As can be seen from Fig. 4, the density of slip lines increases gradually as the location approaches the impacted surface. The density of slip lines at the very surface becomes so high that individual slip lines can no longer be distinguished. This is especially true for the 180-minute processed sample the very surface of which not only shows no distinguishable individual slip lines but also contains no visible grain boundaries. In contrast, the very surface of the 70-

minute processed sample still contains visible grain boundaries, as highlighted by dark arrows. A comparison between Figs. 3 and 4 inevitably leads to a conclusion that the additional microstructural evolution beyond 30-minute impacts, such as the increased slip line density, provides little strengthening.

Based on the XRD peak broadening of the Ni (100) reflection, the grain size at the very surface region as a function of the impacting time has been estimated using eq. (1). The result is summarized in Table 1. Note that the thickness of the surface layer that contributes to the X-ray diffraction is estimated to be less than 9 μm in the present study. Thus, the measurement presented in Table 1 only reflects the average grain size from a top surface layer less than 9 μm thick of the samples. As expected, nanograins have been achieved via the SNH process, and in general, the grain size decreases with increasing the processing time. The continuous grain refinement beyond 20-minute impacting is consistent with the phenomenon that the slip line density in the very surface region increases continuously beyond 70-minute impacting, as revealed in Fig. 4. Both the grain refinement and the increase in the slip line density indicate continued plastic deformation in the very surface region even after prolonged impacts. However, strengthening is saturated at about 30-minute processing beyond which additional plastic deformation and grain refinement contribute only a small amount of strengthening, as evidenced by the small increase in the microhardness in the very surface region (Fig. 3). Thus, it can be concluded that surface strengthening of C-2000® alloy obtained from the SNH treatment (about 150% increase) is mainly due to work hardening, whereas the contribution of grain refinement to strengthening, if any, is small.

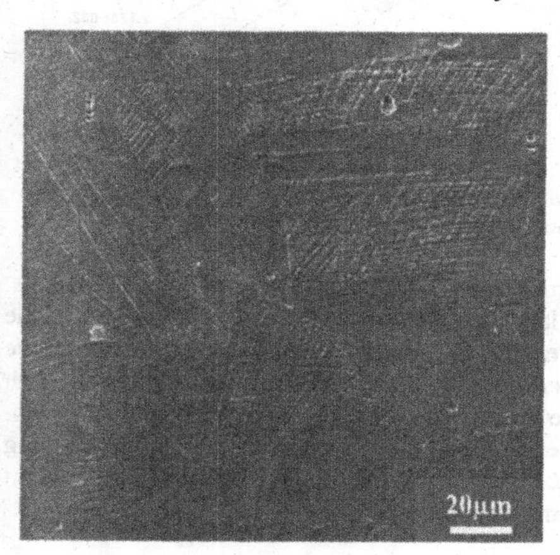

Figure 5: SEM micrograph of C-2000® alloy about 200 μm away from the impacted surface, showing slip lines in the plastically deformed region. The sample was SNH-treated for 70 minutes.

Table I. Grain size of Hastelloy C-2000® alloy SNH-treated as a function of processing time

Processing Time (min)	**20**	**40**	**80**	**120**	**180**
Grain Size (nm)	23.3	21.5	15.5	15.5	16.1

Results of Finite Element Modeling

Figure 6 shows the isosurfaces of the effective plastic strain inside the Hastelloy C-2000® plate after it has been subjected to one and twenty impacts at the same location. Both the size of the plastic zone and the maximum plastic strain value increase with the number of impacts. Further, the shape of the plastic zone resembles a spherical cap. The radius and depth of the spherical cap are 1.17 mm and 0.83 mm, respectively, for one impact, while they become 1.5 mm and 1.1 mm, respectively, for twenty impacts.

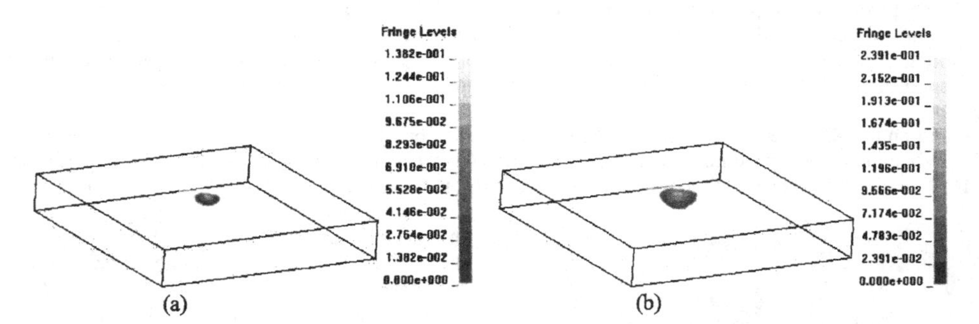

Figure 6: Isosurfaces of the effective plastic strain in the Hastelloy C-2000® plate: (a) subjected to one impact, and (b) subjected to twenty impacts at the same location.

Figure 7 shows the detail of the maximum effective plastic strain as a function of the number of impacts. It can be seen that the first impact gives rise to the largest increase in the plastic strain, while the subsequent impacts only provide a relatively small increase in the plastic strain. The maximum relative work hardening as a function of the number of impacts is shown in Fig. 8. The maximum yield strength used to plot Fig. 8 is computed from eq. (2) based on the maximum effective plastic strain shown in Fig. 7. It can be seen that work hardening beyond fifteen impacts is very small, nearly approaching "strengthening saturation". The trend revealed here is consistent with the microhardness measurement (Fig. 3), which shows that strengthening saturates at 30-minute processing.

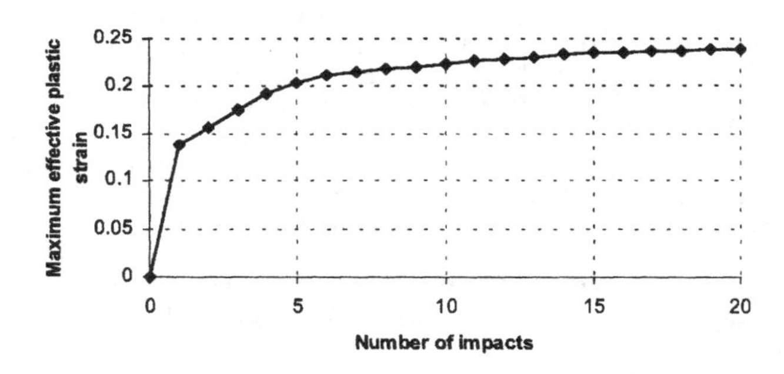

Figure 7: The maximum effective plastic strain as a function of the number of impacts.

The evolution of the effective plastic strain as a function of the distance from the impacted surface and the number of impacts is shown in Fig. 9. The curves of the effective plastic strain are extremely similar to those observed in Fig. 3, showing that a substantial amount of strengthening is achieved in the early stage of impacting, whereas prolonged impacting provides little strengthening because of the strengthening saturation. Since FEM in this study considers only work hardening and the strengthening trends predicted by FEM matches those determined experimentally quite well, it is reasonable to conclude that the simulation from FEM supports the notion that surface strengthening of the C-2000® alloy obtained via the SNH treatment can be accounted for by work hardening and the contribution of grain refinement to strengthening, if any, is small.

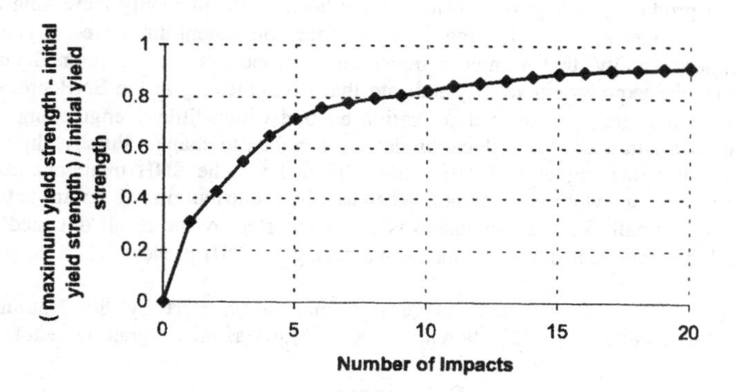

Figure 8: The maximum relative work hardening as a function of the number of impacts.

Finally, it is worthy of mentioning that a quantitative comparison between FEM and experimental results should be made with great caution because of the following two reasons. First, FEM is an idealized case where a single ball repeatedly impacts the same location at a 90-degree angle, whereas the experimental result is obtained from multiple, random, normal and non-normal impacts of five balls. Second, strain rate hardening is not considered in the present FEM. The inclusion of strain rate hardening will increase the value of flow stresses, but will not change the trend discussed above [13]. Furthermore, the consideration of adiabatic heating during impact indicates that the softening due to the temperature rise is negligible [13]. Thus, the conclusion made above will not be changed with a more sophisticated finite element model, which is currently under development [13].

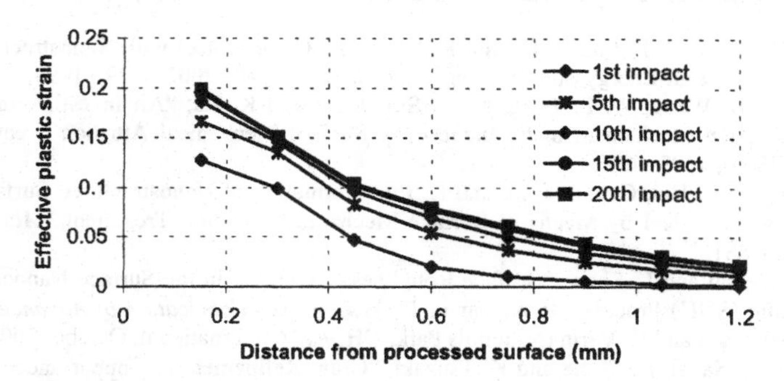

Figure 9: Evolution of effective plastic strain as a function of the distance from the processed surface for a varying number of impacts.

Concluding Remarks

Strengthening and its mechanisms in the surface SPD-based processes have been investigated using a Hastelloy C-2000® alloy. Both experiments and numerical simulations have shown that the SNH process is capable of producing nanograins at the surface layer and appreciably increasing the surface hardness of metallic components, while leaving the core of the component untouched. It is also shown, numerically and experimentally, that a smooth transition in properties can be attained via the SNH process. Furthermore, the experiments clearly indicate that strengthening in the SNH process reaches saturation after a certain degree of plastic deformation beyond which little strengthening is observed even though the microstructure and the dislocation density continue to change. Finally, the experiments suggest that surface strengthening of C-2000® alloy obtained via the SNH treatment (about 150% increase) is mainly due to work hardening, whereas the contribution of grain refinement to strengthening, if any, is small. Such a conclusion is also supported by the result obtained from a 3D finite element model that only considers work hardening during the SNH process.

Acknowledgements - The authors acknowledge the financial support by the National Science Foundation through Grant No. DMR-0207729 with Dr. K. L. Murty as the Program Director.

References

1. H. J. Fecht, E. Hellstern, Z. Fu and W. L. Johnson, "Nanocrystalline Metals Prepared by High-Energy Ball Milling," *Metall. Trans.*, 21A (1990), 2333-2337.
2. C. C. Koch, "The Synthesis and Structure of Nanocrystalline Materials Produced by Mechanical Attrition: A Review," *Nanostruct. Mater.*, 2 (1993), 109-129.
3. L. Shaw, M. Zawrah, J. Villegas, H. Luo and D. Miracle, "Effects of Process Control Agents on Mechanical Alloying of Nanostructured Aluminum Alloys," *Metall. Mater. Trans.*, 34A (1) (2003), 159-170.
4. N. R. Tao, M. L. Sui, J. Lu and K. Lu, "Surface Nanocrystallization of Iron Induced by Ultrasonic Shot Peening," *Nanostruct. Mater.*, 11 (4) (1999), 433-440.
5. X. Wu, N. Tao, Y. Hong, B. Xu, J. Lu and K. Lu, "Microstructure and Evolution of Mechanically Induced Ultrafine Grain in Surface Layer of Al-Alloy Subjected to USSP," *Acta Mater.*, 50 (2002), 2075-2084.
6. G. Liu, S. C. Wang, X. F. Lou, J. Lu and K. Lu, "Low Carbon Steel with Nanostructured Surface Layer Induced by High-Energy Shot Peening," *Scripta Mater.*, 44 (2001), 1791-1795.
7. N. R. Tao, Z. B. Wang, W. P. Tong, M. L. Sui, J. Lu and K. Lu, "An Investigation of Surface Nanocrystallization Mechanism in Fe Induced by Surface Mechanical Attrition Treatment," *Acta Mater.*, 50 (2002), 4603-4616.
8. H. W. Zhang, Z. K. Hei, G. Liu, J. Lu and K. Lu, "Formation of Nanostructured Surface Layer on AISI 304 Stainless Steel by Means of Surface Mechanical Attrition Treatment," *Acta Mater.*, 51 (2003), 1871-1881.
9. J. Villegas, K. Dai and L. Shaw, "Surface Roughness Evolution in the Surface Nanocrystallization and Hardening (SNH) Process," to appear in *Processing and Fabrication of Advanced Materials: XII*, ed. T. Srivatsan and R. Varin (Materials Park, OH: ASM International, October 2003).
10. A. Belyakov, T. Sakai, H. Miura and K. Tsuzaki, "Grain Refinement in Copper under Large Strain Deformation," *Phil. Mag. A*, 81 (11) (2001), 2629-2643.
11. M. Zawrah and L. Shaw, "Microstructure and Hardness of Nanostructured Al-Fe-Cr-Ti Alloys through Mechanical Alloying," Mater. Sci. Eng., A355, 37-49 (2003).
12. H. P. Klug and L. E. Alexander, *X-Ray Diffraction Procedures for Polycrystalline and Amorphous Materials* (London: John Wiley & Sons, Inc., 1954).
13. K. Dai, J. Villegas and L. Shaw, unpublished research.

Processing and Properties of Structural Nanomaterials
Edited by Leon L. Shaw, C. Suryanarayana and Rajiv S. Mishra
TMS (The Minerals, Metals & Materials Society), 2003

Bulk Nanostructured Materials Produced by Mechanical Alloying and Spark Plasma Sintering

J. G. Cabañas-Moreno[1], H. A. Calderón[1], O. Coreño-Alonso[2], M. Umemoto[3], K. Tsuchiya[3], J. R. Weertman[4]

[1]Instituto Politécnico Nacional, ESFM, Depto. de Ciencia de Materiales, Unidad Prof. ALM, Edif..9, Zacatenco, 07338 México D.F., México
[2]Universidad Autónoma del Estado de Hidalgo, Pachuca, Hgo., México
[3]Toyohashi University of Technology, Tempaku-cho, Toyohashi 441-8580, Japan
[4]Nortwestern University, Department of Materials Science, Evanston, IL, USA

Keywords: Nanostructures, Mechanical Alloying, Intermetallics, Co-Ti alloys, NiAl, Cu-Ni alloys

Abstract

For about a decade we have used mechanical alloying processes to produce powder materials which were subsequently consolidated by means of spark plasma sintering. The mechanical alloying process ensured the production of nanostructured powder particles, while the application of plasma sintering was intended to achieve densification in a short time and, concurrently, to avoid excessive coarsening of the microstructures in the sintered materials. Materials produced in this way included intermetallic compounds (Co_2Ti–Co_3Ti, NiAl, and others), as well as different composite materials; in addition, nanostructured copper and Cu-Ni alloys were also produced. As a rule, the combination of mechanical alloying and fast sintering has yielded materials with grain sizes typically below 100 nm. As a result of these refined structures, high mechanical strengths have been observed in most cases in which porosity levels were low. These features of the materials so produced are illustrated here by selected results from two nanostructured intermetallic systems (Co-Ti, NiAl) and from nanocrystalline Cu-Ni alloys.

Introduction

A number of ways have been devised to produce nanostructured materials (grain size below 100 nm) in bulk form and to characterize their mechanical properties. In the pioneering work of J. R. Weertman's group at Northwestern University, nanoparticles were produced by the gas condensation process and subsequently cold or warm consolidated by uniaxial loading using pressures in the GPa range [1,2]. Since then, this experimental approach of powder consolidation has been followed in several different ways, including the use of multiaxial stress states and high homologous temperatures; also, the size of the powder particles has ranged from the nanoscale up to several microns. In some cases, the powder material underwent structural changes during consolidation, perhaps the most remarkable being the crystallization of an originally amorphous structure [3].

The most common concern when using powder consolidation to produce bulk nanostructured materials is clearly the achievement of high densifications without undue coarsening of the grain size. Other concerns include homogeneity, contamination, maximum dimensions of the consolidated pieces, etc. To characterize their mechanical properties, however, adequate densification has been particularly important for nanostructured materials because these properties are usually very sensitive to voids or cracks in high strength materials and more so if specimens are of small size.

In our work, we have used mechanical alloying (MA) to produce nanostructured powder agglomerates which may reach 40-50 μm in size. Advantages of this approach are the production of sizable amounts of powders (tens of grams on a laboratory scale) which frequently can be handled with relative little risk of combustion, contamination and losses, as compared to the case of nanoparticles. The mechanically alloyed powders have been subsequently consolidated by means of the so-called spark-plasma sintering (SPS) process, to obtain specimens with maximum sizes of 2-3 cm in diameter and several mm in thickness. The SPS process allowed very fast heating rates and minimized the time required to attain high densification levels. Overall, the combination of MA and SPS has yielded high densification levels in nanostructured specimens, including intermetallic compounds, metal-matrix composites and simple metallic materials [4]. The present paper summarizes the main results regarding the microstructure and properties of nanostructured Co-Ti and NiAl intermetallic compounds, as well as those from nanostructured Cu-Ni alloys. Issues concerning MA processing can be found elsewhere [3-6].

Experimental Procedures

The intermetallic phases investigated correspond to the systems Co-Fe-Ti and Ni-Al; in addition, pure Cu and Cu-Ni alloys (20 and 30 at.% Ni) were also produced. The alloyed powders were produced by mechanical alloying using attritor or shaker (SPEX) mills. Milling conditions are described in previous publications [3-6]. The phases found in the as-milled powders before sintering are indicated in Table 1. Powder consolidation was performed in a SPS-1030 Dr. Sinter machine (Sumimoto Coal Mining Co.) operated under a partial vacuum of about 20 Pa. Sintering conditions (uniaxial pressure, nominal temperature and time) used in the experiments are also indicated in Table1. The microstructure and mechanical properties of the as-sintered materials were characterized by SEM, TEM and XRD techniques. The phases detected in the sintered materials are also indicated in Table 1.

Table 1. Constituent Phases and Sintering Conditions.

System	Major Phases in As-Milled Products	Sintering Conditions			Major Phases in Sintered Products
		Temp. Range (°C)	Pressure (MPa)	Time (s) at Temperature	
Co-Fe-Ti	amorphous phase + elements (nanograins)	800 - 1100	0 - 60	300	$Co_3Ti + Co_2Ti$
Ni-Al	NiAl (nanograins)	700 - 1500	50	300 - 1800	NiAl
Cu-Ni	Cu Cu-Ni solid sol'n.	200 - 500	100	300	Cu Cu-Ni solid sol'n.

Results

Co-Fe-Ti Alloys

The composition of the sintered alloys varied within the following ranges (in at.%): Co (55–67%); Fe (8–13%); Ti (28–35%); Cr (1–2.5%). Examples of their microstructures are given in Figs. 1(a,b). They consist of two main constituents, the one in higher proportion resembling "grains" of a light contrast in the SEM micrographs of Fig.1, while the second one can be observed either as "grains" of striped (light-dark) contrast or localized regions at the "boundaries" of the first constituent. Boundaries such as those observed in Figs. 1(a,b) actually correspond to the former surfaces of the powder particles, as revealed by comparison of the "grain" size in Fig. 1 (a,b) with the particle size in Fig. 2.

Figure 3 shows a bright-field TEM image from the interior of one of the light-gray "particle-grains" displayed in Fig. 2. It can be seen that the microstructure in these materials is mostly nanocrystalline. The size range of the (true) grains shown in Fig. 3 is about 20–150 nm. In turn, electron diffraction revealed the existence of two phases: Co_3Ti and hexagonal Co_2Ti. Rietveld analysis of x-ray diffraction data [7] yielded average crystallite sizes between 30 and 85 nm, depending on the sintering temperature.

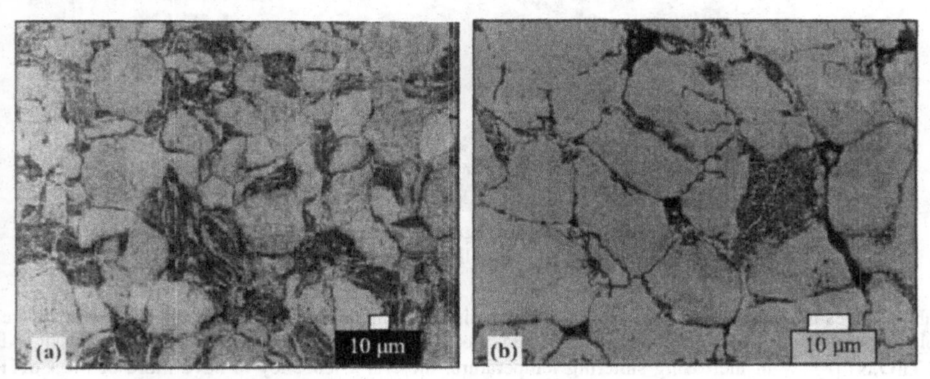

Figure 1. Backscattered SEM micrographs from sintered Co-Fe-Ti specimens. Average compositions: (a) $Co_{55}Fe_9Ti_{34}Cr_2$; (b)) $Co_{67}Fe_4Ti_{28}Cr_1$.

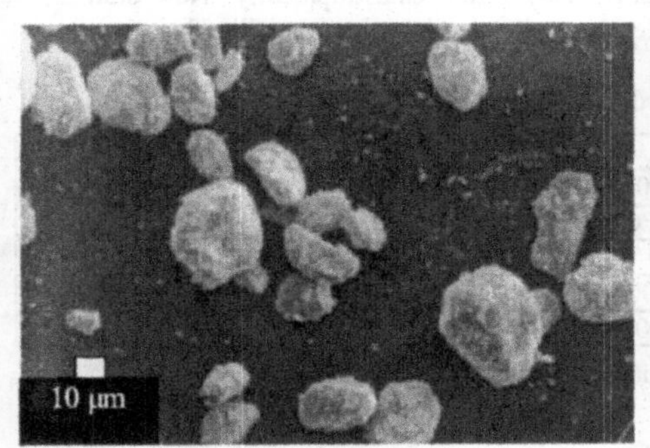

Figure 2. SEM image from as-milled powders

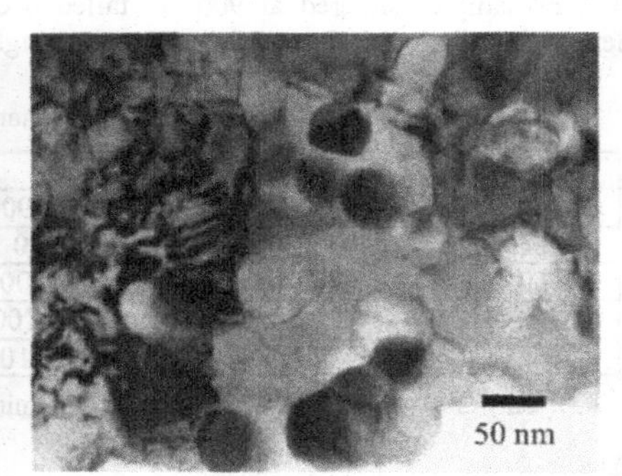

Figure 3. TEM image from sintered material.

Figures 4(a-c) show the microstructure in the darker constituent. The SEM images in Figs. 4(a,b) indicate that this microconstituent consists of at least two different phases. Electron diffraction in the TEM gave patterns compatible with the structure of Co_3Ti and (hexagonal) Co_2Ti, but also revealed some amorphous regions, appearing as elongated bands in the microstructure, as the one delineated by the pair of white lines in Fig. 4c. In addition, relatively high oxygen and nitrogen contents were found in these regions by EDS microanalysis; in some cases, it was possible to associate these elements to fine Ti(O,N) particles, like the ones encircled in Fig. 4c. In general, it was observed that the microstructure of the Co-Fe-Ti alloys was an intimate two-phase mixture of Co_3Ti and Co_2Ti, but some compositional inhomogeneities remained even in specimens sintered at 1100 °C.

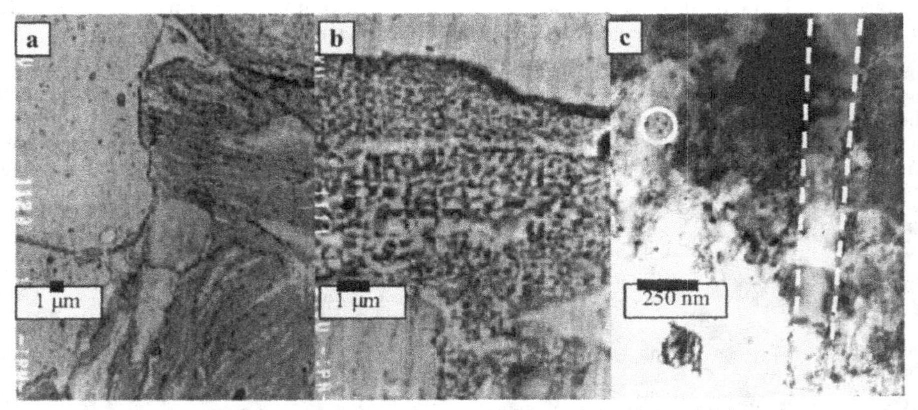

Figure 4. (a,b) . Backscattered SEM micrographs from sintered Co-Fe-Ti specimens.
(c) TEM image from material in (b); white lines indicate location of amorphous band.

Table 2 gives the values of microhardness and porosity determined for sintered Co-Fe-Ti materials. The volume fraction of Co_3Ti and Co_2Ti are indicated; the balance to 100% (not indicated) has been assigned to Ti(O,N) [8]. With increasing sintering temperature, there is a tendency to have more of the Co_3Ti phase at the expense of Co_2Ti. The small grain sizes found in specimens fully consolidated at 900 °C are the most likely reason for the high hardness values reported in Table 2. A few attempts at compression testing of samples sintered at 900 °C failed because they broke catastrophically. New studies are underway in order to produce and characterize single-phase alloys of both intermetallic phases.

Table 2. Porosity and Microhardness of Sintered Co-Fe-Ti alloys.

Material / Major Phases (vol.%)	Porosity (%)	Microhardness (GPa)
Co_3Ti (22.6%)– Co_2Ti (71.1%)§, sintered at 900 °C	< 1	8.3 – 10.4
Co_3Ti (74.5%)– Co_2Ti (18%) , sintered at 800 °C*	24	6.5
Co_3Ti(94.2 %)– Co_2Ti (4.7%) , sintered at 900 °C	< 1	7.1 – 9.5
Co_3Ti(71.4%)– Co_2Ti (23.8%) , sintered at 1000 °C*	< 1	7.0
Co_3Ti(98.4 %)– Co_2Ti (1.2%) , sintered at 1100 °C*	Not detected	5.0

* only one specimen tested $^§Co_{67}Ti_{33}$ nominal composition $Co_{75}Ti_{25}$ nominal composition

NiAl

Bulk specimens of the NiAl intermetallic compound were produced by SPS of mechanically alloyed powders having nearly equiatomic contents of Ni and Al, as well as about 2 at.% Fe [5,8]. The highest densification level attained was 93 % using sintering temperatures in the range 900–1500 °C. X-ray diffraction reported only the presence of the NiAl phase. A Warren-Averbach analysis of the diffraction data led to the cumulative crystallite size distributions shown in Fig. 5. According to these data, crystallite sizes remained small (average values < 15 nm) even for material sintered at 1500 °C, but it was evident from TEM observations that the grain size was becoming highly heterogeneous with increasing sintering temperature [8]. Figure 6 illustrates the high proportion of nanograins still found in specimens sintered at high temperature (1300 °C in this case). A bimodal grain size existed in samples sintered at temperatures of 1000 °C or higher; one component having sizes between 10 and 50 nm (Fig. 5) and the other between 0.5 and 2 µm.

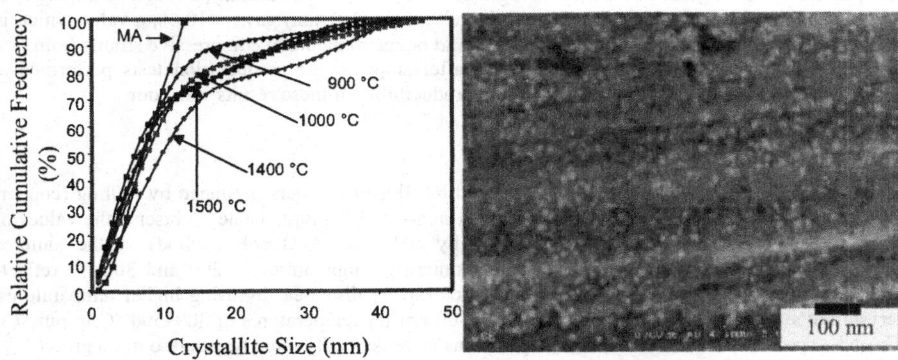

Figure 5. Crystallite size distribution (cumulative) in NiAl sintered at different temperatures.

Figure 6. High resolution SEM micrograph from NiAl specimen sintered at 1300 °C

Despite lacking full densification, it was possible to perform compression tests at temperatures up to 500°C. No ductility was observed at room temperature, but some ductility was detected in tests performed at and above 300 °C. As frequently observed with mechanically alloyed materials, the strength of the sintered materials was very high. The values of yield or fracture stresses in compression were higher than 1.0 GPa at all testing temperatures. The dotted lines in Fig. 7 indicate the strength levels of NiAl specimens sintered at 1000, 1300 and 1500 °C, and compares them with strengths of equiatomic NiAl and particle-reinforced NiAl [9-12]. As shown, the strength of the equiatomic NiAl material produced from mechanically alloyed powders is far superior to that from conventional material and about the same as the strength reported for NiAl having large volume fractions of dispersoids.

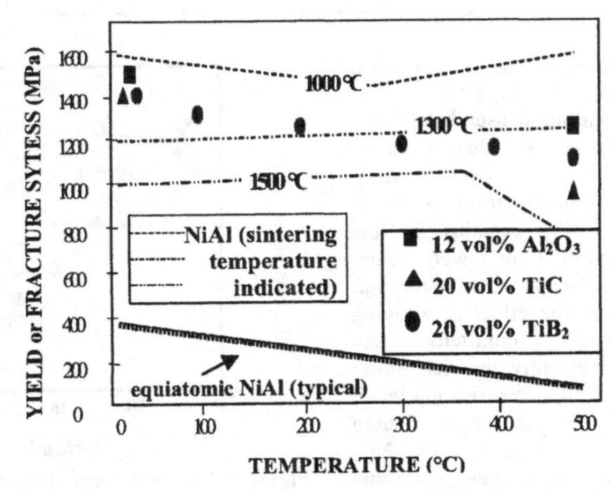

Figure 7. Compressive strength of NiAl from MA (dotted lines), conventional NiAl and particle-reinforced NiAl (from refs. [10-13]) as a function of temperature.

Since second-phase contents of 10 – 20 vol% were certainly not found in our NiAl specimens [8,9], the only explanation left for their high strengths lies in their nanocrystalline structure. From the published data [13] on the effect of grain size on the strength of NiAl (Hall-Petch relation), a rough estimate [8] of a likely upper limit for the volume fraction of "large" grains (2–3 μm) gave ~ 10 %, a value which is consistent with the experimental observations. It should be mentioned that compressive strains from 5 to 21% were measured directly on some specimens after stopping the compression tests performed at temperatures of 1000 °C or higher. However, the reproducibility of these results was poor.

Cu-Ni Alloys

Nanocrystalline Cu and Cu-Ni alloys were produced by SPS of powders produced by milling (copper) and MA (Cu-Ni). The sintering conditions are given in Table 1. In turn, Table 3 presents the values of densification and crystallite size (as determined by a Warren-Averbach method) in the sintered specimens. The wide range of density values for sintering temperatures of 200 and 300 °C reflects inhomogeneities in their state of consolidation, which mostly disappear by using higher temperatures. The crystallite size, however, becomes larger than 100 nm for temperatures of 400–500 °C in pure Cu and $Cu_{70}Ni_{30}$ specimens, while the $Cu_{80}Ni_{20}$ specimens were generally more resistant to grain growth.

Table 3. Densification Levels and Crystallite Sizes in Cu-Ni Specimens.

Composition (at.%)	Sintering Temperature (°C)	Densification (%)	Crystallite Size (nm)
Cu (100)	200	80-86	30
	300	70-83	55
	400	91-92	55
	500	95	> 100
Cu-Ni (80/20)	200	73-77	27
	300	66-84	23
	500	95	52
Cu-Ni (70/30)	400	78-80	41
		94-95	> 100

Figure 8 compares hardness data for pure Cu from several sources, including our results. The material prepared by milling is usually harder – at similar densification levels – than materials produced by other processes. This probably means that some of the deformation hardening induced by milling stills remains in the specimens sintered at 400 °C or lower. Again, strengths of Cu and Cu-Ni specimens were high in compression testing. The effect of alloying with Ni on the yield stress at room temperature is displayed in Fig. 9. Evidently, the yield stress is strongly influenced by the densification level; however, it also seems clear that the hardening effect by solid solution is greater in $Cu_{80}Ni_{20}$ than in $Cu_{70}Ni_{30}$ in these nanocrystalline materials.

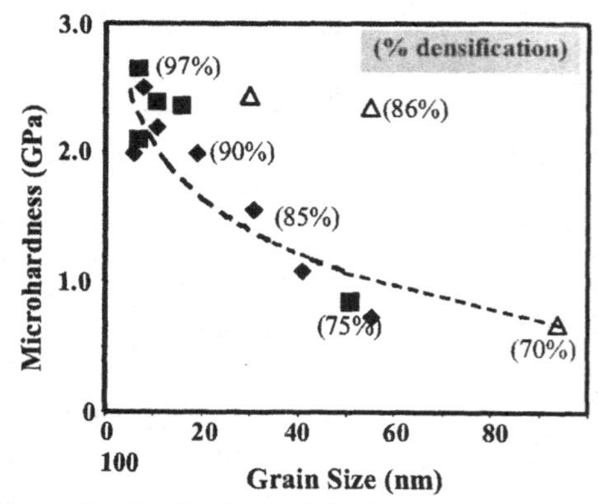

Figure 8. Microhardness of bulk nanocrystalline Cu. Filled symbols from Refs. [1,2,14]; open symbols, from present work.

Specimens having densification levels of 80% or more showed considerable ductility, as illustrated in Fig. 10a, but their fracture usually occurred by propagation of multiple cracks and the specimens broke into several pieces. The test of the Cu specimen shown in Figure 10a was arrested before the point of fracture; Fig. 10b shows an enlarged image of the originally flat, lateral surface after a height reduction of more than 50%. It is evident here that some extrusion-like features are present on the surface of this specimen, and that they run more or less parallel to the direction of loading. It is still unknown what kind of deformation mechanism is operating during deformation of these nanocrystalline materials.

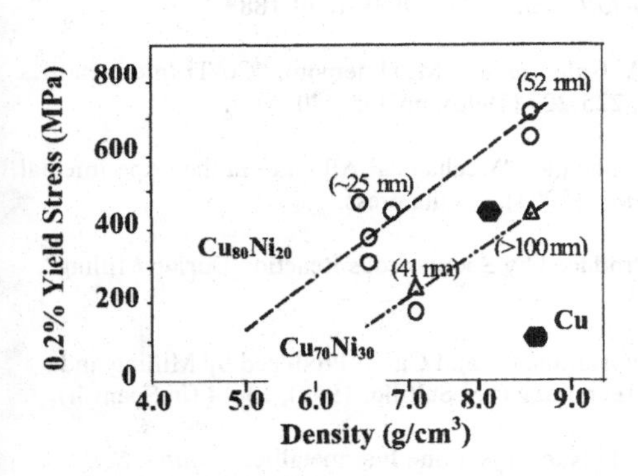

Figure 9. Measured yield stresses as a function of density for Cu and Cu-Ni alloys.

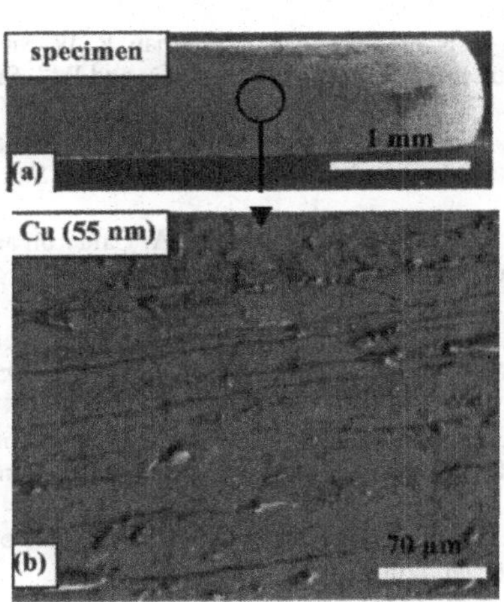

Figure 10. Micrographs from deformed Cu specimen. (a) Overall view. (b) Detail from previously polished surface.

Final Remarks

In producing bulk nanostructured materials by the fast consolidation of mechanically alloyed (or highly milled) powders, two aspects have been very important: homogeneity and contamination of the powders. They are interrelated, since homogeneity refers to both microstructure (e.g., grain and particle size) and composition; even when some contamination is allowed, it should be kept in mind that this occurs at random during the milling operation and there may not be enough milling time left to make uniform the contamination in the powder charge. Compositional variations at the scale of the particle size could possibly be eliminated if long sintering times were applied, but this of course increases the chances to induce too much grain growth. These effects are probably exacerbated when working with small powder charges; therefore, MA in relatively large batches (at least 50 – 100 g) should be prefered even in work at the laboratory scale. After all, one of the main advantages of MA is precisely the capability of producing large quantities of nanostructured materials.

Acknowledgements

Some of the work presented was derived from the thesis dissertations of Dr. R. Martínez-Sánchez (CIMAV, México) and Ms. A. Pulido-Rodríguez. The work on Cu and Cu-Ni alloys was done in collaboration with Prof. J. R. Weertman (Northwestern Univ., USA) under the NSF-CONACYT research program. JGCM and HAC are recipients of COFAA-IPN fellowships. Their work has been funded for several years by CONACYT and CGPI-IPN.

References

1. G. W. Nieman, J. R. Weertman and R. W. Siegel, "Mechanical Behavior of Nanocrystalline Cu and Pd", *J. Mater. Res.*, 6 (1991), 1012-1025.

2. G. E. Fougere, J. R. Weertman, R. W. Siegel and S. Kim, "Grain Size Dependent Hardening and Softening of Nanocrystalline Cu and Pd", *Scripta metall. mater.*, 26 (1992), 1879-1883.

3. R. Martínez Sánchez, J.G. Cabañas Moreno, H.A. Calderón and M. Umemoto, "Co-Ti Intermetallics Made by Mechanical Alloying", *Mater. Sci. Forum*, 225-227 (1996), pp.435-440.

4. J. G. Cabañas-Moreno, H. A. Calderón and M. Umemoto, "Mechanical Alloying in the Experimental Production of Structural Materials", *Mater. Sci. Forum*, (2003) – in the press.

5. O. Coreño Añonso et al., "Al-Ni Intermetallics Produced by Spontaneous Reaction During Milling", *Mater. Sci. Forum*, 343-346 (2000), 290-295.

6. A. Pulido-Rodríguez, "Characterization of Nanocrystalline Cu and Cu-Ni Produced by Milling and Plasma-Assisted Sintering" (M.Sc. thesis, Instituto Tecnológico de Saltillo, 1998), 13-14 (in Spanish).

7. F. Cruz-Gandarilla et al., "XRD Studies of Co-Fe-Ti Nanocrystalline Intermetallics", *Mater. Sci. Forum*, (2003) – in the press.

8. O. Coreño-Alonso et al., "Microstructure and Properties of NiAl Intermetallic Compound Produced by Mechanical Alloying and Consolidated by Spark Plasma Sintering", Advances in Technology of Materials and Materials Processing Journal, 4 (2002), 1-6.

9. O. Coreño-Alonso, "Fabrication of Nanocrystalline NiAl by Mechanical Alloying and its Consolidation" (Ph. D. thesis, Instituto Politécnico Nacional, 2001), 45-60 (in Spanish).

10. K. Xu and R. J. Arsenault, "High Temperature Deformation of NiAl Matrix Composites", *Acta mater.*, 47 (1999), 3023-3030.

11. Z. P. Xing et al., "Microstructure and Mechanical Behavior of the NiAl-TiC *in situ* Composites", *Metall. Mater. Trans. A*, 28A (1997) 1079-1088.

12. M. Dollar et al., "The Role of Microstructure on Strength and Ductility of Hot-Extruded Mechanically Alloyed NiAl", *Metall. Trans.* A, 24A (1993), 1993-2008.

13. D. B. Miracle, "The Physical and Mechanical Properties of NiAl", *Acta metall. Mater.*, 41 (1993), 649-666.

14. R. W. Siegel, "Cluster-Assembled Nanophase Materials", *Ann. Rev. Mater. Sci.*, 21 (1991), 559-578.

Processing and Properties of Structural Nanomaterials
Edited by Leon L. Shaw, C. Suryanarayana and Rajiv S. Mishra
TMS (The Minerals, Metals & Materials Society), 2003

The Effects of Equal Channel Angular Pressing on Microstructure and Tensile Properties of Spray Deposited Al-Cu-Mg Alloy

Kyung H. Chung*, Dong H. Shin and Enrique J. Lavernia***

**Department of Chemical Engineering and Materials Science, University of California, Davis, Davis, CA 95616*
***Department of Metallurgy and Materials Engineering, Hanyang University, Ansan, Kyunggi-do, Korea 425-791*

Keywords: Spray deposition, equal channel angular pressing, nanocrystalline material, mechanical properties, microstructure

Abstract

The ECAP process was applied for spray deposited Al-4.4Cu-0.8Mg alloy to produce fine-grained bulk material without the limitation over cross-sectional dimensions. The variation of the density and tensile properties of spray deposited Al-Cu-Mg alloy was investigated and the microstructures were analyzed through scanning electron microscopy (SEM) and transmission electron microscopy (TEM) observations. The ECAP of the spray deposited Al-Cu-Mg alloy is successfully performed at 200 °C. After a single pass, the density becomes nearly full and the remaining pore size is reduced to less than 3 μm from about 20 μm. At the same time, an elongated structure develops with an average width of 200 nm and ~ 1 μm length. With repeated ECAP passes, the microstructure becomes more equiaxial and homogeneous. The grain size reduces to the 100 ~ 250 nm range after four passes. After ECAP, the room temperature strength of the spray deposited Al-Cu-Mg alloy is increased up to 110%, and the strength increase is attributed to the reduced amount of pore, the increased dislocation density and the reduced grain size.

1. Introduction

Recently, various severe plastic deformation (SPD) processing techniques have been used to obtain bulk ultrafine-grained (UFG) materials, which reportedly exhibit enhanced strength and superplastic properties [1]. Among these techniques, equal-channel angular pressing (ECAP) has been one of the most attractive methods due to its effectiveness in producing bulk UFG materials without any residual porosity. Compared to other conventional forming methods, the ECAP process has a distinct advantage: it produces UFG materials without reducing the available dimensions for working parts. In this regard, successful applications have been reported for various materials such as Al alloys and their composites [2,3], Cu [4], Mg alloys [5], Ni [6], Zn alloys [7] , Fe [6], and low carbon steels [8,9].

Spray deposition process has been proposed as an alternative process in obtaining the UFG materials due to their rapid cooling rates (~1×10^4 K/sec) [10,11]. Moreover, spray deposition is reported to exhibit physical attributes that are typical of rapid solidification, namely: increased solid solubility, non-equilibrium phases and the absence of macro-segregation [10-12]. However the presence of approximately 3% porosity is another characteristic of spray deposited bulk materials. Therefore, the post-processes after spray deposition, e.g., extrusion and rolling, are typically needed to increase their density for structural applications [13]. These post-processes after spray-deposition cause a dimensional limitation due to the reduction of the cross-sectional area during the processes and the microstructural changes due to the relatively high temperature of its operation, e.g., the extrusion temperature of Al-Mg alloy being 400°C [13]. The application of spray deposited materials as structural parts has been limited accordingly.

The current investigation examines the possibility of ECAP process as a post-process for spray deposited materials without the limitation over cross-sectional dimensions. The variation of the density and tensile properties of spray deposited Al-4.4Cu-0.8Mg alloy was investigated during the ECAP process. In addition, the microstructures of spray deposited Al-4.4Cu-0.8Mg alloys subjected to ECAP with severe plastic strain were analyzed through scanning electron microscopy (SEM) and transmission electron microscopy (TEM) observations.

2. Experimental Procedure

The Al-4.4Cu-0.8Mg alloy was heated to 800 °C and atomized using N_2 gas jets at a dynamic pressure of 1.20 MPa. The distribution of droplets was then directed toward and deposited upon rotating, water cooled Cu substrate. Table 1 shows the composition of the alloy used in this work after spray deposition. Details of the experimental procedure are available in the literature [21]. After machining the cylindrical samples, with size of F 18 ×110 mm, from the spray deposited billet, up to four passes of ECAP were carried out on the samples at 200 °C. A detailed description of the ECAP apparatus used in this study can be found elsewhere [8]. In brief, the ECAP die was designed to yield a shear strain of ~ 1.83 during each pass [14]. It contained an inner contact angle and arc of curvature at the outer point of contact between the channels of 90° and 20°, respectively. The densities of the samples were measured with a precision balance with an accuracy of ±0.1 mg, Model AG204 by Mettler Toledo Co. (Columbus, OH), following ASTM B3-11. At each condition, three samples were examined and the density of each sample was measured five times. Specimens for TEM were prepared through mechanical thinning of samples to ~40 μm followed by a twin jet polishing technique with a solution of 25% nitric acid and 75% methanol under an applied potential of 25 V at -30°C. TEM images and corresponding selected area diffraction (SAD) patterns of each sample were obtained by a JEOL JEM 2010 machine operating at 200kV. Tensile tests for each of the conditions were performed according to ASTM E8-95 on samples machined from the samples as the spray deposited and after each ECAP pass. The tests were conduced at room temperature, 200 and 300°C.

Table 1. The chemical composition of the alloy used in this paper after the spray deposition. (wt. %)

Element	Cu	Mg	Mn	Al
Composition	4.42	0.82	0.69	Bal.

3. Results and Discussion

Structure of pores and density

Figure 1 presents the change of pore structures during the ECAP processes. The sample before the ECAP process (Fig. 1(a)) shows the distribution of the irregular shaped pores in the range of 1 to 20 μm in size. Usually several individual pores are gathered closely to make a group, and those pore groups are surrounded by needle-shaped precipitates or tangled with each other. Between the pore groups, the pore density is relatively low, and the spacing between groups is about 10 to 50 μm. When ECAP is applied, Fig. 1 (b) and (c), the pore size is reduced to less than 3 μm after one pass and reduced further to less than 1 μm after four passes as well as the amount of pore. The pore grouping is diminished gradually during the repeated ECAP processes. As shown in Fig. 1 (c), the pore grouping has disappeared after four passes of ECAP in the spray deposited Al-Cu-Mg alloy.

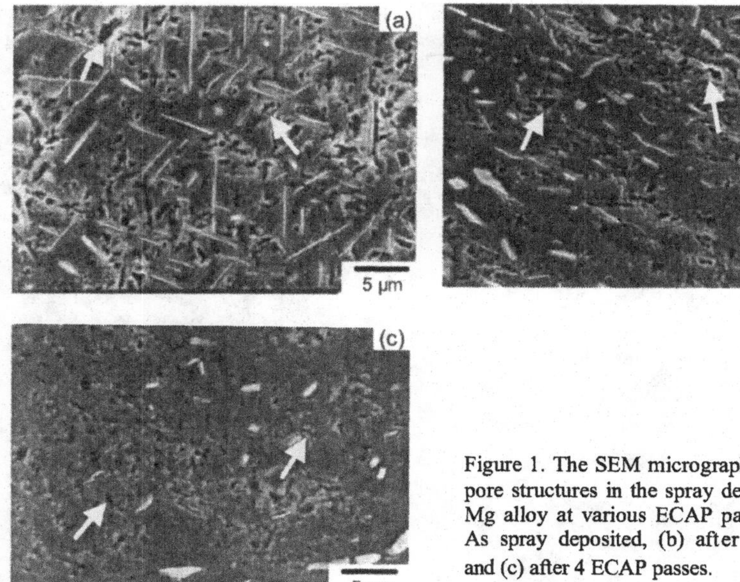

Figure 1. The SEM micrographs showing the pore structures in the spray deposited Al-Cu-Mg alloy at various ECAP pass number; (a) As spray deposited, (b) after 1 ECAP pass and (c) after 4 ECAP passes.

Figure 2 shows the density variation of the spray deposited Al-4.4Cu-0.8Mg alloy during the ECAP process. The alloy density instantly reaches near to its fullest after the first ECAP pass, then the density remains at its fullest during repeated ECAP passes. Similar behavior was observed in the variation of hardness during the ECAP of pure aluminum[15]: there was an immediate increase in the hardness after the first pass, a minor additional increase up to the second pass, and then essentially no change in the hardness with subsequent passes.

The behavior of pore structure during the ECAP of spray deposited Al-Cu-Mg alloy can be categorized into two regions. In the early stage, the reduction of the amount of pores is dominant and the density reaches full density after the first ECAP pass. At the same time, the breakdown of pore group structure is initiated. In the later stage, the pores are redistributed into a uniform distribution, and an accompanying decrease in the pore size gradually proceeds through the repeated ECAP passes.

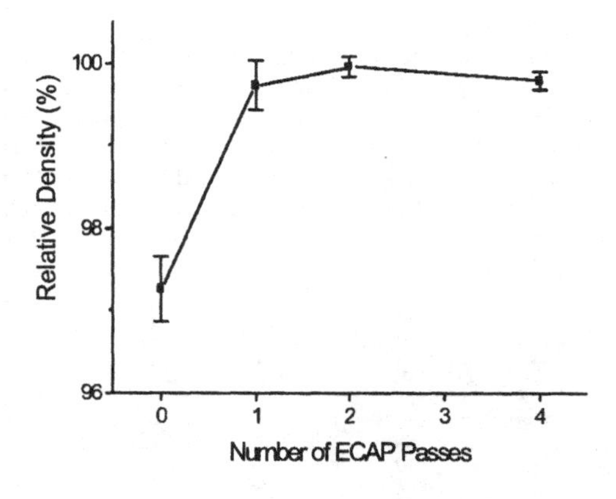

Figure 2. The density variation of spray deposited Al-Cu-Mg alloy with varying the number of ECAP passes experienced.

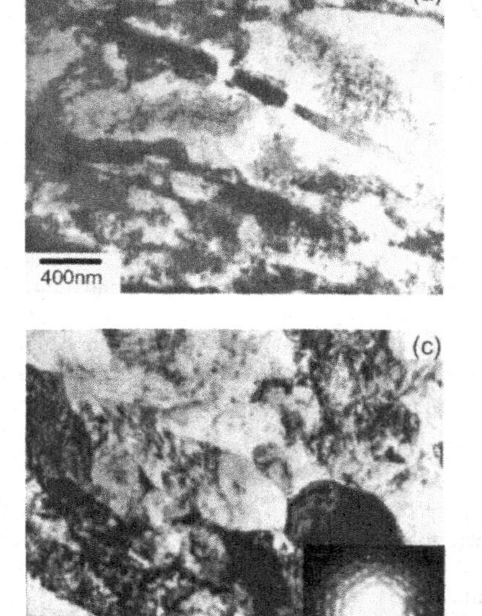

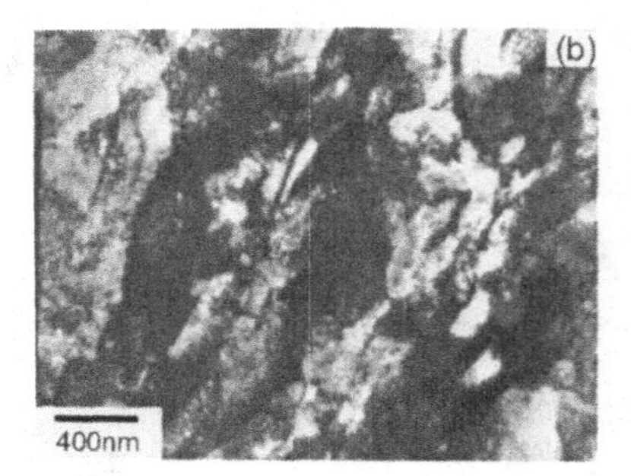

Figure 3. The TEM micrographs showing the grain structures of the spray deposited Al-Cu-Mg alloy at various ECAP pass number; (a) After 1 ECAP pass, (b) after 2 ECAP passes and (c) after 4 ECAP passes.

Microstructure

Fig. 3 shows TEM micrographs of spray deposited Al-Cu-Mg alloy after various ECAP passes. Fig. 3 (a) shows, after single pass, a heavily deformed structure which mainly consists of parallel bands of elongated grains, having a length of ~ 1 μm and a width of ~ 200 nm. As shown in Fig. 3 (b), the grain size is reduced to a range of 150 to 300 nm after two passes of ECAP, and the grain morphology is still elongated, but the lengths of the grains are much shorter than those of the Al-Cu-Mg alloy after single pass. After four passes of ECAP (Fig. 3 (c)), the elongated structures are almost eliminated and the equiaxed grains, with an average grain size of 100 to 250 nm, are formed. This microstructure evolution is reported to be typical in ECAPed aluminum alloys [16]. In addition, SAD yields a diffused ring pattern, which implies the existence of grain boundaries having high angles of misorientation.

Mechanical Properties

Figure 4 shows the variation of the tensile strength, yield strength and elongation with varying numbers of ECAP passes at various temperatures. The yield strength and tensile strength at room temperature, Fig. 4(a), are drastically increased about 110 and 70 %, respectively, after the first ECAP pass. Then both of the strength values are saturated and remain at a similar level during the following ECAP passes. The elongation also remains in the range of 10 to 15 %.

The drastic increases in the strengths of the spray deposited alloys after ECAP were attributed to the combined effects of the elimination of pores, the increased dislocation density and the refined grain size, which were caused by ECAP. The relationship between the strength and amount of pore is usually suggested as follows [17]:

$$\frac{\sigma}{\sigma_F} = e^{(-k \cdot P)} \quad (1)$$

Where, s is the strength of a porous material, s_F is the strength of material with full density, k is a constant which is generally 7 to 9 for most metals [17] and P is the amount of pore in the material. From our investigation, the relative density of the as-spray-deposited Al-Cu-Mg alloy was 97.5 % and the full density was obtained after the ECAP process. Therefore, it is expected that the strength would increase by 20 % after the ECAP pass by the elimination of pore in the materials. The increase of dislocation density, already observed by TEM as seen in Fig. 4, and the grain size reduction from 50 μm to 200 nm took place at the same time during the ECAP pass. The strength increase from the grain refinement and the increased dislocation density after the ECAP process are typical characteristics, which are observed in variety of materials [17-19]. For the ECAP of spray deposited material, the densification occurred and contributed to the strength increase by ECAP, in addition to the effects of dislocation density and grain refinement, which resulted in the drastic strength increase after ECAP.

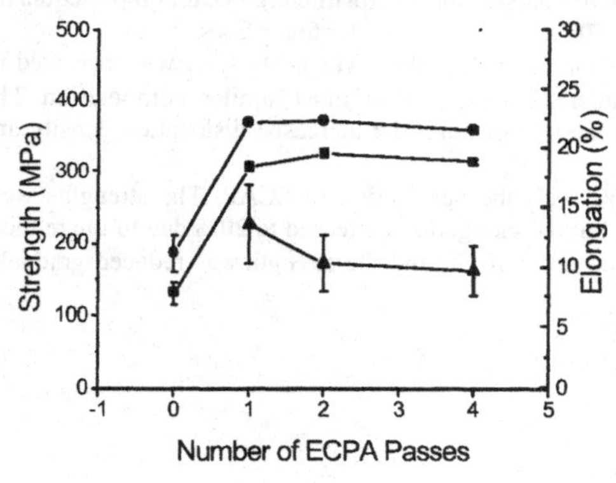

Figure 4. The variations of the ultimate tensile strength (●), yield strength (■) and elongation (▲) values with varying the number of ECAP pass of spray deposited Al-Cu-Mg alloy tested at room temperature.

The strength of the spray deposited Al-Cu-Mg alloy after ECAP has been compared with those of the conventional 2014 Al alloy, which has a composition similar to that of the present alloy, as shown in Table 2. The spray deposited alloy before ECAP shows strength similar to that of 2014Al alloy without heat treatment (2014 Al-O), with little elongation. But after ECAP, the strength level increases to a level comparable with that of heat-treated 2014Al, which is the strongest form of commercially available 2014 Al [20]. It is worth noting that one pass of ECAP yielded a strengthening effect on the Al alloy similar to that attained after an extrusion and a heat treatment, which takes longer and costs more.

Table 2. The comparison of the tensile properties of the spray deposited Al-Cu-Mg alloy and commercial 2014 Al.

Material	Process	Yield Strength (MPa)	Tensile Strength (MPa)	Elongation (%)	Reference
Al-4.4Cu-0.8Mg	As spray deposited	132	183	N/A	This Study
	Spray deposition + ECAP (4 passes)	303	395	15	
2014 Al	O-treatment	97	186	14	[21]
	T6 heat treatment	365	415	7	
	Extrusion + T6 heat treatment	380	450	8	

4. Conclusions

1. The ECAP of the spray deposited Al-4.4Cu-0.8Mg alloy was successfully performed at 200 ℃. After a single pass, the density became full and the remaining pore size was reduced to less than 3 μm from about 20 μm. At the same time, an elongated structure developed with an average width of 200 nm and ~ 1 μm length. With repeated ECAP passes, the microstructure became more equiaxial and homogeneous. The grain size reduced to the 100 ~ 250 nm range after four passes.
2. After ECAP, the room temperature strength of the spray deposited Al-Cu-Mg alloy was increased up to 110%, which is comparable to the commercial alloy having most similar composition. The strength increase is attributed to the reduced amount of pore, the increased dislocation density and the reduced grain size.
3. At 200 ℃, the strength improved about 20% with the application of ECAP. The strengths were slightly reduced with repeated ECAP passes, but the elongation increased to 20% due to the reduced grain size. At 300 ℃, the elongation increased up to 90 %, and the strength was reduced gradually with repeated ECAP passes.

Acknowledgements

The authors acknowledge the financial supports provided by the Army Research Office (Grant No. DAAD19-01-1-0627) and the Office of Naval Research (Grant Nos. N00014-00-1-0109 and N00014-01-1-0882). One of the authors (DHS) also gratefully acknowledges financial support provided by 'The 21st Century New Frontier Research and Development Program'.

References

[1] R.Z. Valiev, R.K. Islamgaliev and I.V. Alexandrov, Prog. Mater. Sci. **45**, 103(2000)
[2] Y. Iwahashi, M. Furukawa, Z. Horita, M. Nemoto and T. G. Langdon, Metall. Mater. Trans., **29A**, 2245(1998)
[3] R. Valiev, R. Islamgaliev, N. Kuzmina, Y. Li and T. G. Langdon, Scripta Mater., **40**, 117(1999)
[4] R. Valiev and I. Alexandrov, NanoStruct. Mater., **12**, 35(1999)
[5] R. Valiev, A. Korznikov and R. Mulyokov, Mater. Sci. Eng., **A168**, 141(1993)
[6] V. Segal, Mater. Sci. Eng., **A197**, 157(1995)
[7] D. H. Shin, W. Kim and W. Y. Choo, Scripta Mater., **41**, 259(1999)
[8] D. H. Shin, B.C. Kim, K.T. Park and W. Y. Choo, Acta Mater., **48**, 3245(2000)
[9] D.H. Shin, Y.S. Kim and E.J. Lavernia, Acta Mater., **49**, 2387(2001)
[10] E. J. Lavernia, J. A. Ayers and T. S. Srivatsan, Inter. Mater. Rev., **37**, 1(1992)
[11] H. Jones, J. Mater. Sci., **19**, 1043(1984)
[12] E. J. Lavernia, Inter. J. Rapid Solidif., **5**, 47(1989)
[13] P. Lengsfeld, J. A. Juares-Islas, W. A. Cassada and E. J. Lavernia, Inter. J. Rapid Solidif., **8**, 237(1995)
[14] Y. Iwahashi, J. wang, Z. Horita, M. Nemoto and T.G. Langdon, Scripta Mater., **35**, 143(1996)
[15] K.Nakashima, Z. horita, M. Nemoto and T.G. Langdon, Mater. Sci. Eng., **A281**, 82(2000)
[16] P.B. Berbon, N.K. Tsenv, R.Z. Valiev, M. Furukawa, Z. Horita, M. Nemoto and T.G. Langdon, Metall. Mater. Trans., **29A**, 2237(1998)
[17] A. Squire, Trans. Metall. Soc. AIME, **171**, 485(1942)
[18] Y. Iwahashi, Z. Horita, M. Nemoto and T. G. Langdon, Acta Mater., **45**, 4733(1997)
[19] J. Wang, M. Furukawa, Z. Horita, M. Nemoto, R. Z. Valiev and T. G. Langdon, Mater. Sci. Eng., **A216**, 41(1996)
[20] ASM Handbook Committee, Metals Handbook 9th ed., vol. 2 "Properties and Selection: Nonferrous Alloy and Pure Metals", ASM, Metals Park, OH, 73(1979)
[21] L. Del Castillo and E. J. Lavernia, Metall. Mater. Trans., **31A**, 2287(2000)

Processing and Properties of Structural Nanomaterials
Edited by Leon L. Shaw, C. Suryanarayana and Rajiv S. Mishra
TMS (The Minerals, Metals & Materials Society), 2003

Elevated Temperature Deformation Behavior of Nanostructured Al-Ni-Gd Alloys

X. L. Shi[1], R. S. Mishra[1] and T. J. Watson[2]

[1]Department of Metallurgical Engineering, University of Missouri, Rolla, MO 65409
[2]Materials and Processes Engineering, Pratt & Whitney, East Hartford, CT 06108

Keywords: Aluminum alloy; deformation behavior; dispersion strengthening; nanostructure; threshold stress

Abstract

Crystallization of amorphous alloys provides opportunities to produce bulk nanostructured materials with high volume fraction of second phase particles. In this study, the elevated temperature deformation behavior of $Al_{87}Ni_7Gd_6$ alloys in which rod-like nano-crystalline particles are dispersed along the grain boundaries was characterized in the temperature range of 523 to 673 K. The results, which cover four orders of magnitude in strain rate, show an increase in apparent stress exponent with decreasing temperature. The introduction of a threshold stress into the analysis leads to stress exponent of ~5 and a true activation energy of 136 kJ/mol in $Al_{87}Ni_7Gd_6$ alloy. The normalized threshold stress exhibits strong temperature dependence. The operative deformation mechanism is discussed in terms of dislocation climb.

1. Introduction

Development of high specific strength aluminum alloys that are microstructurally and mechanically stable at temperature up to 573 K has been pursued for the last two decades for elevated temperature applications [1]. For high temperature applications, the classical precipitate strengthening has the disadvantage of losing it strengthening effect due to the coarsening/dissolution of precipitates. So thermally stable dispersoids, such as oxides, carbides and intermetallic compounds, are useful to maintain the strengthening at high-temperatures. Recently, development of the high-strength Al-based alloys via non-equilibrium processing has received significant attention. In particular, crystallization of amorphous alloys provides opportunities to produce bulk nanostructured materials with high volume fraction of second phase particles. The phase transformation from amorphous state has some unique features: (1) homogenous nucleation, (2) high nucleation frequency, (3) low growth rate, and (4) nanoscale inter-particle spacing [2]. All of these are beneficial for developing high-strength dispersion-strengthened alloys.

In this paper, the elevated temperature deformation behavior of as-extruded dispersion-strengthened $Al_{87}Ni_7Gd_6$ (at %) alloy was evaluated at various temperatures and strain rates. In addition, the observed deformation behavior is compared to results of a number of dispersion- strengthened aluminum alloys with metallic/intermetallic dispersoids.

2. Experimental procedure

The atomized $Al_{87}Ni_7Gd_6$ powders were consolidated at 422 K, and extruded into 22.9 mm (0.9 inch) diameter rods at 623 K with an extrusion ratio of 11:1. Mini-tensile specimens (gage length 1.3mm and gage width 1.0mm) were electro-discharge machined from the as-extruded rods in the longitudinal direction. The tensile specimens were polished to the final thickness of ~0.5 mm and 1 μm finish. Tension tests were performed on a custom-built, computer-controlled mini-tensile tester in the temperature range of 523 to 673 K and initial strain rates range of $5x10^{-5}$ to $1x10^{-2}$ s^{-1}.

The microstructure of the extruded rod was examined using a Philips EM430 transmission electron microscope (TEM) at 300 kV. TEM specimens were prepared using ion-milling.

3. Results

Figure 1 shows a TEM micrograph of as-extruded $Al_{87}Ni_7Gd_6$ alloy. Rod-like particles with an average length of ~180 nm and diameter of ~30 nm are dispersed along the grain boundaries. The average grain size of aluminum matrix is ~200 nm.

Fig 1. Bright-field TEM micrograph of as-extruded $Al_{87}Ni_7Gd_6$ alloy.

The variation of tensile strength with the temperature at strain rate$1x10^{-3}s^{-1}$ is shown in Fig. 2. It can be noted that the tensile strength of this alloy was 445MPa at 523 K. True stress-strain curves at various strain rates at 673 K are shown in Fig. 3. At 673 K, the as-extruded alloy exhibited good ductility at all strain rates. However, there is no apparent relationship between strain rate and failure strain. The maximum failure strain of ~ 0.32 was obtained at strain rate of $1x10^{-4}$ s^{-1}. The variation of flow stress with strain rate is shown in Fig. 4. The apparent stress exponent increases with decreasing temperature.

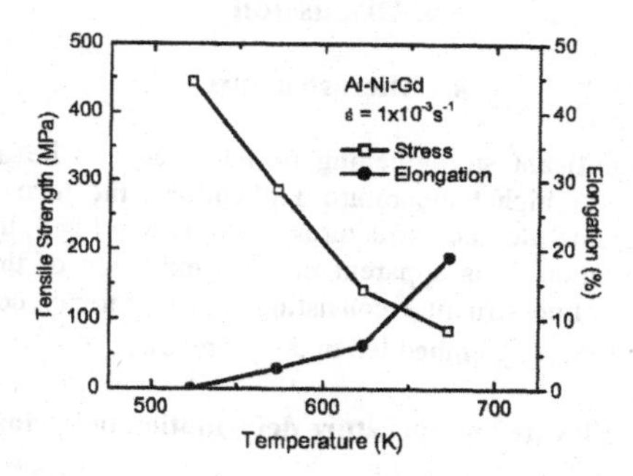

Fig.2. Variation of tensile strength with temperature at initial strain rate of $1x10^{-3}s^{-1}$ for $Al_{87}Ni_7Gd_6$ alloy.

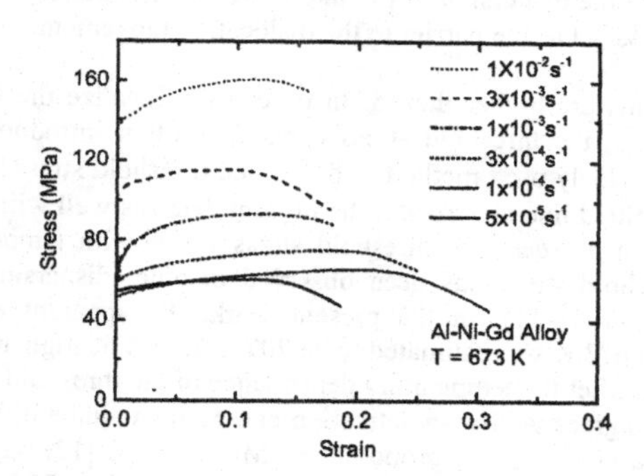

Fig.3. Stress-strain behavior of $Al_{87}Ni_7Gd_6$ alloy at 673 k for various initial strain rates.

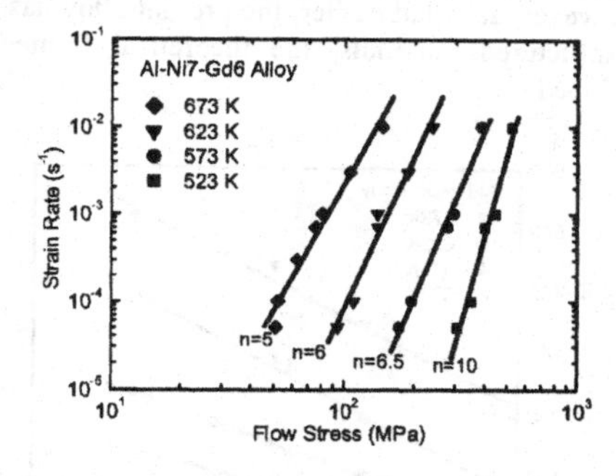

Fig.4. Double logarithmic plot of strain rate versus flow stress at different temperatures.

4. Discussion

4.1 Microstructure

It is well recognized that significant strengthening of Al-based can be achieved by utilizing non-equilibrium processing [3-4]. For high temperature applications, the focus is to produce alloys with dispersion strengthening and nanoscale microstructures. As shown in Fig.1, the intermetallic compounds do not exhibit preferred orientation. It is apparent that hot extrusion of the amorphous powders can result in the formation of ultra-fine structure consisting of fine ternary compound ($Al_{15}Ni_3Gd_2$) and binary compound M_3Gd (M-Al, Ni) [5], embedded in the Al matrix.

4.2 Elevated temperature deformation behavior

The tensile strength of this alloy was 445 MPa at 523 K at strain rate of $1x10^{-3}$ s^{-1}. The good elevated temperature strength is due to the dispersion of the intermetallic compounds which are thermally stable and insoluble in the matrix and act as the barrier to the dislocation movement.

The observed stress exponents are higher than 5. In order to rationalize the high stress exponents in dispersion-strengthened alloys, the threshold stress concept is often introduced [6]. Lagnegorg and Bergman [7] introduced the widely used method to determine threshold stress by plotting $\dot{\varepsilon}^{1/n}$ versus σ and extrapolating the linear fitted line to zero $\dot{\varepsilon}$. The present data fits well with the true stress exponent of 5, shown in Fig. 5. It can be seen that threshold stress varies with temperature. Similar trend of temperature-dependent threshold stress has been observed in other dispersion-strengthened Al-based alloys, as is shown in Fig. 6 [8-11]. In the present work, the apparent activation energy in the temperature ranges of 573 to 673 K was estimated to be 207 kJ/mol. The high apparent activation energy is rationalized to 136 KJ/mol after the temperature dependence of the threshold stress is considered, as is shown in Fig. 7. This value agrees well with activation energy for volume diffusivity in pure Al (Q_L = 142 KJ/mol). The threshold stress model proposed by Mishra et al [12] suggests that the origin of threshold stress is due to the attractive dislocation-particle interaction. This model as well as other models for threshold stress for dislocation creep is based on interaction of lattice dislocations with particles within the grain. However, as noted earlier, the present alloy has particles mostly on the grain boundaries. For such nanostructured materials, the theoretical framework for dislocation-particle interaction has not been developed.

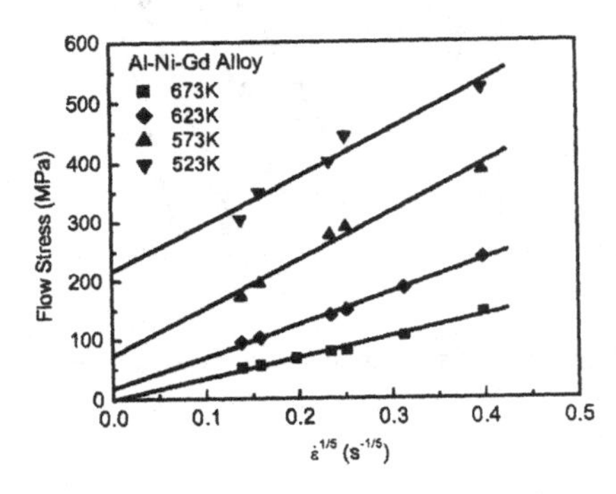

Fig.5. The variation of flow stress as a function of $\dot{\varepsilon}^{1/5}$ in $Al_{87}Ni_7Gd_6$ alloy.

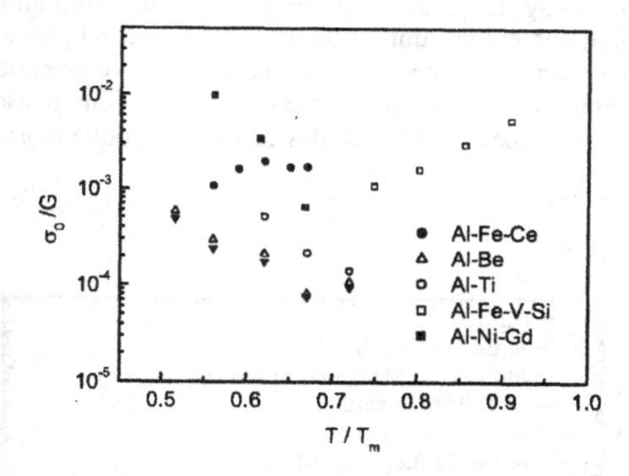

Fig.6. Threshold stress (normalized with respect to the shear modulus) as a function of homologous temperature for dispersion-strengthened alloys

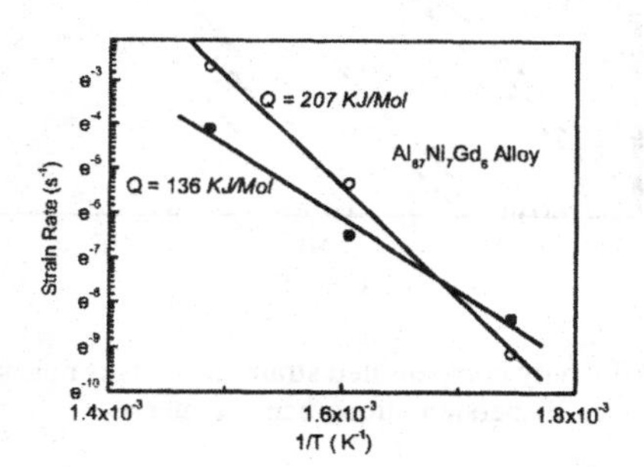

Fig. 7. An Arrhenius plot of strain rate against inverse of temperature

The deformation behavior of a material at elevated temperatures can be represented by the constitutive equation which incorporates threshold stress as given by [13, 14]

$$\dot{\varepsilon} = (AGbD_L)/(kT)\ ((\sigma-\sigma_0)/G)^n$$

where $\dot{\varepsilon}$ is the strain rate, D_L the lattice self-diffusivity, G the shear modulus, b the Burgers vector, k the Boltzmann's constant, T the absolute temperature, σ the applied stress, σ_0 the threshold stress, A the dimensionless constant, n the true stress exponent. The normalized strain rate, $\dot{\varepsilon}\, kT/DGb$, is plotted against the normalized effective stress $(\sigma-\sigma_0)/G$, on double logarithmic scales in Fig. 8. For the data analysis, values of the lattice self-diffusivity were obtained using D_L [M^2/s] = 1.71 x 10^{-4} exp (-142/RT) [15], and G [MPa] = 3.0 x 10^4-16T [16], where R is the universal gas constant. It can be seen that the slope of the fitted line is 5, which indicates the dislocation-climb controlled deformation. For comparison, a number of dispersion-strengthened alloys with metallic/ intermetallic dispersoids and pure Al are also included in this plot [9-12,17]. The nanostructured alloy in the present work exhibits good elevated temperature strength, which is superior to that for conventional Al-based alloys as well as for

Al-based alloys developed by rapid solidification processing. The achievement of good elevated temperature strength of this alloy is presumably because of the formation of the finely mixed nanostructure consisted of high volume fraction of thermal stable second phase particles homogenously distributed in the Al matrix, which cannot be produced by the conventional thermo-mechanical processing. In this case, the high thermal stability of the nanoscale second phase particles is presumably due to the absence of the internal defects in the particles due to the unique fabrication route [2] and slow diffusivity of Gd and Ni.

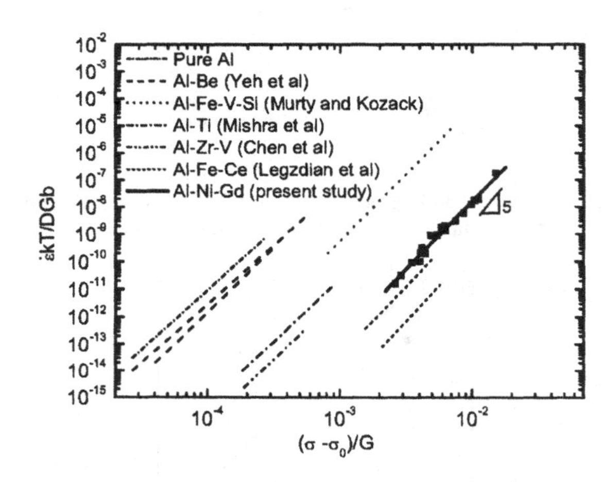

Fig .8 Temperature and diffusivity compensated strain rate versus normalized effective stress for dispersion-strengthened materials.

5. Conclusions

- Homogenous dispersion of fine intermetallic compounds can be obtained by the use of the amorphous precursor. Nanostructured $Al_{87}Ni_7Gd_6$ alloy exhibits good elevated temperature strength.
- The apparent stress exponent was higher than 5 and the apparent activation energy was 207 KJ/mol. The $Al_{87}Ni_7Gd_6$ alloy exhibits temperature-dependent threshold stress.
- The Deformation mechanism is likely to be dislocation-climb mechanism.

6. Acknowledgment

The authors gratefully acknowledge the financial support of DARPA through contract number F336115-01-2-5217.

References

1. Y-W. Kim and W.M.Griffth (eds), *Dispersion strengthened aluminum alloys* (TMS-AIME, Warrendale, PA, 1988), 31
2. A. Inoue, H.M. Kimura, "Fabrication and Mechanical Properties of Bulk Amorphous, Nanocrystalline, Nanoquasicrystalline Alloys in Aluminum-based System" *J. Light Metals*, 1 (2001), 31-41
3. J.S. Benjamin and M.J. Bomford, "Dispersion Strengthened Al Made by Mechanical Alloying" *Metal. Trans. A*, 8A (1977), 1301-1305.
4. F.H. Froes, C.M. Ward-Close and W. Baeslack, "Advanced Synthesis of Light-Weight Metals" *J. Adv. Mats.*, 25(1)(1993), 20.
5. M.C. Gao and G.J. Shiflet, "Devitrification Phase Transformations in Amorphous Al sub 85 Ni sub 7 Gd sub 8 Alloy" *Intermetallics*, 10(11-12)(2002), 1131
6. J.H. Gittus, "Theoretical Equation for Steady-state Dislocation Creep in a Material Having a Threshold Stress," *Proc. Roy. Soc. A*, 342(1975), 279-287.
7. R. Lagneborg, and B. Bergman, "Stress/Creep-Rate Behavior of Precipitation-Hardened Alloys" *Metal. Sci.*. 10(1976), 20.
8. D. Legzdina, T.A. Parthasarathy, "Deformation Mechanisms of a Rapidly Solidified Al--8.8Fe--3.7Ce Alloy" *Metall. Trans. A*, 18A (10)(1987), 1713
9. Y.H. Yeh, N. Nakashima, H. Kurishita, S. Goto, H. Yoshinaga, "Threshold Stress for High-Temperature Creep in Particle Strengthened Al--1.5 vol.% Be Alloys" *Mater. Trans, JIM*, 31(4)(1990), 284
10. R.S. Mishra, A.G.Paradkar, K. N. Rao, "Steady State Creep Behavior of a Rapidly Solidified and Further Processed Al--5 wt.% Ti Alloy" *Acta Metall.Mater.*41 (1993), 2243
11. G.S. Murty, M.J. Koczak, "Rate Sensitivity in the High Temperature Deformation of Dispersion Strengthened Al--Fe--V--Si Alloys" *J. Mater. Sci.*, 24(2)(1989), 510
12. R.S. Mishra, and T.K. Nandy, "The Threshold Stress for Creep Controlled by Dislocation--Particle Interaction" *Phil. Mag. A*, 69 (1994), 1097-1109
13. K.T. Park, E. J. Lavernia, F.A. Mohamed, "High Temperature Creep of Silicon Carbide Particulate Reinforced Aluminum" *Acta Metall.Mater.*38 (1990), 2149
14. K.T. Park, E.J. Lavernia, F.A. Mohamed, "High-temperature Deformation of 6061 Aluminum" *Acta Metall.Mater.*42 (1994), 667
15. J.E. Bird, A.K. Mukherjee, J.E. Dorn, in: D.G. Brandon, A. Rosen (Eds.), *Quantitative Relation between Properties and Microstructure* (Israel University Press, Jerusalem, 1969), 225
16. T. S. Lundy, J. F. Murdock, *J. Appl. Phys.* 33(1962), 1671
17. Y.C. Chen, M.E. Fine, J.R. Weertman, "Microstructural Evolution and Mechanical Properties of Rapidly Solidified Al--Zr--V Alloys at High Temperatures" *Acta Metall.Mater.*38 (1990), 771

Processing and Properties of Structural Nanomaterials
Edited by Leon L. Shaw, C. Suryanarayana and Rajiv S. Mishra
TMS (The Minerals, Metals & Materials Society), 2003

Mechanical Properties of Ultrafine-Grained Titanium Aluminide/Titanium Silicide Composites Prepared by High Energy Milling

Thomas Klassen, Rainer Bohn, C. Suryanarayana,* Georg Fanta, and Rüdiger Bormann

GKSS Research Center Geesthacht GmbH, Institute for Materials Research, Department of Powder- and Nanotechnologies, Max-Planck-Strasse, D-21502 Geesthacht, Germany

*Permanent address: Department of Mechanical, Materials and Aerospace Engineering, University of Central Florida, Orlando, FL 32816-2450, USA

Abstract

This study was undertaken to systematically investigate the powder-metallurgical synthesis and mechanical behavior of intermetallic/ceramic composites with grain sizes in the nano- and submicron range. γ-TiAl with 0.5 to 60 vol.% Ti_5Si_3 content was chosen as a model system. The resulting microstructure was either a fine dispersion of Ti_5Si_3 in the γ-TiAl matrix or intermixed grains of γ-TiAl and Ti_5Si_3, with comparable grain sizes. Four objectives were pursued: (1) measurement of truly grain size-dependent mechanical properties, not affected by any processing flaws, (2) coverage of a wide range of grain sizes, extending from several micrometers down to the nanometer-range, thus ensuring that the results may be related to the data available for materials with conventional grain sizes, (3) clarification, whether the mechanical behavior is governed by chemistry and/or microstructure, and (4) evaluation of the results with respect to the underlying mechanisms. Dense composite material was obtained by high-energy milling and subsequent hot isostatic pressing. At room temperature, the grain size dependence of hardness and yield strength was described by the well-known Hall-Petch relationship. Contrary to the behavior of conventional alloys, the ductility of submicron-grained composites dropped if the grain size was further reduced. This may be attributed to the increasing difficulties for deformation based on single dislocation glide mechanisms. In the high temperature range, the flow stress is strongly reduced. Superplastic deformation became feasible at temperatures $\leq$800 °C, allowing for easy forming of parts. Silicide particles impede grain growth, but they also promote cavitation during tensile straining. Higher silicide content did not change the deformation behavior substantially, indicating that the mechanical behavior is mainly controlled by microstructure rather than chemistry. The mechanisms of deformation are similar to those established for coarse-grained materials, though at significantly reduced temperatures.

Introduction

The mechanical properties of a material strongly depend on its microstructure, which can be controlled by the processing conditions. With the introduction of novel production- and processing-methods like high-energy-milling, inert gas condensation, electrodeposition or severe plastic deformation, it became possible to synthesize materials with very small grain sizes in the submicron- or nanometer range [1-7]. The very high density of two-dimensional defects like grain- and phase-boundaries should significantly influence the mechanical properties of ultrafine-grained materials. However, up to now, there is no complete understanding of the mechanisms that govern the mechanical behavior of these materials. Generally, rules or equations describing the correlation between mechanical data and microstructural parameters of conventional grained materials may either continue or break down, if the grain size falls markedly below 1 μm. Deviations may arise from the fact, that size parameters gradually reach the dimensions of some characteristic lengths related to the original deformation mechanism [8,9]. As an example, applicability of the Hall-Petch relation [9,10], characterizing the correlation between yield stress and grain size at low temperatures, is still controversial. It is unclear, whether dislocation-based mechanisms still dominate in materials with such small grain sizes or if diffusional processes become more significant due the close meshed net of grain boundaries. The existing results are quite inconclusive (e.g. [11,12]). Several reasons might be responsible for that nonuniformity: The complex and often not well characterized preparation methods, processing flaws, or simply the small quantities of available test material can lead to questionable results. Therefore, the purpose of this study is (a) to measure truly grain-size-dependent mechanical properties, unaffected by any processing flaws, (b) coverage of a wide grain size range, in order to ensure that the results may be related to the data available for materials with conventional grain sizes and (c) evaluation of the results in terms of the underlying deformation mechanisms.

The alloy system Ti-Al-Si was chosen because of its technological potential as a structural material for high temperature applications in engines and turbines. However, up to now, γ-TiAl-based compounds suffer from brittleness at low temperatures, high deformation resistance at hot-working temperatures and insufficient creep strength in the high temperature application area of about 700 – 800°C. In the micrometer range, ductility could be slightly improved by alloying additions, processing control and grain refinement [13,14]. Consequently, it would be interesting to observe whether this tendency can be continued in the submicron regime. Silicon was added in order to cause a fine dispersion of second phase particles inside the γ-TiAl matrix. Two intentions were pursued by incorporating these silicides: (i) Grain growth should be minimized during preparation and processing of the material. (ii) With regard to high temperature application properties, nano- and submicron-grained compounds with a fine silicide dispersion might be regarded as a suitable precursor for the development of a creep resistant material.

Experimental

Prealloyed gas-atomized powders of composition Ti-48.9Al and Ti-37.5Si (in at.%) as well as pure silicon powders were mixed in different proportions and high-energy-milled in an 8 l attrition mill (Zoz GmbH) for 72 h. To avoid atmospheric impurities, storage and transport of the powders were conducted under vacuum or inert gas conditions. Milling was performed in an atmosphere of flowing Ar–5 vol.% H_2 to decrease the cold welding between the powder particles and grinding tools. Before consolidation, the powder was degassed by vacuum annealing at 440°C. Fully dense samples were obtained by hot isostatic pressing at temperatures between 750°C and 1150°C, and a pressure of 200 or 300 MPa for 2 hours; the higher temperatures being used for powders containing higher volume fractions of the silicide phase. Full details of the powder processing methods used in these investigations may be found in references 13 and 14.

Microstructural investigations were performed using X-ray diffraction (Siemens D 5000), scanning (Zeiss DSM 962) and transmission electron microscopy (Philips CM 200). Both microscopes were equipped with an energy-dispersive X-ray (EDX) detector. The average grain size of the consolidated samples was determined by the linear intercept method. Microhardness was measured on metallographically polished samples under a load of 1 N, using a depth-sensing micro-indentation instrument (Fischerscope H100) with a resolution of 1 nm, equipped with a Vickers indenter. The apparatus was able to record the time-dependent displacement of the indenter, thus allowing the low temperature creep behavior to be investigated. The mechanical behavior of the consolidated samples was examined both in compression and tension testing. Compression testing was carried out on spark-eroded and ground specimens of cylindrical shape with 4 mm diameter and 8 mm height. The Instron 1195 machine was equipped with a three-zone furnace to ensure constant and uniform heating of both the specimen and the pistons. The compression tests were run in a flowing 99.998% pure argon gas atmosphere to minimize the influence of oxidation. The compression strain was recorded via 3 alumina rods, pressed directly against the specimen or the upper die and connected to an LVDT with an accuracy of 5 μm, which is positioned outside the heating zone of the furnace. The maximum deformation detectable was limited to 1 mm, corresponding to a strain of 12.5%. The compression tests were conducted at different strain rates varying from 4×10^{-5} s^{-1} to 4×10^{-2} s^{-1} and in the temperature range of 900 to 1100 °C. Tensile testing of the samples was conducted on plate-shaped samples with a gauge length of about 22 mm, width of 2 mm and thickness of 1.2 mm. The samples were prepared from the HIP consolidated specimens by electro-discharge machining. Tensile tests were performed at temperatures between 800°C and 1000°C in air.

Results and Discussion

The microstructure of the γ-TiAl intermetallic/ξ-Ti_5Si_3 ceramic composite powders processed by high-energy milling and HIP are very homogeneous. Details about the microstructural features of the composites containing up to 32 vol.% of the silicide phase were described in our earlier publications [15-22]. The mechanically alloyed Ti-31.6Al-21.6Si (at.%) powder HIPed at the highest temperature of 1150 °C contains a volume fraction of the ξ phase of about 60%. The ξ-Ti_5Si_3 grains are completely intermixed with the γ-TiAl grains, with approximately similar grain sizes. In this sense, the microstructure of this alloy is similar to those observed in the other alloys with a high ξ-Ti_5Si_3 content [16,17]. The microstructural data of the composites containing up to 32 vol.% of the silicide phase were recently published in [19,22] along with their impurity contents. Considering the multistep powder-metallurgical processing of the samples, it should be noted that the amount of contamination by interstitial atoms like O (≈ 0.3 at.%) and N (≈ 0.1 at.%) is very low. Nevertheless, the material does not reach the purity of ingots. As the O and N contents probably exceed the solubility limit of the γ-TiAl phase, small amounts of oxides and nitrides may be present within the microstructure.
At low temperatures, submicron-grained alloys based on γ-TiAl are characterized by high values of hardness and compressive yield strength (Fig. 1) [19]. As reported previously [19], this behavior can be explained by the classical Hall-Petch relationship, resulting in an inverse square-root dependence of hardness, H (or yield strength, σ) on grain size, d, according to the equation:

$$\sigma\,(H) = \sigma_o\,(H_o) + Kd^{-1/2} \qquad (1)$$

with the constant K = 1.4 MPa√m and H_o = 2,200 MPa in the case of the hardness data and 1.0 MPa√m and σ_o = 125 MPa for the yield strength data. Consequently, further grain refinement leads to a continued increase of hardness and strength. Figure 1 shows the variation of microhardness with inverse square root of the mean grain size for a series of differently processed γ-TiAl-based alloys and composites over a wide range of grain sizes (obtained by a number of conventional and non-equilibrium processes, including high-energy milling, as in the present investigation). The constants calculated above for K and H_o (σ_o) also fit the data for conventional coarse-grained materials. Hence, it suggests that it is mainly the grain size that is determining the room temperature hardness (and yield strength) of these materials. Similar to observations of Jang et al. on mechanically alloyed, oxide-dispersion-strengthened Ni_3Al [21], the influence of chemistry is – within certain limits – reflected rather indirectly by controlling the microstructure, i.e. the grain size.
On closer inspection of the compression curves shown in Fig.1(b), several further characteristics may be noticed:

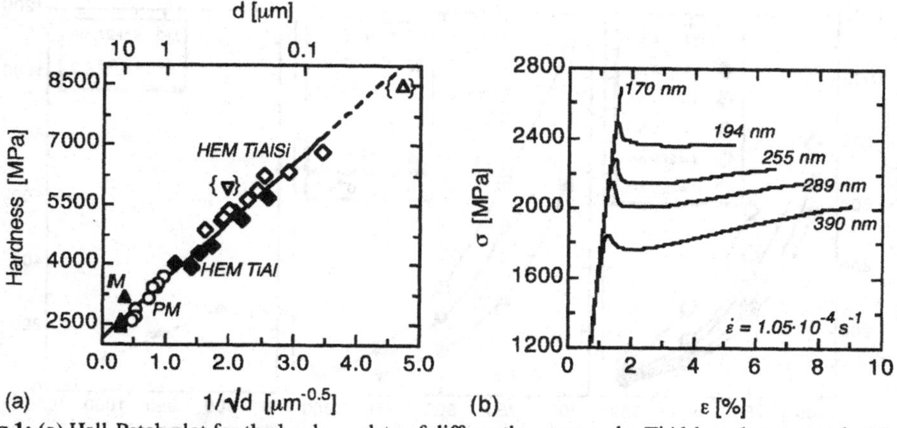

Fig.1: (a) Hall-Petch plot for the hardness data of differently processed γ-TiAl-based compounds: High-energy milled (HEM), cast (IM) and gas-atomized (PM, from [19]); (b) Compression stress-strain curves of Ti-45Al-2.4Si with various mean grain sizes as indicated; the{lower yield stress/grain size}-data couples also show the Hall-Petch-type behavior (see also [19]).

(a) All samples show discontinuous yielding. This phenomenon may be partly caused by the few number of dislocations initially present. Furthermore, the few existing dislocations may be pinned by small particles (oxides, nitrides), dissolved elements like Si and Fe (0.5 at.% Fe originating from wear of the milling tools) and – to a lesser degree – also by low soluble interstitials like O and N.
(b) Strain hardening is reduced with decreasing grain size and increasing strain.
(c) Grain refinement within the submicron range causes a continuous drop of fracture strain.
TEM analysis shows that deformation at room temperature is accomplished by a mixture of dislocation glide and mechanical twinning [19]. However, as the grain size is further reduced, twinning becomes more and more prevailing. In most grains only one twinning system is operative. Therefore, the absence of strain hardening and loss of ductility are quite comprehensible; the parallel twins hardly produce any interaction forces. Additionally, the von Mises criterion of at least five independent slip systems to ensure reasonable plasticity is not met by a deformation solely based on mechanical twinning. This means that dislocation glide and mechanical twinning as the prevailing deformation mechanisms are hampered by the fine-meshed net of grain-boundaries present in submicron-grained alloys.

In the high temperature range above 500°C, these relations change completely, i.e. ultrafine-grained materials become very soft. As demonstrated in Figure 2, there is a grain size-dependent inversion of yield strength, which means that the smaller the grain size of the specimen, the stronger and sharper the drop in yield strength upon raising the temperature. Silicon-containing γ-TiAl-based compounds with a fraction of up to 60 vol.% $Ti_5(Si,Al)_3$ show a similar behavior as the binary TiAl alloys (Figure 3). Grain refinement causes a reduction of flow stress, suggesting favorable conditions for hot-working of ultrafine-grained Ti-Al-Si compounds.

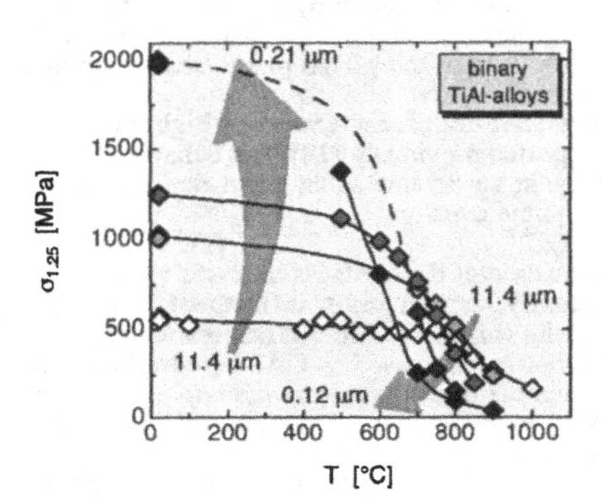

Fig. 2: Compressive yield strength, $\sigma_{1.25}$ (after 1.25% plastic strain) as a function of temperature for binary TiAl-alloys with different grain sizes.

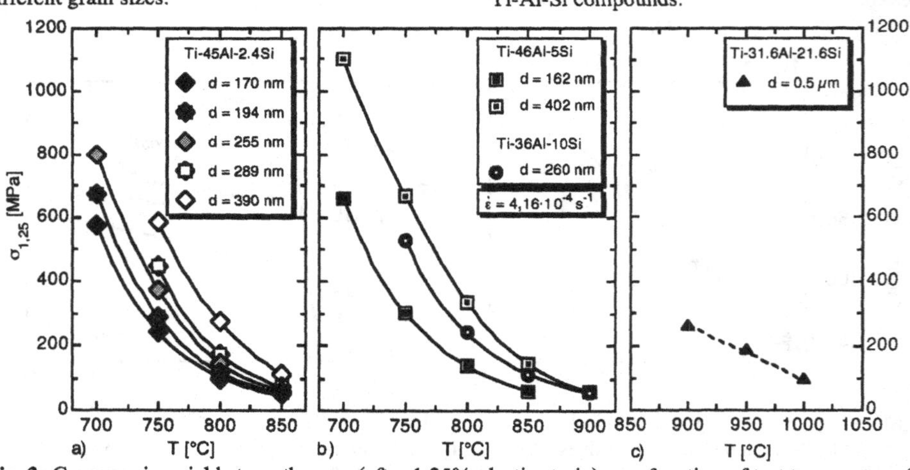

Fig. 3: Compressive yield strength, $\sigma_{1.25}$ (after 1.25% plastic strain) as a function of test temperature for different grain sizes and compositions. (a) Ti-45Al-2.4Si, (b) Ti-46Al-5Si, and Ti-36Al-10Si, and (c) Ti-31.6Al-21.6Si.

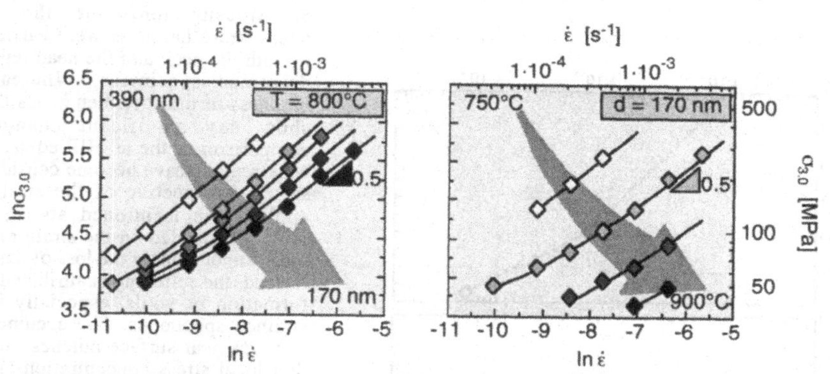

Fig.4: (a) Strain rate dependence of flow stress for compounds of the type Ti-45Al-2.4Si with grain sizes varying between 390 and 170 nm at 800°C (a) or with temperatures increasing from 750°C to 900°C for a grain size of 170 nm(b); $\sigma_{3.0}$ refers to the flow stress after 3 % of true plastic deformation.

Figure 3 (c) shows the compressive yield strength, $\sigma_{1.25}$ (after 1.25% plastic strain) as a function of test temperature for the Ti-31.6Al-21.6Si (at.%) alloy containing 60 vol% of silicides. Similar to the other compositions, the yield strength decreases continuously with increasing test temperature from 262 MPa at 900 °C to 94 MPa at 1000 °C.

All submicron-grained Ti-Al-Si compounds are characterized by a marked strain rate sensitivity at temperatures ≥ 500°C. In Figure 4, the flow stress of Ti-45Al-2.4Si is depicted as a function of the applied compression rate at a constant temperature of 800°C for different grain sizes (a) or at different temperatures for a constant grain size of 170 nm. Both, refining the microstructure or increasing the test temperature lead to softening, i.e. higher strain rates at a given stress or lower stresses at a given strain rate. In the present case, reduction of the mean grain size by a factor of two admits about a fivefold increase of the strain rate. Thus, grain refinement is comparable to a temperature increase with respect to the deformation behavior. This fact is of great technological interest, as it allows hot-working of submicron-grained TiAl alloys at reduced temperatures [18]. The compressive stress vs. strain rate plots at different strain rates for the Ti-31.6Al-21.6Si alloy are shown in Figure 5 and show a similar behavior. For all the compositions investigated, strain rate sensitivity, m, which is the slope of the respective curves, varies between 0.3 and 0.55, suggesting the possibility of superplastic deformation behavior in these alloys.

The strong strain rate dependence of submicron-grained Ti-Al-Si compounds has to be kept in mind if the temperature dependence of the flow stress (Figures 1 and 2) is to be properly interpreted. Especially, from a high flow stress (under the strain rate conditions applied), it does not automatically follow that a good creep resistance will be observed [25].

In order to actually demonstrate the potential of submicron-grained γ-TiAl-based alloys for superplastic deformation at comparatively low temperatures and to verify the results obtained by compressive loading, tensile tests were performed. The results for Ti-45Al-2.4Si are summarized in table 1, and may be described as follows:

1. At strain rates $\dot{\varepsilon} \leq 3.2 \times 10^{-3}\ s^{-1}$ elongations $\varepsilon \geq 225$ % were reached (Figure 6). Larger deformations could not be managed due to the limited length of the furnace. By increasing the strain rate to $\dot{\varepsilon} = 6.4 \times 10^{-3}\ s^{-1}$ the specimens fail after elongations of 103 or 113 %, respectively.
2. The coefficients of the strain rate sensitivity are comparable to those obtained by compression testing. Increasing the strain rate leads to higher strain rate sensitivities.
3. Concurrent with an increase of $\dot{\varepsilon}$ and m, the fracture strain drops. However, necking of the samples is not observed.
4. The microstructure has conserved its equiaxed grain character after deformation. Contrary to compression loaded specimens, limited grain growth occurs. However, within the strain rate range investigated, increasing the deformation rate leads to a reduction of grain growth.

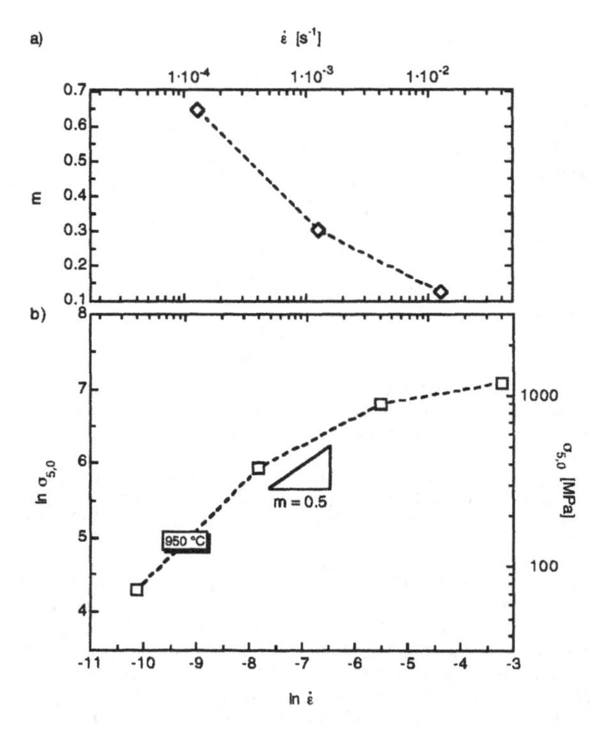

Fig.5: Strain rate dependence of compressive flow stress $\sigma_{5.0}$ for Ti-31.6Al-21.6Si and respective strain rate exponent m.

5. Measurements of the room temperature hardness were carried out in both the neck and the head region of deformed specimens. Whereas the hardness of the specimen heads did not show any significant changes in comparison to the as-HIP'ed state, the neck regions have become considerably softer, irrespective of the strain rate applied. As mentioned above, grain growth depended on the strain rate and thus cannot explain the loss of hardness. Instead, the softening is attributed to the formation of voids, especially in fast strained specimens. The accumulation of pores near surface notches indicates that local stress concentrations largely contribute to cavitation.

In order to elucidate, whether high temperature deformation is also governed by microstructure rather than chemistry, tensile tests on the Ti-31.6Al-21.6Si specimen containing 60 vol% of silicides were performed. In this case, a 3d-continuous network of the silicide phase can be expected. Nevertheless, total extensions of up to 250% of the initial value were achieved (Fig. 7). Together with the strain rate exponent around 0.5 (Fig. 5 (a)) and the equiaxed microstructure after testing [26], potential superplasticity is thus confirmed for high silicide content alloys also.

sample	ε_1	ε_2	m	$\varepsilon_{fracture}$[%]	porosity [%] head / neck
F	$3.2\cdot10^{-3}\,s^{-1}$	$6.4\cdot10^{-3}\,s^{-1}$	0.48	103	0.15 / 1.10
E				113	
D	$1.6\cdot10^{-3}\,s^{-1}$	$3.2\cdot10^{-3}\,s^{-1}$	0.45	≥ 175	
C	$8\cdot10^{-4}\,s^{-1}$	$1.6\cdot10^{-3}\,s^{-1}$	0.43	≥ 175	
B	$4\cdot10^{-4}\,s^{-1}$	$8\cdot10^{-4}\,s^{-1}$	0.40	≥ 175	0.06 / 0.41

Table 1: Data of tensile tests for Ti-45Al-2.4Si.

Based on a stress exponent of $n = 1/m \approx 2$ and a grain size exponent of $p = 2.23$ [17], we suggest superplastic deformation as general deformation mechanism for all compositions of submicron-grained Ti-Al-Si compounds. Deformation is accomplished by grain-boundary sliding, which is accommodated by diffusional processes within the grains. Such a mechanism would not essentially differ from the deformation modes acting in coarse-grained material. However, it is effective at significantly lower temperatures.

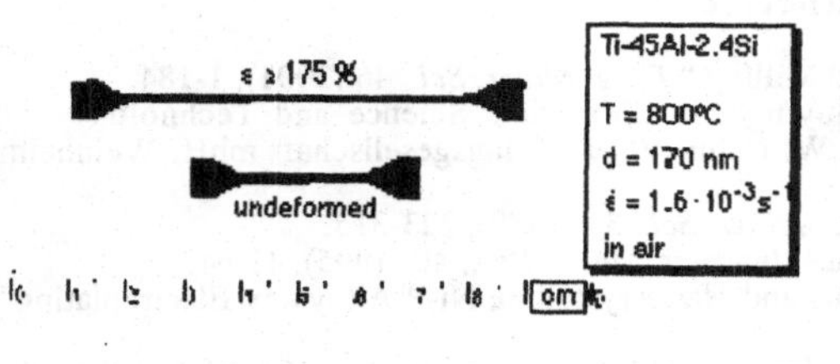

Fig.6: Alloy Ti-45Al-2.4Si, tensile-strained at 800°C in air. The test was stopped as the zone of constant temperature in the furnace was exceeded.

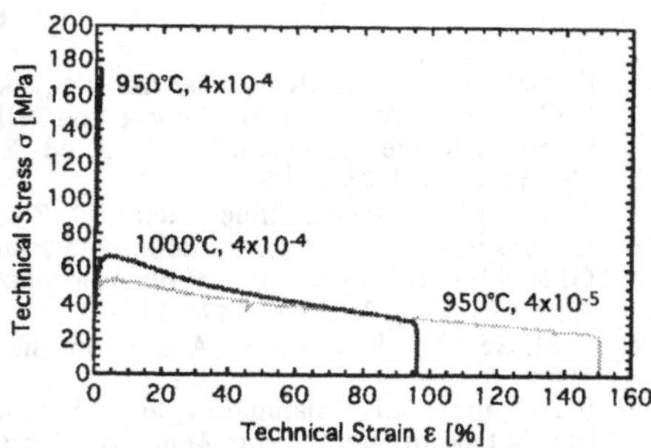

Fig.7: Alloy Ti-31.6Al-21.6Si, tensile-strained at different temperatures in air until fracture.

Summary and Conclusions

1. Multiphase γ-TiAl-based compounds with a very fine-grained microstructure were produced by high-energy milling and hot isostatic pressing. The grain size of the γ-TiAl matrix varied between about 160 and 480 nm, whereas the size of precipitated ξ- $Ti_5(Si,Al)_3$ particles varied between 80 and 190 nm, depending on the HIP conditions.
2. At room temperature, hardness and yield strength follow the Hall-Petch-Relation, i.e. dislocation glide and mecha-nical twinning as the prevailing deformation mechanisms are hampered by the fine-meshed net of grain-boundaries present in submicron-grained alloys.
3. At temperatures above 500°C, these alloys become very soft. The smaller the grain size, the strongeris the alloy and faster is the drop of flow stress. Deformation occurs without any strain hardening.
4. During deformation at high temperatures, the finely dispersed silicide particles prevent grain growth as long as the HIP temperature is not exceeded.
5. The flow stress is rather sensitive to the applied strain rate. Superplasticity becomes feasible at 800°C, allowing elongations ≥ 225 %. In connection with a stress exponent of around $n = 2$ and a grain size exponent of $p = 2.2$, it is concluded that superplastic deformation is accomplished by grain boundary sliding accommodated by diffusional processes inside the γ-TiAl grains. Such a mechanism does not differ from the deformation modes described for conventionally grained materials deformed at temperatures ≥ 1000°C.
6. At low silicide content, the silicide phase is not directly involved in the process of deformation. In particular, Ti-36Al-10Si compounds with about 32 vol.% of the silicide phase reveal the same activation energy as alloys of composition Ti-45Al-2.4 Si with only 9.6 vol.% $Ti_5(Si,Al)_3$. In both cases only the γ-TiAl phase seems to contribute to the overall deformation.
7. At high silicide content of about 60 vol%, the silicide phase forms a continuous network. Nevertheless, similar superplastic behavior is found at higher temperatures, indicating that the deformation behavior at small grain sizes is controlled mainly by the microstructure rather than the chemical composition or phase fractions.

Acknowledgments

This work was supported by the German Science Foundation (Deutsche Forschungsgemeinschaft, DFG) within the scope of Collaborative Research Center (Sonderforschungsbereich, SFB) 371. Thanks are due to Dr. R. Gerling for providing the powders for high energy milling. The experimental help of U. Lorenz and R. Behn is gratefully acknowledged. C. Suryanarayana is grateful to the German Science Foundation (Deutsche Forschungsgemeinschaft, DFG) for award of a fellowship to work at the GKSS Research Center during part of the summers of 2002 and 2003.

References

1. C. Suryanarayana, „Mechanical Alloying and Milling," *Prog. Mater. Sci.,* 46 (2001), 1-184.
2. C.C. Koch, „Mechanical Milling and Alloying", In „Materials Science and Technology – A Comprehensive Treatment", vol. 15, ed. R.W. Cahn, VCH Verlagsgesellschaft mbH, Weinheim, Germany, 1991, 193-245.
3. H. Gleiter, „Nanocrystalline Materials", *Prog. Mater. Sci.*, 33 (1989), 223-315.
4. C. Suryanarayana, „Nanocrystalline Materials", *Internat. Mater. Rev.,* 40 (1995), 41-64.
5. G. McMahon, and U. Erb, „Bulk Amorphous and Nanocrystalline Ni-P Alloys by Electroplating", *Microstr. Sci.*, 17 (1989), 447-457.
6. V.M. Segal, V.I. Reznikov, A.E. Drobyshevskiy, and V.I. Kopylov, *Russian Metall.*, 1 (1981), 99-105.
7. R.Z. Valiev, R.K. Islamgaliev, and I.V. Alexandrov, „Bulk Nanostructured Materials from Severe Plastic Deformation", *Prog. Mater. Sci.,* 45 (2000), 103-189.
8. E. Arzt, „Size Effects in Materials due to Microstructural and Dimensional Constraints: A Comparative Review," *Acta Mater.*, 46 (1998), 5611
9. A. Lasalmonie, J.L. Strudel, „Influence of Grain Size on the Mechanical Behaviour of Some High Strength Materials", *J. Mater. Sci.*, 21 (1986), 1837-1852.
10. E.O. Hall, „The Deformation and Ageing of Mild Steel: III. Discussion and Results",*Proc. Phys. Soc.*, B64 (1951), 747-753.
11. D.G. Morris, „Mechanical Behavior of Nanostructured Materials", Vol. 2 of Materials Science Foundations, Trans Tech Publications, Zurich, Switzerland, 1998.
12. R.W. Siegel and G.E. Fougere, „Mechanical Properties of Nanophase Materials", In "Nanophase Materials: Synthesis-Properties-Applications", eds. G.C. Hadjipanayis and R.W. Siegel, Kluwer, Dordrecht (1994), pp. 233-261.
13. F.H. Froes, C. Suryanarayana, and D. Eliezer, „Synthesis, Properties and Applications of Titanium Aluminides", *J.Mater. Sci.*, 27 (1992), 5113-5140.
14. Y.-W. Kim, Ordered Intermetallic Alloys, Part III: Gamma Titanium Aluminides," *JOM,* 46 (1994), 30-39.
15. T. Klassen, M. Oehring, R. Bormann, „Microscopic Mechanisms of Metastable Phase Formation During Ball Milling of Intermetallic TiAl Phases," *Acta Mater.*, 45 (1997), 3935-3948.
16. R. Bohn, G. Fanta, T. Klassen, R. Bormann, "Submicron-Grained Multiphase TiAlSi Alloys: Processing, Characterization and Microstructural Design", *J. Mater. Res.*, 16 (2001), 1850-1861.
17. R. Bohn, T. Klassen, and R. Bormann, "Mechanical Behavior of Submicron-Grained γ-TiAl based Alloys at Elevated Temperatures," *Intermetallics,* 9 (2001), 559-569.
18. G. Fanta, R. Bohn, M. Dahms, T. Klassen, R. Bormann, „The Effect of Ultrafine Grained Microstructures on the Hot-Workability of Intermetallic/Ceramic Composites based on γ-TiAl," *Intermetallics,* 9 (2001), 45-49.
19. R. Bohn, T. Klassen, and R. Bormann, „Room-Temperature Mechanical Behaviour of Silicon-Doped TiAl Alloys with Grain Sizes in the Nano- and Submicron Range," *Acta Mater.,* 49 (2001), 299-311.
20. R. Bohn, T. Klassen, R. Bormann, „Compaction of High-Energy-Milled TiAlSi Powders by HIP: A Simple Estimation of the Finest Grain Size Achievable," *Adv. Eng. Mat.,* 3 (2001), 238-242.
21. R. Bohn, M. Oehring, T. Pfullmann, F. Appel, and R. Bormann. In "Processing and Properties of Nanocrystalline Materials", eds. C. Suryanarayana et al., TMS, Warrendale, PA, 1996, pp. 355-366.
22. R. Bohn, G. Fanta, T. Klassen and R. Bormann, „Microstructural Characterization and Mechanical Behavior of Nano- and Submicron-Grained Titanium Aluminide/Titanium Silicide Composites Prepared by High Energy Milling," in:"Powder Materials: Current Research and Industrial Practices", eds. F.D.S. Marquis, N. Thadhani, and E.V. Barrera, TMS, Warrendale, PA, 2001, 107-126.
23. M. Oehring, F. Appel, T. Pfullmann, R. Bormann, „Mechanical Properties of Submicron-Grained TiAl Alloys Prepared by Mechanical Alloying", *Appl. Phys. Lett.,* 66 (1995), 941-943.
24. J.S.C. Jang, C.C. Koch, „The Hall-Petch Relationships in Mechanically Alloyed Ni_3Al with Oxide Dispersoids", *Scripta Metall.*, 22 (1988), 677-682.
25. G. Fanta, R. Bohn, M. Dahms, T. Klassen, R. Bormann, "Advanced Processing and Application Properties of γ-TiAl by Microstructure Design", Proceedings of Materials Week 2000 München, Werkstoff-Informationsgesellschaft mbH, Frankfurt, Germany (2002), on CD-ROM.
26. C. Suryanarayana, T. Klassen, R. Bormann, unpublished.

Processing and Properties of Structural Nanomaterials
Edited by Leon L. Shaw, C. Suryanarayana and Rajiv S. Mishra
TMS (The Minerals, Metals & Materials Society), 2003

Mechanical Properties/Microstructure Correlation in a Devitrified Fe-Based Metallic Glass

N.A. Mara[1], A.V. Sergueeva[1], A.K. Mukherjee[1]

[1]University of California, Davis, One Shields Avenue, Davis, CA 95616

Keywords: mechanical properties, nanocrystalline, devitrified metallic glass

Abstract

One of the greatest challenges of testing the high-temperature mechanical properties of nanocrystalline materials is to find a single-phase material that is stable at test temperature. Tensile tests of different microstructures arising from different heat treatments of Fe-based metallic glass are presented. Most interestingly, a single phase α-Fe microstructure with equiaxed, randomly oriented 15 nm grains is produced through the devitrification of Vitroperm ($Fe_{73.5}Cu_1Nb_3Si_{15.5}B_7$) metallic glass. This microstructure was stable during tensile testing at 600°C, showing a strain rate exponent correlating to grain boundary sliding (m=0.5), but little ductility, and strengths to 1250 MPa. It is suggested that the lack of dislocation activity at the small grain sizes can cause the observed brittle behavior. At temperatures up to 725°C, grain growth occurs, leading to elongations as large as 65% at flow stresses of 250 MPa. This investigation is supported by NSF, Division of Materials Research, grant NSF-DMR-0240144.

Introduction

In recent years, increasing amounts of investigation has been conducted in the area of room temperature mechanical properties of nanostructured materials. However, due to the inherent instablility of the extremely fine microstructures at test temperatures, investigation of mechanical properties at elevated temperatures is still rare. The constitutive relation that dictates the dependence of strain rate on stress, grain size and temperature is given by the Mukherjee, Bird, Dorn (MBD) equation: [1].

$$\dot{\varepsilon} = A[(DGb/kT)](b/d)^p(\sigma/G)^n \quad (1)$$

where $\dot{\varepsilon}$ is the steady-state strain rate, D is the appropriate diffusivity (volume, grain boundary, or chemical interdiffusivity), G is the shear modulus, b is the Burger's vector, k is Boltzmann's constant, T is the test temperature, d is the grain diameter, p is the grain size exponent, σ is the applied stress, and n is the stress exponent (n=1/m, m=strain rate sensitivity of the flow stress). Of these values, the stress exponent, diffusivity, and grain size exponent are all mechanism-dependent, and their values can give good insight to the dominant deformation mechanism for a given set of test conditions.

Based on findings in the literature, the characteristic strain rate exponent values can range widely, as can the activation energy for deformation, which will usually vary from lattice diffusion to grain boundary diffusion of certain species responsible for controlling the rate of deformation, depending on the deformation mechanism at hand.[2-6] This range of scatter is typical in the available experimental data, and shows that simply having a fine grain size does not necessarily mean that a material will deform at elevated temperatures according to the classical superplasticity-type mechanisms that were developed for microcrystalline solids where finer grain size generally yielded better superplastic ductility. There are clearly other possibilities for high-temperature deformation mechanisms at diminishing length scales. One of the great difficulties in characterizing the deformation of nanomaterials is the number of variables in the test matrix. In the literature, not only is the grain size varied, but also the chemistry, grain size distribution, number of phases, etc. As such, the literature data is still not complete enough to make a definitive model for the mechanisms of high temperature plasticity at the nanoscale. To best fill in some of these knowledge gaps, what is ideally needed is a specimen that has uniform chemistry, that is relatively stable at high test temperatures, and where a method for varying the grain size from a few nanometers to a few microns would be available. Among the many methods of producing nanostructured materials, devitrification of amorphous solids offers some of these attractive advantages. For example, with the correct chemistry, devitrification allows for a wide range of microstructures to be evolved by simply changing annealing conditions without the problems of contamination or residual porosity that are inherent to other processing methods such as electrodeposition or inert-gas condensation. With the right type of chemical composition, as long as the annealing is carefully carried out, the retention of unwanted amorphous or other phases can be minimized, and grain sizes as low as 20 nm can be generated. [7] Alloys of composition close to that of FINEMET iron-based metallic glass show great promise in this area, although the original FINEMET [8] alloy with a composition of $Fe_{73.5}Cu_1Nb_3Si_{13.5}B_9$ can encounter a primary crystallization process that results in a multiphase structure consisting of differing types of iron borides ($(FeNb)_2B$, FeNbB) in addition to an α-iron/silicon solid solution phase. [9] Complex microstructures such as these are usually not conducive to studies of plasticity due to the inherent brittleness of the crystalline phases attributable to the lack of slip systems needed to accommodate dislocation motion. However, certain changes in alloy chemistry developed later in this family of alloys such as seen in Vitroperm ($Fe_{73.5}Cu_1Nb_3Si_{15.5}B_7$), can lead to a product with the ability to be devitrified to a fully one-phase material with an extremely fine microstructure.

In this paper, a commercially available metallic glass (Vitroperm, $Fe_{73.5}Cu_1Nb_3Si_{15.5}B_7$) was crystallized into a single phase structure of α-Fe with a grain size of approximately 15 nm. This crystallized structure was then subjected to constant strain rate tensile testing in an Argon atmosphere at different temperatures and strain rates. Analysis of the microstructure before and after testing using analytical transmission electron microscopy (TEM), high-resolution TEM (HRTEM), and X-ray diffraction (XRD) reveals a change in microstructure and sheds some light on the rate-controlling resulting deformation mechanism at test temperatures.

Experimental Procedures

Strips of melt-spun Vitroperm ribbon approximately 2 mm wide, ~20 microns thick, and with a gage length of 4 mm were used in this investigation. Tensile testing was carried out at a variety of strain rates using a custom-built mini tensile tester with displacement resolution of 5 microns and load resolution of ~1 gram that is capable of testing materials with a gage length and width as small as 1 mm, and thicknesses down to 5 microns. All testing was carried out in a high-purity (99.999%) argon atmosphere to minimize effects of oxidation. A 600°C anneal for 1 hour was performed *in-situ* in the tensile tester prior to tensile testing to give a microstructure consisting of randomly oriented, equiaxed 15 nm α-Fe grains.

Microstructural analysis was conducted using a Topcon 002B TEM at 190 kV for conventional bright/dark field/diffraction microscopy, and a Philips CM-200 at 200 kV for analytical (EDS, EELS) and high-resolution microscopy. X-ray analysis was performed using a Scintag XDS 2000 X-ray diffractometer with a 3 axis goniometer and Cu K_α radiation. DSC scans were made on a Perkin-Elmer

DSC 7 from ambient to 725°C at a heating rate of 40°C/minute. This heating rate was chosen because it mimics the heating rate encountered in the tensile testing furnace.

Results

As-received microstructure:

The as-received microstructure of Vitroperm is not completely glassy. It contains a number of α-Fe crystallites, as evidenced in Figure 1a.

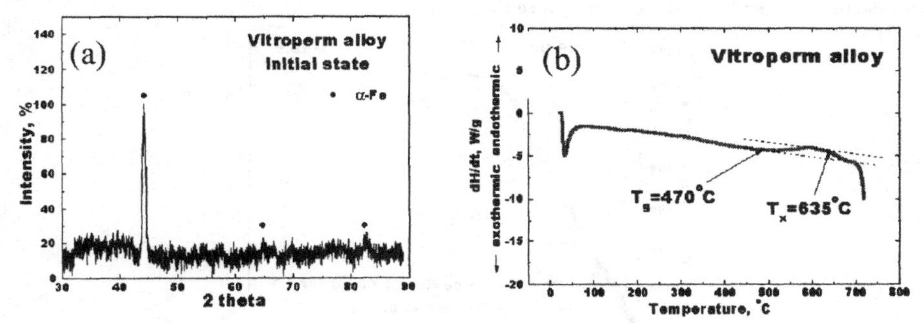

Figure 1: (a) XRD scan of as-received Vitroperm material. Note the presence of α-Fe crystallites. (b) DSC trace for as-received Vitroperm at a heating rate of 40°C/min. Note the values of the crystallization temperature T_x and the glass transition temperature T_g.

By coupling the information seen in the above figure with a DSC scan seen in Figure 1b, it is clear that since nanocrystallites are already present in the amorphous matrix, a short anneal below the crystallization temperature T_x can lead to the production of a devitrified structure. An anneal at 600°C for 1 hour was carried out in order to fully crystallize the sample into a single phase microstructure without inducing crystallization of a second phase that may have a different composition than the existing α-Fe nanocrystals. This simple anneal produces the microstructure seen in Figure 2.

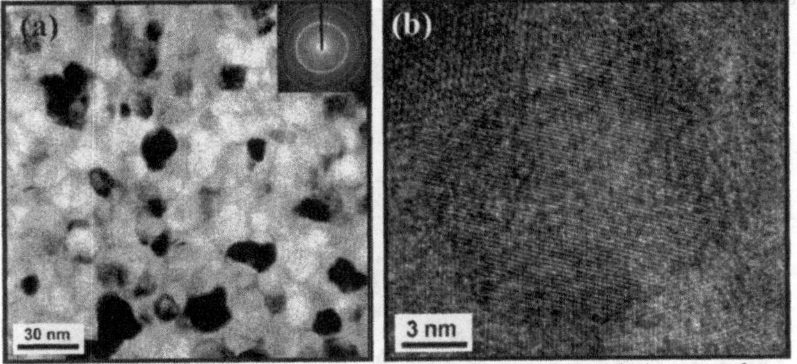

Figure 2: (a) Microstructure and Selected Area Diffraction (SAD) pattern of Vitroperm after annealing at 600°C for 1 hour.
(b) HREM image of Vitroperm after annealing at 600°C for 1 hour.

The structure is single-phase α-Fe, as seen in the electron diffraction pattern, and verified by XRD (not shown). All of the alloying elements are in solid solution in the α-Fe matrix, and the ~15 nm grains are very equiaxed and randomly oriented. In order to verify that there was no residual amorphous phase at the grain boundaries, HRTEM investigation was carried out. Figure 2(b) depicts a typical micrograph. All of the boundaries and triple points investigated showed no evidence of a residual amorphous phase and appeared clean of contamination by other phases. This 15 nm equiaxed, randomly oriented, single phase microstructure seen above becomes the basis for the following investigation of its mechanical properties in tension at high temperature. When tested at 600°C at varying strain rates, the material shows the limited ductility and high strength shown in Figure 3.

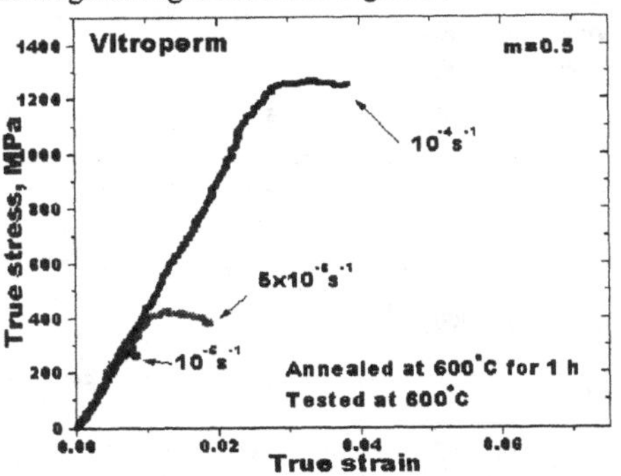

Figure 3: True stress/True strain curve for Vitroperm after 600°C 1h anneal, tested at 600°C.

It is to be noted that the strain rate sensitivity value, m for these testing conditions is 0.5. As will be discussed later, this value usually corresponds to grain boundary sliding related elevated temperature deformation mechanism like superplasticity. However, there is hardly much plasticity revealed in the plot.

A look at the structure of the 600°C sample after testing at strain rate 10^{-4} shows little grain growth, and retention of the original one phase microstructure. (Figure 4)

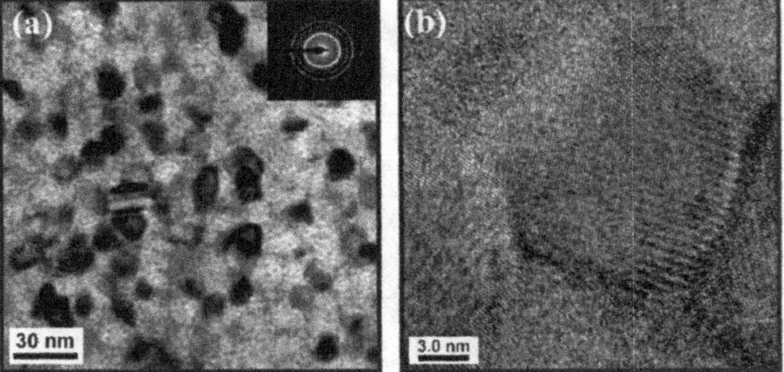

Figure 4: (a) Microstructure of the 600°C, strain rate=10^{-4} sample after tensile testing. Single phase α-Fe structure is retained. **(b)** Corresponding high resolution TEM image.

In contrast, the behavior at 725°C at the same strain rate is much different. Plasticity is enhanced, steady-state flow is reached, and flow stress values are much lower as seen in Figure 5.

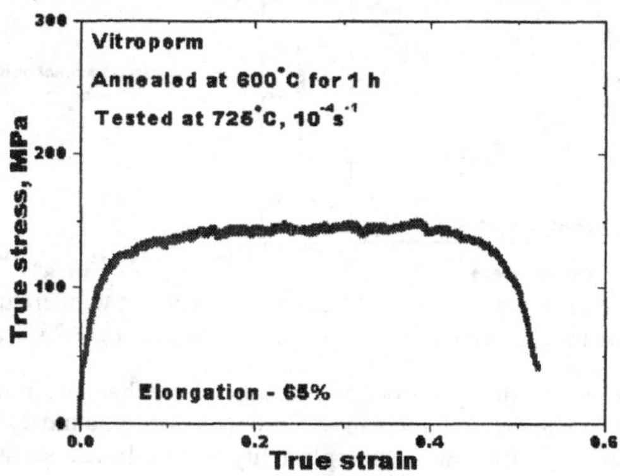

Figure 5: Plot of True Stress/True Strain for Vitroperm after 600°C 1h anneal, tested at 725°C.

As would be expected, the microstructure at the end of this test differs greatly from that found in samples tested at 600°C, as is shown in Figure 8.

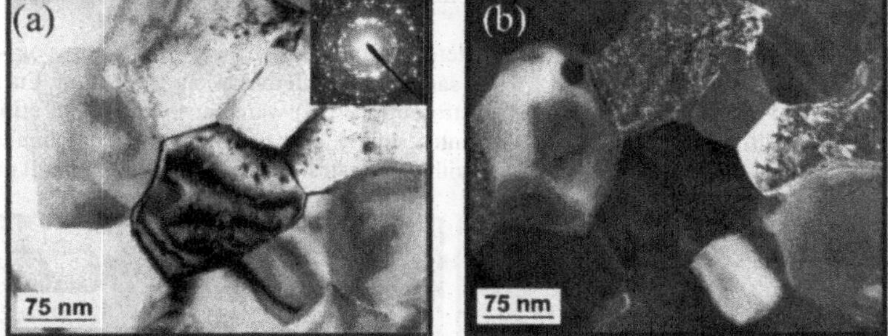

Figure 8: (a) Bright Field image and selected area diffraction pattern of Vitroperm after 600°C 1h anneal, tested at 725°C to a true strain of 0.55. (b) Corresponding dark field image.

From the above selected area electron diffraction pattern and corresponding bright and dark field images, it is clear that a second phase has nucleated at some of the triple points, and the larger grains have grown to ~125 nm in diameter. However, the grains have remained remarkably equiaxed, considering that they have undergone 65% elongation.

Through the use of several isotemperature strain-rate jump tests (not shown), a logarithmic true stress vs. logarithmic true strain plot can be constructed as depicted in Figure 9a. Figure 9b shows the Arrhenius plot derived from Figure 9a, where strain rate vs. temperature values were taken at an isostress condition at 170 MPa across three test temperatures (675-725°C) and two orders of magnitude of strain rate.

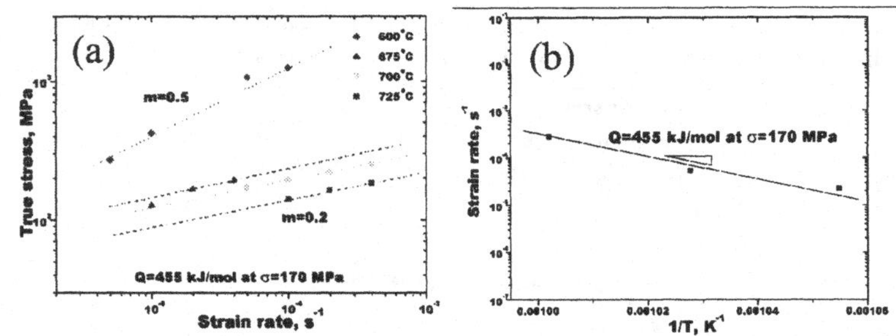

Figure 9: (a) Plot of results for strain-rate jump tests at different test temperatures. (b) Arrhenius plot revealing activation energy value for tensile deformation at 675°C and above.

As is evident in the above figure, there is a definite change in mechanism from the tests at 600°C and those at 675°C and above. The slope that gives m=0.5 points to grain boundary sliding as a deformation mechanism, which is usually associated with superplasticity, but in these tests at 600°C, the deformation behavior is mostly elastic, with minimal plasticity. For the ductile specimens above 675°C, the slope yields a strain rate exponent of 0.2 that is usually tied to power-law creep; a mechanism which depends on dislocation climb and lattice diffusion to accommodate the plastic flow, and may result in elongation of the deformed grains in the absence of grain rotation. We do not see elongation of grains that could correspond to 65% ductility in the post-tested TEM micrographs.

Crucial to the investigation of the rate-controlling deformation mechanism at temperatures of 675°C and above is the composition of the grains in the sample after deformation. Electron Energy Loss Spectroscopy (EELS) and Energy Dispersive Spectroscopy (EDS) were carried out in an effort to find the components of the final microstructure. Most interestingly, an EELS map based on Niobium shows the segregation of Nb to some of the large grains, with the greatest concentration in the small particles at the triple points in Figure 10.

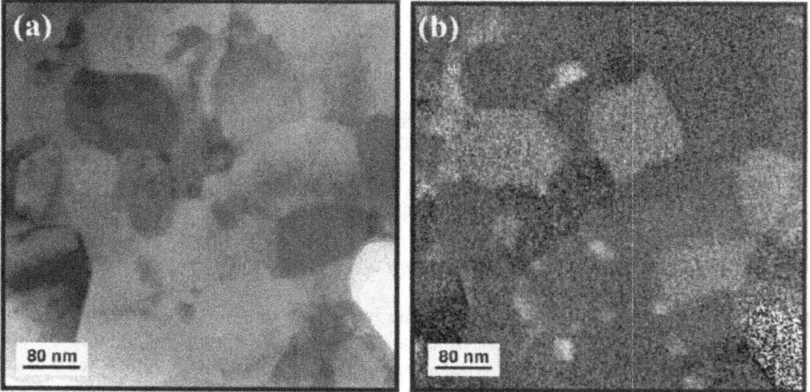

Figure 10: (a) shows the bright field microstructure after testing at 725°C at strain rate 10^{-4}; (b) is the EELS map of Nb, with the lighter areas corresponding to greater Nb concentration.

Further investigation using EDS confirms the EELS results, as well as reveals segregation of copper to the grain boundaries (not shown).

Discussion

For the 600°C tests, the microstructure consists of a one-phase, 15-20 nm grain size material throughout the duration of the test. Usually, for mechanisms with m=0.5, grain boundary sliding occurs, and at stress concentrations such as triple points, dislocations are emitted and slip and climb their way to another grain boundary to accommodate the deformation. This model assumes that the grain size is sufficient to permit dislocations to form and propagate through the grain interior or along grain boundaries. At grain sizes as low as 15 nm, dislocations may not exist, so there is no way to accommodate the strain after sliding has occurred, unless by diffusion-based mechanisms, which are probably not working fast enough in this case to avoid fracture. This results in a stress concentration at triple points that cannot be relieved, so a void nucleates and quickly grows, leading to early fracture. The strain rate exponent of m=0.5 corresponds to the actual sliding of the grains as the rate-limiting step for deformation, since no other mechanism, especially those operated by dislocations, is present.

For the 675°C to 725°C tests, there is evolution of another phase at the triple points of the original grains. If clean, deformation-free grains are formed at stress concentrations such as triple points through the diffusion of Nb-containing phases, the growth of these phases could explain the accommodation of the strain seen in the tensile tests. If it is assumed that grain boundary sliding occurs, which is reasonable since grains remain equiaxed after deformation, stress concentrations such as triple points and particles at grain boundaries can help to emit dislocations. Once emitted, the dislocations must move their way through the matrix to aid in accommodation of the resulting strain. In this case, it is possible for either grain boundary sliding, dislocation glide, or dislocation climb to be the rate-limiting step. With a strain rate sensitivity value of m=0.2, this rate limiting step should be the same as seen in power-law creep—the climb of dislocations within the grain interior. In some of the Fe-rich grains (See Figure 8), there is some sort of particle-like diffraction contrast viewable in the center of the grains which has not been yet identified, and may account for the obstacles needed to justify the above mechanism of deformation.

The Arrhenius plot in Figure 9(b) shows an activation energy value of 455 kJ/mol that is higher than any literature values for lattice self-diffusion or the other present species in α-Fe. This value reflects the complexity of diffusion-based processes in FINEMET alloys that have been found in other studies [10]. Due to the large number of alloying elements contained in the α-Fe initial structure, and the formation of secondary phases during tensile testing, it is difficult to attribute this activation energy to any particular diffusion mechanism at this time. Analysis of the secondary phases is to be conducted in future work, and this may shed some light on the significance of this value.

Other authors have found that in FINEMET metallic glasses, Nb tends to reside at the grain boundaries in the as-crystallized structure and will pin the boundaries, leading to a structure that is quite stable at high temperatures. [11] This coincides with the above results, in that the grain size is very stable up to a certain point, and once Nb is segregated to its own phase away from the α-Fe grain boundaries, grain growth of all phases quickly proceeds. Furthermore, if the 15 nm grains are too small to have dislocation activity to accommodate the deformation at triple points, it is logical that once the grains have grown sufficiently to allow dislocation motion that the strength should drop, and plasticity should increase. This is in fact exactly the trend that is exhibited in the data. An important question that remains to be answered is: "At what critical grain size do dislocations begin to take part in deformation?" Experiments are planned to tailor the grain size to try to find an answer to this question.

Conclusions

A single-phase α-Fe matrix with grain size of ~15 nm was produced through the devitrification of Vitroperm metallic glass. The grains are stable during tensile testing at 600°C, and show great strength, but little ductility. However, an analysis of strain rate sensitivity shows a strain rate exponent of 0.5, which usually corresponds to grain boundary sliding as the rate limiting step. Testing at higher temperatures up to 725°C reveals formation of at least one other phase, grain growth to approximately 150 nm, 65% elongation at much lower flow stresses, and a transition to a possible dislocation-climb based deformation accommodation mechanism, although no grain elongation was observed. TEM investigation shows that Nb has segregated to new grains mainly located at triple points of the original structure. Nucleation and growth and the associated grain boundary migration of these new phases is expected to play an important role in deformation accommodation processes.

Acknowledgements

The authors would like to acknowledge the National Science Foundation, Division of Materials Research for its financial support under grant NSF-DMR-0240144.

References

[1] J.E. Bird, A.K. Mukherjee, J.E. Dorn, "Correlations Between High-Temperature Creep Behavior and Structure" Int. Conf. Proc., Israel Univ. Press (1969) 255.

[2] R.S. Mishra, A.K. Mukherjee, "Superplasticity in nanomaterials" Superplasticity and Superplastic Forming 1998. Proceedings of a Conference held as part of the TMS Annual Meeting. TMS. (1998) 109.

[3] A.V. Sergueeva, V.V. Stolyarov, R.Z. Valiev, A.K. Mukherjee, Scripta Materialia, 45 (2001) 747.

[4] H. Hahn, R.S. Averback, Journal of the American Ceramics Society, 74 (1991) 2918.

[5] K. Taketani, A. Uoya, K. Ohtera, T. Uehara, K. Higashi, A. Inoue, T. Masumoto, Journal of Materials Science, 29 (1994) 6513.

[6] K. Higashi, "Positive exponent superplasticity in advanced aluminum alloys with nano or near-nano scale grained structures". Materials Science and Engineering A, 166, (1993) 118.

[7] Xue-Dong Liu, Met. Trans., 39 (1998) 783.

[8] Y. Yoshizawa, S. Oguma, K. Yamauchi, Journal of Applied Physics. 64 (1988) 6044.

[9] N.I. Noskova, A.P. Potapov, and V.A. Lukshina, "Nanocrystallization of Metallic Glasses Under Different Conditions", Materials Science Forum, 307 (1999) 125-134.

[10] E. Illekova, "FINEMET-type nanocrystallization kinetics", Thermochemica Acta, 387 (2002) 47.

[11] T. Kulik, Journal of Non-crystalline Solids, 287 (2001) 145.

Processing and Properties of Structural Nanomaterials
Edited by Leon L. Shaw, C. Suryanarayana and Rajiv S. Mishra
TMS (The Minerals, Metals & Materials Society), 2003

Grain Size Effects in Nanocrystalline Electrodeposits

U. Erb[1], G. Palumbo[2], D. Jeong[1], S. Kim[1], K. T. Aust[1],

[1]University of Toronto, Dept. of Materials Science and Engineering, Toronto, Canada, M5S 3E4
[2]Integran Technologies Inc., 1 Meridian Road, Toronto, Canada, M9W 4Z6

Keywords: *Nanocrystalline Electrodeposits, Properties, Grain Size, Structural Applications*

Abstract

Over the past decade electrodeposition as a low cost synthesis method for fully dense nanostructured materials has attracted considerable interest in academia and industry. Several industrial applications have been developed for these materials based on i) their unique properties and ii) design concepts which use grain size of these materials as a microstructural tool to optimize overall system performance. The key issue in any application of structural nanomaterials is a thorough understanding of the grain size dependence of various physical, chemical and mechanical properties of these materials. In this paper recent advances in the understanding of grain size effects in various electrodeposited nanocrystalline materials will be summarized. Properties which are either relatively independent or strongly dependent on grain size will be presented and specific examples of structural applications of electrodeposited nanostructures will be discussed.

Introduction

One of the earliest large scale applications of nanostructured materials, specifically designed as a structural nanomaterials technology, was the in-situ electrosleeve nuclear steam generator tubing repair technology developed in the early 1990's and in continuous use since 1994 in both Canadian CANDU and US pressurized water reactors [1, 2]. In this process, steam generator tubes (e.g. Alloys 600 or 400) whose structural integrity was compromised by localized degradation phenomena (e.g. intergranular stress corrosion cracking, intergranular corrosion, pitting) were repaired by coating the inside of the tube with a thick nanocrystalline Ni-P microalloy to restore a complete pressure boundary. A local electrodeposition cell was established in areas of structural degradation. The probe designed for the formation of the electrodeposition cell consisted of two inflatable seals to localize the area to be plated, supply lines for electrolyte flow, heating and electrical power and a non-consumable anode, all fitted inside the narrow tube (10-16 mm diameter) and remotely controlled, about 30-50 ft away from the actual repair side. During the electrosleeving process the degraded host tube became the cathode and the electroplating process was continued until the required thickness (typically 1 mm) of the sleeve was deposited.

The grain size of the electrosleeve material (Ni-P microalloy) was adjusted to be in the 50-100 nm range by controlling the electrodeposition parameters (e.g. bath composition, pH, temperature, current density, on and off times during pulsed current electrodeposition) to give the desirable combination between strength and ductility of the material. Some of the important mechanical properties of the electrosleeve materials are compared with conventional polycrystalline nickel in Table 1.

Table 1: *Summary of the mechanical properties of nanocrystalline Electrosleeve Ni.*

Property	Conventional Ni*	Electrosleeve
Yield Strength, 25°C (77°F)	103MPa (14.9Kpsi)	667MPa (96.7Kpsi)
Yield Strength, 350°C (662°F)	-	492MPa (71.3Kpsi)
Ultimate Tensile Strength, 25°C (77°F)	403MPa (58.5Kpsi)	855MPa (124.0Kpsi)
Ultimate Tensile Strength, 350°C (662°F)	-	714MPa (103.5Kpsi)
Elongation, 25°C (77°F)	50%	15-23%
Modulus of Elasticity, 25°C (77°F)	207GPa (30.0Mpsi)	204GPa (29.6Mpsi)
Modulus of Elasticity, 350°C (662°F)	-	179GPa (26.0Mpsi)

* ASM Handbook, Desk Edition, ASM International, Materials Park, Ohio (1998).

Other important property requirements for the electrosleeve material included good matching of the thermal expansion coefficient with that of the host tubes, excellent fatigue performance in the 0.5 – 25 Hz frequency range, excellent corrosion resistance and ferromagnetic properties which make them suitable for non-destructive inspection. The material was fully approved by both Canadian and US nuclear regulatory bodies under ASME Code-Case N-569.

Since there is still the widespread misconception that electroplating can only be used to produce thin nanocrystalline coatings on various substrates it is worth noting in the context of structural applications that electrodeposition is one of the most versatile techniques for the synthesis of fully dense nanostructured materials (pure metals, alloys, composites). Many different product forms can be made by this technique including thin and thick films, freestanding sheet, foil, plate and wire, powders and composites. For example, Fig. 1 shows a section of a free standing plate of fully dense nanocrystalline (15 nm grain size) Ni-Fe alloy, plated to a thickness of about 4 mm and shown here without any substrate.

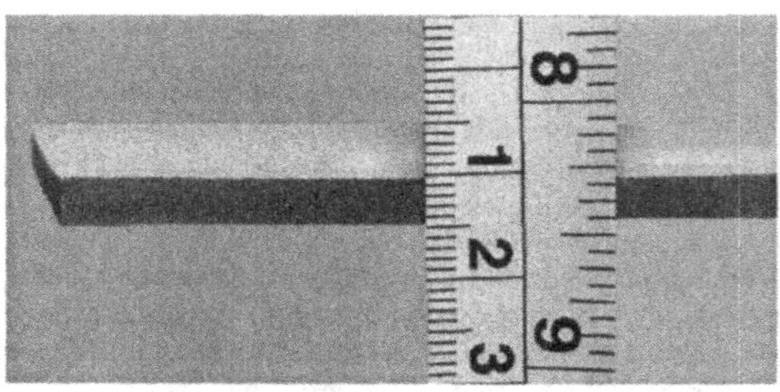

Figure 1: *Example of a thick freestanding section of nanocrystalline (grain size: 15 nm) nickel-iron electrodeposit with a thickness of 4 mm (Courtesy of G. D. Hibbard, Integran Technologies Inc.).*

Effect of Grain Size on Properties

The design of any application of nanocrystalline materials requires a good understanding of various physical, chemical and mechanical properties as a function of grain size. It has been shown that different properties scale differently with grain size. In extensive studies on nanocrystalline materials produced by electrodeposition over the past 15 years [e.g. 3-5], we have demonstrated that there are basically two groups of properties. The first group shows strong grain size dependence, whereas the second group of properties are nearly independent of grain size. In the following two sections some of the important property versus grain size graphs are summarized for nickel. Typically, these properties were measured on fully dense bulk electrodeposits in free-standing sheet form (i.e. without any substrate) with thicknesses in excess of 0.1 mm. Each point on these graphs represents the average of several (typically 3-10) measurements. It should be noted that other materials produced by electrodeposition show directionally similar property changes.

Grain Size Dependent Properties

For a better understanding of some of the property charts it is important to consider changes in the volume fractions of intercrystalline defects (i.e. grain boundaries and triple junctions) as a function of grain size.

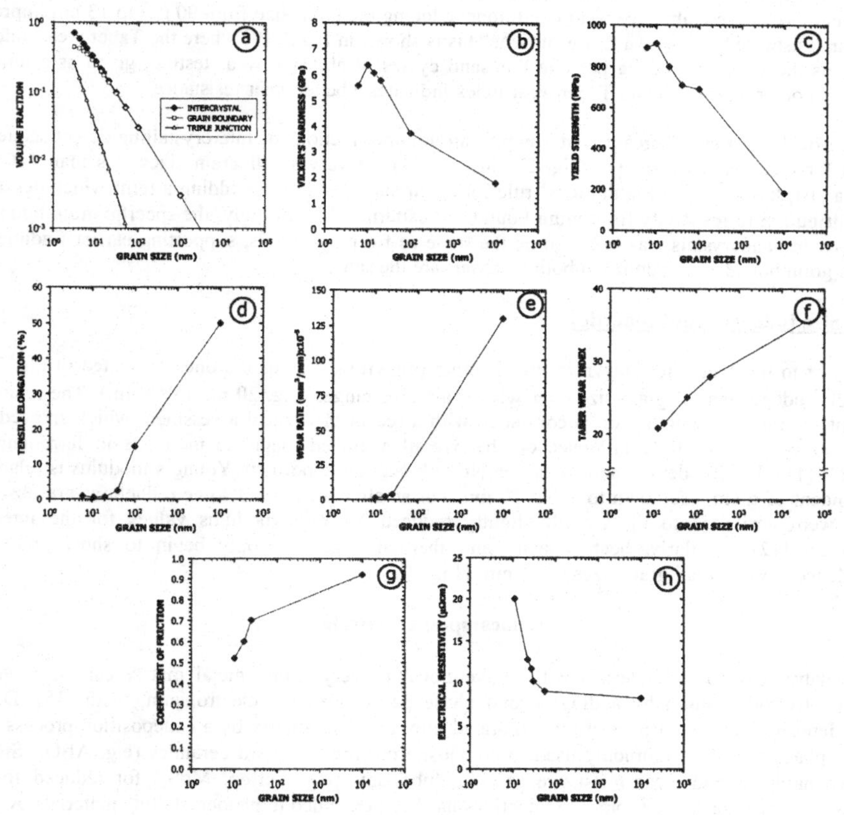

Figure 2: *Volume fractions of intercrystalline defects (a) and grain size dependent properties for nickel (b-h).*

Fig. 2a shows these volume fractions based on the assumptions of a 14-sided regular tetrakaidecahedron as the grain shape and a grain boundary thickness of 1 nm [6]. As expected from Hall-Petch strengthening, the hardness (Fig. 2b), yield strength (Fig. 2c) and tensile strength of nickel increase considerably when the grain size is reduced from 10μm to 10 nm. Both hardness and yield strength show some softening at the smallest grain sizes (< 10 nm) consistent with the inverse Hall-Petch behavior observed for many other nanocrystalline materials [e.g. 4]. Tensile elongation (Fig. 2d) is greatly reduced for grain sizes less than 100 nm. However it should be pointed out that tensile elongation is probably not the best indicator for the overall ductility of nanocrystalline materials. Generally, for Ni we found much larger ductility in bending [2] and during cold rolling [7]. In fact, the latter study showed that over 700% deformation can be obtained for nanocrystalline Ni with an average grain size of about 50 nm and a relatively broad grain size distribution. Pin-on disk wear testing (Fig. 2e) showed that grain size reduction results in an improvement of the adhesive wear resistance by a factor greater than 100 times [8]. At the same time the coefficient of friction is reduced by a factor of 2 (Fig. 2g). Even under severe abrasive wear conditions reducing the grain size from 90 μm to 13 nm improves the wear resistance by almost a factor of two. This is shown in Fig 2f [9] where the Taber wear index is defined as the wear loss in mg per one thousand cycles of abrasive wear testing, given as a unitless number. In other words, a lower Taber wear index indicates a better wear resistance.

The electrical resistivity increases with increasing volume fractions of intercrystalline defect scattering centers for electrons as shown in Fig. 2h, in particular at very small grain sizes less than 100 nm. McCrea [10] has extended Matthiesens' rule for electrical resistivity by adding a term which describes the contribution to resistivity from grain boundary scattering. Interestingly, the specific grain boundary resistivity in nanocrystals was found to be the same as for polycrystals, supporting earlier findings [4] that the grain boundary structures in both materials are the same.

Grain Size Independent Properties

In contrast to the properties shown in Fig. 2, other properties such as the ones presented in Fig. 3 are relatively independent of grain size over wide grain size ranges (e.g. 20 nm to 10 μm). The relatively constant saturation magnetization is consistent with a recent theoretical assessment which showed that grain boundaries have little influence on the overall averaged magnetic moments in ferromagnetic materials [11]. In fully dense nanocrystalline Ni with negligible porosity Young's modulus is relatively independent of grain size down to about 20 nm. For smaller grain sizes some reductions are observed which were attributed to the overall slightly reduced Young's modulus values for the interface component [12]. Similarly, heat capacity and thermal expansion only begin to show grain size dependence at very small grain sizes (< 20 nm) [13, 14].

Nanocomposite Materials

Metal matrix composite materials with conventional polycrystalline metal matrix can be produced through electrodeposition by adding second phase particles to the electroplating bath [15]. During electrodeposition these particles are incorporated into the metal matrix by a codeposition process [15]. Second phase particles commonly used in composite coatings are hard ceramics (e.g. Al_2O_3, SiC) to improve hardness and wear resistance or solid lubricants (e.g. Teflon, MoS_2) for reduced friction applications. The concept of composite coatings has been extended to nanocrystalline materials over the past several years [16]. Figure 4 shows an example of property improvements that can be achieved when reducing the crystal size to the nanocrystalline range in the metal matrix [17]. The properties of conventional pure nickel electrodeposits (10 μm grain size) and commercially available Ni + 7% SiC composite coatings are shown in Figs. 4a and 4b, respectively. For these materials substantial improvements in hardness yield, strength and ultimate tensile strength can be achieved by adding SiC, however at considerable loss in tensile ductility. Grain size reduction to 11 nm in pure nickel results in even higher mechanical strength values (Fig. 4c) which can be further improved by incorporating about 1.8% SiC with an average particle size of 200 nm in a Ni matrix with a grain size of 13 nm (Fig. 4d).

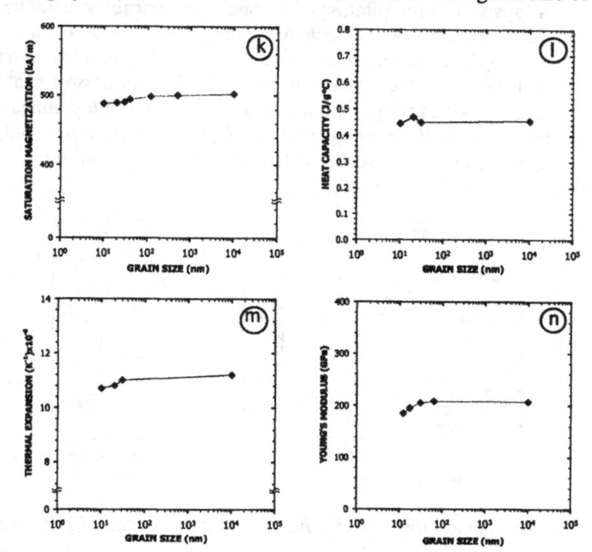

Figure 3: *Examples of grain size independent properties for nickel.*

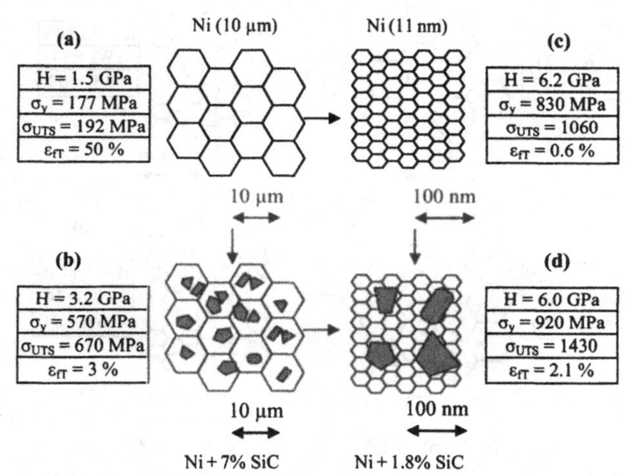

Figure 4: *Hardness (H), yield strength (σ_y), tensile strength (σ_{UTS}) and ductility in tension (ε_{fT}) for polycrystalline nickel and nickel silicon carbide composites (a, b) and nanocrystalline pure nickel and nickel-silicon carbide composites (c, d).*

Heat Treatable Nanocrystalline Alloys

Nanocrystalline alloys produced by electrodeposition can be plated as supersaturated solid solutions [18]. Upon heat treating, these alloys show precipitation of second phase particles in addition to grain growth. By choosing appropriate heat treatments, the mechanical properties can be enhanced considerably by controlling final grain size as well as second phase particle size and volume fraction in these "in-situ" composite materials. Examples of the effect of heat treatment on the hardness of heat treatable Co-P and Ni-P alloys are shown in Fig. 5. For example, the hardness of Ni-1% P alloy annealed at 400^{0}C for 15 minutes is almost twice the hardness of the nanocrystalline (10 nm) starting material. Longer annealing times or higher annealing temperatures have the effect of overaging resulting from excessive grain growth and second phase particle coarsening [19].

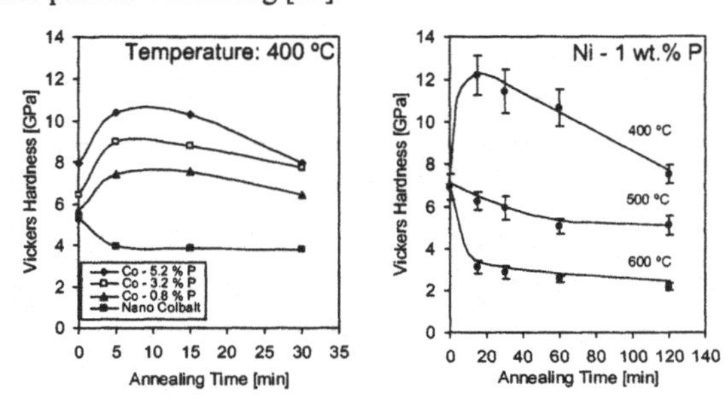

Figure 5: *Vickers hardness vs. annealing time for heat treatable Co-P and Ni-P alloys.*

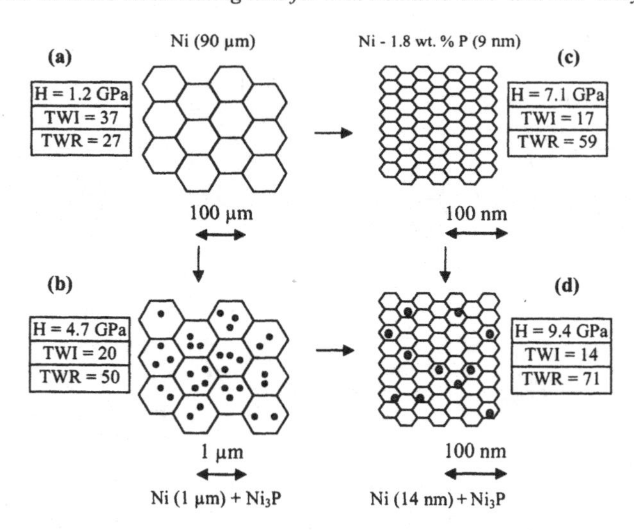

Figure 6: *Comparison of abrasive wear property improvements in conventional polycrystalline (a, b) and nanocrystalline (c, d) Ni-P alloys through heat treatment. (TWI = Taber wear index; TWR = Taber wear resistance; H: Hardness).*

Figure 6 demonstrates the property enhancements for heat-treatable Ni-P alloys in terms of hardness, Taber wear index and Taber wear resistance, the latter two being commonly used to express a materials resistance to abrasive wear [20]. Taber wear index was defined in the section on "Grain Size Dependent Properties". Taber wear resistance is the reciprocal of Taber wear index times 1000, also given as a unitless number. Submicron-sized Ni_3P precipitates in conventional nickel matrix already increase the Taber wear resistance by a factor of 2 (Fig. 6b) compared with conventional Ni (Fig. 6a). A slightly better wear resistance is achieved simply by reducing the grain size in Ni (Fig. 6c) without any second phase particles. The highest wear resistance was obtained for a precipitation hardened material consisting of a nanocrystalline Ni-matrix (14 nm) with nano-sized (< 10 nm) Ni_3P precipitates (Fig. 6d).

Designing Structural Nanomaterials Applications

Using the property charts shown in Figs. 2 and 3 and an Ashby type approach [21] performance indicators for some important properties in structural applications for nanocrystalline materials can be readily assessed. Table 2 shows that remarkable improvements can be achieved in specific strength (σ_y/ρ), elastic energy storage capacity (σ_y^2/E) and thermal shock resistance ($\sigma_y/E\cdot\alpha$) simply by grain size reduction from 10 μm to 10 nm. Of course for any specific application, other property changes with grain size must also be considered. For example, for some applications the simultaneous increase in electrical resistivity for grain sizes less than 100 nm may be either beneficial (e.g. for reduced eddy current losses) or detrimental (e.g. when high conductivity is required). The important point to make here is that with a good understanding of the effect of grain size on all properties (including the properties not specifically addressed in this paper such as thermal stability [e.g. 22], corrosion resistance [e.g., 23]) of importance for a particular application, grain size becomes an important design parameter in efforts to achieve optimum system performance. In the past, the success of this approach has been clearly demonstrated for the case of the electrosleeve technology, with the initially installed sleeves being in continuous operation for nearly 10 years. For this particular application, the grain size range of 50 – 100 nm was chosen (rather than 10 nm, for example, for maximum strength) because the total required property set (in terms of strength, ductility, thermal stability, etc.) provided optimum system performance for this grain size range.

Table 2: *Relative improvements in performance indicators for Ni by grain size reduction (normalized with respect to 10 μm grain size).*

Performance Indicator \ Grain size	10 μm	1 μm	100 nm	10 nm
Specific strength (Maximize σ_y/ρ)	1	2.2	3.4	4.6
Elastic energy storage (Maximize σ_y^2/E)	1	4.8	11.7	21.5
Thermal shock resistance (Maximize $\sigma_y/E\cdot\alpha$)	1	2.2	3.5	4.9

σ_y: yield strength; E: Young's modulus; ρ: density; α: thermal expansion coefficient.

Conclusions

Structural applications for nanocrystalline electrodeposits can be readily developed through microstructural design and optimization considerations which include i) grain size dependence of various properties, ii) effects of second phase particles in composite materials and iii) careful analysis of typical performance indicators rather than looking only at some individual properties. This approach has been proven successful for the application of nanostructured materials under very severe conditions (i. e. the nuclear industry) and continues to be applied in a variety of other applications.

Acknowledgements

Financial support from the Natural Sciences and Engineering Research Council of Canada (NSERC) and Integran Technologies Inc., and contributions by Dr. C. Cheung and Mr. J. Victor-Beale are gratefully acknowledged.

References

1. F. Gonzalez, A. M. Brennenstuhl, G. Palumbo, U. Erb, W. Shmayda and P. C. Lichtenberger, Mat. Sci. For., 225-227 (1996) 831.
2. G. Palumbo, F. Gonzalez, A. M. Brennenstuhl, U. Erb, W. Shmayda and P. C. Lichtenberger, Nanostr. Mat., 9 (1997) 737.
3. K. T. Aust, U. Erb and G. Palumbo, in Mechanical Properties and Deformation Behavior of Materials Having Ultrafine Microstructures, M. Nastasi et al., (eds.), Kluwer Academic Publ. (1993).
4. U. Erb, G. Palumbo, R. Zugic and K. T. Aust, in Processing and Properties of Nanostructured Materials, C. Suryanarayana et al., (eds.), TMS, Warrendale, PA (1996) 93.
5. U. Erb, K. T. Aust, G. Palumbo, J. L. McCrea and F. Gonzalez, in Processing and Fabrication of Advanced Materials IX, T. S. Srivatsan et al., (eds.), ASM International, Materials Park, OH (2001) 253.
6. G. Palumbo, S. J. Thorpe and K. T. Aust, Scripta Metall. Mater., 24 (1990) 1347.
7. J. Brassard, M.Sc. Thesis, University of Toronto, Ontario, Canada (2003).
8. A. M. El-Sherik and U. Erb, in Nickel-Cobalt '97: Volume IV, Applications and Materials Performance, F. N. Smith et al., (eds.), The Metallurgical Society of CIM, Montreal, PQ (1997).
9. D. H. Jeong, F. Gonzalez, G. Palumbo, K. T. Aust and U. Erb, Scripta Mater., 44 (2001) 493.
10. J. L. McCrea, Ph.D. Thesis, University of Toronto, Ontario, Canada (2001).
11. B. Szpunar, U. Erb, G. Palumbo, K. T. Aust and L. J. Lewis, Phys. Rev. B, 53 (1996) 5547.
12. Y. Zhou, U. Erb, G. Palumbo and K. T. Aust, Z. Metallkunde, (2003) accepted.
13. T. Turi and U. Erb, Mat. Sci. Eng., A204 (1995) 34.
14. B. Szpunar, L. J. Lewis, I. Swainson and U. Erb, Phys. Rev. B, 60 (1999) 10107.
15. W. H. Safranek, The Properties of Electrodeposited Metals and Alloys, American Electroplaters and Surface Finishers Society, Orlando, FL. (1986).
16. U. Erb, A. M. El-Sherik, G. Palumbo and K. T. Aust, Nanostr. Mat., 2 (1993) 383.
17. A. F. Zimmerman, G. Palumbo, K. T. Aust and U. Erb, Mat. Sci. Eng., A328 (2002) 137.
18. U. Erb, A. M. El-Sherik, US Patent # 5,352,266; U. Erb, A.M. El-Sherik, C. Cheung and M. J. Aus, US Patent # 5,433,797.
19. U. Erb, G. Palumbo and K. T. Aust, in Nanostructured Films and Coatings, G. M. Chow et al., (eds.), NATO Science Series, Kluwer Academic Publ., Dordrecht, The Netherlands (2000) 11.
20. D. H. Jeong, U. Erb, K. T. Aust and G. Palumbo, J. Metastable and Nanocrystalline Materials, 15-16 (2003) 635.
21. M. F. Ashby, Materials Selection in Mechanical Design, Pergamon Press, Oxford (1992).
22. G. Hibbard, U. Erb, K. T. Aust, U. Klement and G. Palumbo, Mat. Sci. For., 386-388 (2002) 387.
23. S. H. Kim, K.T. Aust, U. Erb, G. Ogundele and F. Gonzalez, AESF SUR/FIN 2002 Proceedings, American Electroplaters and Surface Finishers Society, Orlando, Fl. (2002) 225.

Processing and Properties of Structural Nanomaterials
Edited by Leon L. Shaw, C. Suryanarayana and Rajiv S. Mishra
TMS (The Minerals, Metals & Materials Society), 2003

Processing-Controlled Mechanical Properties and Microstructures of Bulk Cryomilled Aluminum-Magnesium Alloys

David Witkin[1] and E.J. Lavernia[2]

[1]Department of Chemical Engineering and Materials Science, University of California, Irvine, CA 92697
[2]Department of Chemical Engineering and Materials Science, University of California, Davis, CA 95616

Keywords: Aluminum Alloys, Structural Materials, Nanostructured Materials

Abstract

We have been successfully realizing bulk nanostructured aluminum alloys via a process of cryomilling, consolidation and extrusion. The resulting extrusions have consistently demonstrated greatly enhanced strength relative to conventional materials while the grain size after processing is generally in the range of 100 to 300 nm. Within the nanostructured matrix we find grains that are on the order of several hundred nm's and often elongated in the extrusion direction. The suggestion that these grains are responsible for increasing the ductility of the materials inspired an effort to engineer ductility via the concept of multiple length scales. We present mechanical property data for cryomilled Al-Mg alloys that were processed under different consolidation and extrusions conditions, as well as materials engineered for higher ductility through the addition of unmilled powder to the cryomilled prior to consolidation. In the latter case, the presence of 15, 30 and 50 weight percent unmilled powder enhanced the ductility relative to the all-cryomilled samples without sacrificing the strength derived from the grain-size refinement and cryomilling-related dispersoid strengthening in the matrix.

Introduction

One of the presumed benefits of nanostructured materials is the additional strength to be gained from a refinement of grain size. Based on the empirical Hall-Petch relationship, decreasing the grain size should lead to an increase in the strength of the material, according to the relationship

$$\sigma_y = \sigma_0 + kd^{-1/2} \tag{1}$$

where σ_y is the yield strength, σ_0 is the friction stress (interpreted as the stress necessary to move one dislocation), d is the grain size and k is a constant. There have numerous research efforts that have successfully demonstrated this presumption in different metallic and alloy systems [1].

There is a difference, however, between synthesizing materials appropriate to the interrogation of fundamental physical and mechanical properties and producing samples that are of sufficient size and dimension to be considered prototype or proxy for a structural material. Techniques such as inert gas condensation and mechanical attrition take opposite approaches to achieving nanocrystalline dimensions. In the former, a rapid change in phase from vapor to solid yields particles of nanometer-scale and thus nanocrystalline character, while in the latter grain size refinement is the result of repeated plastic deformation. The result is typically metallic powder, which requires additional processing, such as high-pressure, low-temperature compaction to produce a sample amenable to testing. A recent paper has demonstrated that the process of mechanical attrition can produce agglomerated powders that can then be compressed to a flat coupon from which tensile samples can be cut [2].

In many cases, the mechanical properties of the nanocrystalline materials have been measured using very small samples, because the samples themselves did not support a standard sized testing specimen. In scaling up from small experimental quantities to batches or continuous outputs, economics or unexpected material properties may thus emerge as obstacles to the production of structures based on nanomaterials.

In the present work, the mechanical properties and microstructures are described for bulk nanostructured aluminum-magnesium alloys produced by cryomilling. Cryomilling is a low-energy mechanical attrition technique that is performed in a cryogenic medium, liquid nitrogen. The cryomilled powders are subsequently consolidated and extruded using standard powder metallurgy techniques, and the resulting extrusions are of sufficient size to render standard tensile specimens. The strength of the cryomilled materials derives both from grain refinement and from the formation of dispersed aluminum nitrides during cryomilling [3]. These nitrides form as platelets with dimensions on the order of several nanometers long and several atomic layers thick, and act as agents of dispersoid strengthening [4]. The particles also impart considerable thermal stability to the microstructure of the material, allowing for some flexibility in processing the material without sacrificing the strength gained from grain size refinement. In addition, processing conditions, specifically consolidation and extrusion, may be varied to produce varying mechanical properties.

Experimental

The materials described herein were all consolidated from cryomilled pre-alloyed powders of spray atomized Al 5083 (a widely used commercial alloy) and Al-7.5Mg. The cryomilling process was performed in two different attritors that varied according to size: a smaller, laboratory Szegvari attritor in which one-kilogram powder charges were milled, and a larger, commercial scale attritor, in which a single 20-kg batch of powder was milled. Other processing aspects of the milling were essentially identical: the milling medium was liquid nitrogen, the milling ball to charge weight ratio was 32:1, and stearic acid was added as a process control agent to prevent excessive cold welding of the aluminum particles. The milling balls were ¼' diameter 440 C stainless steel. Milling conditions consisted of eight hours of milling with the attritor operating at 180 rpm with liquid nitrogen levels maintained at a level to allow continual milling in a liquid nitrogen/stainless steel ball/powder slurry.

After cryomilling, the powders were consolidated and extruded following two different processing routes. In the first, the cryomilled powders were transferred to aluminum cans fitted with valved stems and degassed at a temperature of 400 °C for up to 24 hours, whereupon the stems were welded shut. Powders from the laboratory-scale attritor were put in cans with a capacity of approximately 80 grams of powder, while the large-scale batch was canned *in toto* in a single large can. The canned powder was consolidated by hot isostatic pressing (HIPping) under varying temperature and pressure conditions. The single large-batch HIP can was consolidated using a proprietary HIP cycle in a commercial press at a higher temperature and lower pressure than the laboratory HIP cycles. Prior to extrusion, the HIP cans were machined from the outside of the laboratory-scale billets. The resulting cylinders were typically 2.5 cm in diameter and to 8 cm long. These billets were either extruded at a 9:1 ratio on a 400-ton press at varying temperatures (300, 325 and 350 °C) or extruded at UES Corp., Wright-Patterson Air Force Base, OH again under proprietary conditions which included a 6.5:1 extrusion ratio and a temperature that was considerably lower than the other extrusions. In the case of the large-scale HIPped billet, 3-inch diameter billets were harvested using electro-discharge machining and extruded at UES Corp. at a 6.5:1 extrusion ratio and temperatures of 200 and 225 °C.

In the second processing route, the cryomilled powders were first cold isostatic pressed (CIPped) at room temperature and approximately 400 MPa. The CIPped compact was degassed at 400 °C for 24 hours. After degassing, the billets were extruded at approximately 500 °C using the Dyna-Pak process, a high-energy, high-strain rate extrusion process. The resulting extrusions were approximately ¾ in diameter and included the CIP can. The extrusions were swaged to an overall diameter of ½ to produce rods commensurate with size requirements for tensile samples.

An overview of the processing history is given in Table I. Round tensile samples were machined from the extrusions and room-temperature tensile tests were conducted in accordance with ASTM Method E 8 at strain rates of approximately 10^{-3}. The microstructures described herein are based on examination using transmission electron microscopy (TEM).

Table I. Processing Conditions for Cryomilled Al 5083

Sample No.	HIP Conditions			Extrusion Conditions		Extrusion Size
	Temp. (°C)	Pressure (MPa)	Cycle Time (Mins.)	Ratio	Temp. (°C)	Diam. x Length (mm)
1	325	185	205	9:1	300	8 x 300
2[a]	325	185	205	9:1	325	8 x 200
3[a]	325	185	205	9:1	350	8 x 200
4	350	173	205	6.5:1	*	13 x 275
5	300	173	205	6.5:1	*	13 x 275
6	*	*	*	6.5:1	200	33 x 500
7	*	*	*	6.5:1	225	33 x 500
	CIP Conditions			Dyna-Pak Extrusion		Dimensions after hot Swaging (mm)
8	310 MPa			525 °C		13 x 600

[a]HIPped billet was split to use in two separate extrusions.
*Processing variables are considered proprietary and not available for publication.

Results and Discussion

Room Temperature Mechanical Properties

The yield stress, ultimate tensile strength, and elongation to fracture for the extrusions are shown in Table II. Several obvious trends relating mechanial properties and processing histories emerge from the room-temperature tensile tests. The primacy of extrusion temperature as the determining processing variable is suggested by the results for samples 1, 2 and 3. Each of these samples was extruded at the same ratio on a laboratory press. There is a strong correlation between extrusion temperature and mechanical behavior; specifically, there is an inverse relationship between extrusion temperature and strength and the expected relationship between strength and ductility.

The mechanical properties of samples 4 and 5 adhere to this same trend, although it should be noted that in order to extrude the material at the lower temperature, a different press and extrusion ratio were used. In this case, the yield stress of the sample HIPped at 350 °C and extruded at the lower temperature (around 200 °C) is nearly 135 percent higher than the material HIPped at 325 °C and extruded at 300 °C. The difference in the yield stresses and ultimate strengths for Samples 4 and 5 may not be simply attributable to the difference in HIPping cycle temperatures and may be an artifact of the particular materials.

Samples 6 and 7 again show the influence of extrusion temperature on the mechanical properties. In this case, the difference in strength is minor, but the higher extrusion temperature increases the ductility by nearly 35 percent. In comparison, commercial Al 5083 treated in the H112 temper has a yield stress and ultimate stress of 193 and 303 MPa, respectively and a ductility of 16 percent [5].

Finally, sample 8 represents the alternative fabrication process of cold consolidation followed by high-strain rate, high-temperature extrusion. The strength of this material is roughly comparable with Samples 4 and 5, which were HIPped and then extruded under low temperature conditions.

Table II. Room Temperature Tensile Properties of Cryomilled Extrusions

Sample No.	Yield Stress (as 0.2% Offset, MPA)	Ultimate Tensile Strength (MPa)	Elongation at Fracture (%)
1	523	584	3.7
2	453	491	5.9
3	169	199	22.8
4	705	745	1.4
5	660	682	1.4
6	555	636	6.4
7	544	611	8.6
8	676	710	4.0
Conventional Al 5083	145	281	16.3

Microstructures and Microstructural Evolution

Previous investigations have focused on the microstructure and thermal stability of as-cryomilled powders of varying compositions [6,7]. This work has shown that the typical grain size after cryomilling for eight hours under the present conditions for Al and Al-Mg alloys is approximately 20 to 40 nm. Isothermal annealing experiments showed that the grain size of the powders is stable in the nanocrystalline range (up to 100 nm) at temperatures up to 400 °C. During consolidation and extrusion, however, annealing conditions may not be an appropriate comparison, as the application of both temperature and pressure is necessary to deform the powder into a bulk form. The grain size of the bulk extruded material is better described as ultra fine grained or near nanocrystalline, as it is typically 100 to 300 nm.

The stability of this microstructure may be illustrated by comparing two samples processed from cryomilled Al 5083 by very different means. In Figure 1a, a TEM bright field image at low magnification is shown of the longitudinal orientation of Sample 5. In comparison, Figure 1b shows the microstructure of the longitudinal section of an extrusion that was CIPped and then extruded at a high strain rate at 525 °C, the conditions under which Sample 8 was processed. For a final comparison, in Figure 1c, the longitudinal section of Sample 6 is shown, which is the product of a large-scale cryomilling batch, high-temperature HIP cycle, and extrusion as a 3-inch diameter billet at 200 °C.

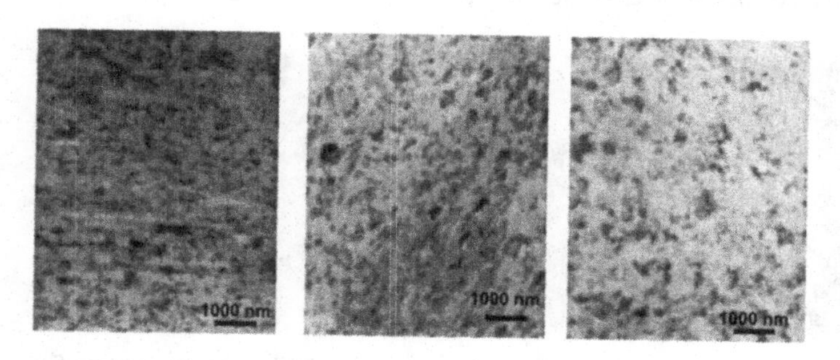

Figure 1a (left), 1b (center) and 1c (right). Microstructures of samples 5, 8 and 6, respectively.

The similarity in grain size for these three examples indicates that the cryomilled material is in some manner stable during thermomechanical processing, regardless of the process conditions. This stability is in accord with previous studies of cryomilled Al and Al 5083 powder, which have demonstrated that nanocrystalline to ultra-fine grain size can be maintained at homologous temperatures as high as 0.8 T_m [6,8]. The stability is attributable to the pinning effects of the cryomilling dispersoids and solute segregation at grain boundaries [8], as well as an increase in the activation necessary for grain growth [6]. Examination of the room temperature tensile properties of these three samples indicates that there must be some other aspects of the extrusions that lead to the variations in strength and ductility.

One explanation is that the microstructures illustrated in Figure 1 do not represent the entire microstructure of the extrusions. Previous investigations of nanocrystalline Al [9], Al alloys [10, 11] and cryomilled Al 5083 [12] have noted the presence of a bimodal or duplex microstructure, in which larger grains are present within the finer-grained matrix. Close examination of the micrographs in Figure 1 reveals that there are grains which are roughly equiaxed and are on the order of 500 to 800 nm, as opposed to the 100 to 300 nm range. While these grains are larger, they remain isolated within the matrix and appear to have a similar relationship (i.e., high-angle grain boundary) to their neighbors as do smaller grains in the matrix.

A more significant feature is the presence of elongated bands of coarse grains, as shown in Figure 2. This micrograph illustrates this feature in a longitudinal section of Sample 5 and thus indicates the extrusion direction. The band running roughly horizontally through the center of the micrograph is approximately 1 ì m wide, and consists of individual grains or sub-grains that are roughly equiaxed.

The origin of these coarse grained bands is suggested by TEM of the as-HIPped precursor of the extruded Sample 5. In Figure 3, one feature of the microstructure of the HIPped billet is shown. In this micrograph, coarse, apparently recrystallized Al appears in roughly a "Y" shape between three different finer-grained matrix regions. The formation of this feature is understood in light of the physical processes that occur during hot isostatic pressing. To achieve consolidation, mass flows from powder particle surfaces under tensile stresses to surfaces under compressive stresses. The result is a filling of the interstices of a packed powder to form a consolidated solid. The "Y" shaped feature in Figure 3 can be interpreted one such filled in volume. Cryomilled powder particles are blocky and sub-spherical in appearance, so the packing density of the powder is likely to be larger than the theoretical packing density of spheres and the overall volume fraction of these features is probably considerably less than would be seen if the powder particles were spherical.

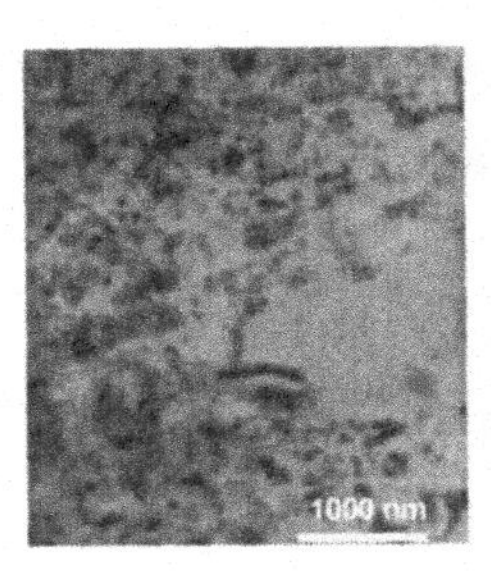

Figure 2. Coarse grained band within fine grained matrix of extruded material.

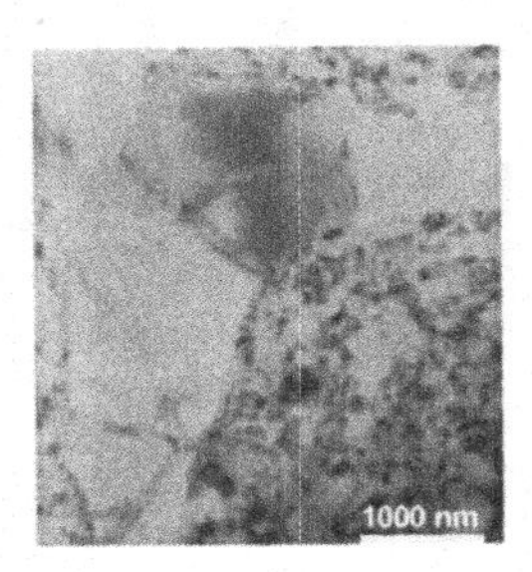

Figure 3. Coarse grains between fine grained regions in HIPped material.

The observed differences in room temperature tensile properties may be rationalized on the basis of these coarse grained bands. The sources of strength in the fine-grained matrix of these extrusions are principally grain size refinement and Orowan strengthening due to the aluminum nitride dispersoids. The degree of strengthening in each extrusion is likely to be similar for many of hese materials because both the grain size and the cryomilling conditions are similar. The overall strength of the extrusion may be dictated by the relative abundance of the coarse grained regions; larger or more numerous coarse-grained regions could be expected to have a deleterious effect on the yield stress and tensile strength.

Another argument for the role of the coarse grained bands in the mechanical behavior can be made in terms of the Hall-Petch equation (Equation 1). For example, in Samples 2 and 3, a change in extrusion temperature of 25 C led to a decrease in yield stress of 284 MPa. Using a value for k of 0.07 MPa· m $^{1/2}$ for Al and a value of 150 MPa for σ_0, to attribute this decrease in strength to a change in grain size alone (bearing in mind that the samples were extruded from the same HIPped material) would require that the grain size of Sample 3 is roughly 250 times the grain size of Sample 2. The actual grain sizes of these two extrusions are roughly 350 nm for Sample 2 and 500 nm for Sample 3, so coarsening the microstructure is clearly not the cause for the drastic change in the tensile properties.

In addition, the coarse-grained regions have a greater capacity for plastic deformation than the fine-grained matrix for two reasons. First, the coarser grain size means a greater distance between grain boundaries, allowing dislocations to glide over longer distances before encountering grain boundaries. Second, if the origin of the coarse grain regions is the inter-particle volume that is filled in during HIPping, these regions should be free of the aluminum nitride dispersoids, which would either be immobile or redissolve into the Al matrix during thermomechanical processing. During shear, mobile dislocations in the coarse grained bands would not interact with dispersoids, so would be less likely to tangle within the grains. The phenomena underlying the observed relationship between strength and ductility for these cryomilled materials may be better explained by recourse to the different behaviors exhibited by the individual constituents of the duplex microstructure rather than as a monolithic material.

Engineered Bimodal Microstructures

These observations of duplex microstructures and the apparent influence of coarse grained regions on the mechanical properties inspired an effort to engineer a material which combined both the strength of cryomilled Al-Mg and the ductility of the conventional material. In this approach, cryomilled Al-7.5Mg powders were combined with unmilled powders of identical composition prior to degassing. These samples were HIPped using the same HIP cycle as Sample 1 in Table I and the same extrusion conditions as Samples 4 and 5 in that table. The room-temperature tensile behavior has been previously reported [13] and is reproduced in Table III. Also shown in Table III are tensile properties that were obtained for a bimodal cryomilled plus unmilled Al 5083 that was processed using CIP and high strain rate processing, as given for Sample 8 in Table I. In this case, the volume fraction of unmilled powder was roughly 50 percent.

Table III. Room Temperature Tensile Properties of Bimodal Cryomilled Al-Mg

Sample Description	Yield Stress (as 0.2% Offset, MPA)	Ultimate Tensile Strength (MPa)	Elongation at Fracture (%)
Al-7.5Mg Extrusions [12]			
All Cryomilled	641	847	3.5
15% Unmilled	630	778	5.0
30% Unmilled	554	734	7.3
Al 5083 Extrusions			
All Cryomilled*	676	710	4.0
50/50 Blend	496	619	7.5

*This is Sample 8 from Tables I and II.

These results are presented graphically in Figure 4, in which the y axis represents the ratio of the yield stress or ultimate tensile strength of the bimodal extrusion to the monolithic, all cryomilled extrusion. In preparing this graph, rather than use the average of three tensile tests for the all cryomilled Al 5083, as is given in Table III, an average of two tests where the yield stress and UTS were very similar was used as the divisor in the ratio.

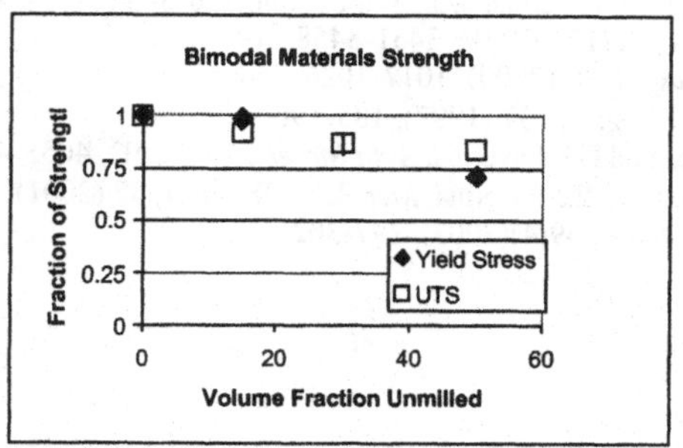

Figure 4. Yield stress and UTS of bimodal extrusions expressed as a fraction of the yield and UTS of all cryomilled extrusions processed the same way. The figure contains results for both Al-7.5Mg and Al 5083.

In comparison to the bimodal materials plotted here, a strain hardened commercial Al 5083 (H112 temper) would plot as approximately 28 percent of the yield stress and 30 to 40 percent of the ultimate strength of either monolithic cryomilled extrusion. The potential of this approach to engineering high-strength Al alloys with increased ductility is ever more apparent in comparison to conventional materials, even after strain hardening treatment.

Conclusions

Cryomilling has been used to successfully produce high-strength Al 5083 extrusions using different post-milling consolidation and extrusions conditions. The extruded microstructures tend to be duplex in nature, consisting of a fine grained matrix and coarser grains that form elongated bands during processing. The exact microstructure, i.e., the abundance and size of the coarse-grained bands, is a function of processing, as changes in processing conditions, particularly extrusion temperature, give rise to different room-temperature mechanical properties. The correlation of the coarse-grained bands with ductility and plastic deformation led to an effort to combine cryomilled and unmilled powder of the same composition together prior to consolidation. These bimodal materials show greater ductility than the all-cryomilled extrusions, but their yield stress and tensile strengths remain much higher than those of conventional Al 5083.

Acknowledgements

This work was supported by the Office of Naval Research and Marine Corps MEFFV Program, Contract Numbers N-00014-01-C0384 and N-00014-03-C013, under the supervision of Rodney Peterson.

References

1. D.G. Morris, "Mechanical Behavior of Nanostructured Materials", *Materials Science Foundations*, vol. 2 (1998), 85 pages.
2. X. Zhang et al., -*Scripta Mater.*, 46 (2002), 661-665.
3. M.J. Luton et al., *Mat. Res. Soc. Symp.*, 132 (1989), 79-86.
4. O. Susegg et al., *Phil. Mag. A*, 68 (1993), 367-380.
5. Metals Handbook, 10th Edition (1990), vol. 2.
6. V.L. Tellkamp et al., *J. Mat. Res.*, 16(4)(2001), 938-944.
7. F. Zhou, R. Rodriguez and E.J. Lavernia, *Mat. Sci. Forum*, 386-388 (2002), 409-414.
8. F. Zhou et al., *J. Mat. Res.*, 16(12) (2001), 3451-3458.
9. M. Legros et al., *Phil. Mag. A*, 80 (2000), 1017-1026.
10. M. Rittner et al., *Mat. Sci. Eng. A*, 237 (1997), 185-190.
11. R.W. Hayes, R. Rodriguez and E.J. Lavernia, *Acta Mater.*, 49 (2001), 4055-4068.
12. V.L. Tellkamp, A. Melmed and E.J.Lavernia, *Met. Mat. Trans. A*, 32 (2001), 2335-2343.
13. D. Witkin et al., *Scripta Mater.*, 49(4)(2003), 297-302.

Processing and Properties of Structural Nanomaterials
Edited by Leon L. Shaw, C. Suryanarayana and Rajiv S. Mishra
TMS (The Minerals, Metals & Materials Society), 2003

Nanocrystallization of Steels by Various Severe Plastic Deformation

M. Umemoto, Y. Todaka and K. Tsuchiya

Toyohashi University of Technology; Toyohashi, Aichi 441-8580, Japan

Keywords: ball mill, ball drop, shot peening, nanocrystalline, microstructure, severe deformation, steel.

Abstract

The formation of nanocrystalline structure (NS) in steels by various severe plastic deformation processes, such as ball milling, a ball drop test, particle impact deformation and air blast shot peening are studied. Layered or equiaxed nanograined region appeared near the specimen surface and dislocated cell structured region appeared interior of specimens. These regions are separated with clearly defined boundaries. The deformation induced nanograined regions have the following common specific characteristics: 1) with grains smaller than 100 nm and low dislocation density interior of grains, 2) extremely high hardness, 3) dissolution of cementite when it exists and 4) no recrystallization and slow grain growth by annealing. It was suggested that the most important condition to produce NS is to impose a strain larger than about 7.

Introduction

Large efforts have been devoted to refine grains of materials since ultrafine-grained materials have often superior mechanical and physical properties to those of coarse-grained counterparts. To obtain nanograined materials (grain size smaller than 100 nm), various methods such as consolidation of ultrafine powders [1], electron beam deposition [2], electrodeposition [3], crystallization of amorphous phase [4], severe plastic deformation [4-13] etc. have been developed. Figure 1 shows representative severe plastic deformation processes to produce nanocrystalline structure (NS): ball milling [4-6], high pressure torsion [7,8], sliding wear [9], a ball drop test [10], ultrasonic shot peening [11] and air blast shot peening [12,13].

In the present study the microstructure evolutions in steels by ball milling, a ball drop test, particle impact test and air blast shot peening are observed. Focusing on iron and steels, the amount of strain, strain rate and other favorable conditions to produce NS are discussed.

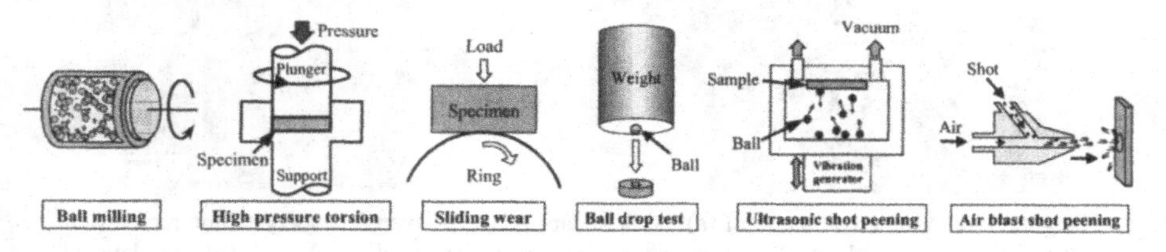

Figure 1: Various severe plastic deformation processes to produce nanograined structure.

Results

Nanocrystallization by Ball Milling

Grain refinement by ball milling to nano-meter range has been reported in various pure metals with bcc [4,5] and hcp [5] and fcc [6] crystal structures.

Figure 2 shows SEM micrographs of the ball milled Fe-0.1C (in mass % hereafter) martensite with initial hardness of 3.2 GPa. From the full view of the cross section of a powder (Fig. 2 (a)) two distinct regions are recognized. One is the dark uniform contrast region with several tens of μm thick observed near the surface of powder and the another is the bright contrast region observed interior of the powder with deformed martensite morphology. The hardness of dark contrast region (8.8 GPa) is much higher than that of bright contrast region (3.9 GPa) as shown in Fig. 2 (b). The boundaries between these two types of regions are clear and sharp. There is no intermediate structure between these two regions. TEM observation (shown below) revealed that the dark contrast area corresponds to nanograined structured region and the bright contrast area corresponds to dislocated cell structured region, respectively. These two types of structures were observed in all the ball milled carbon steels irrespective of the carbon content (0 ~ 0.9 mass%C) or starting microstructure (ferrite, martensite, pearlite or spheroidite).

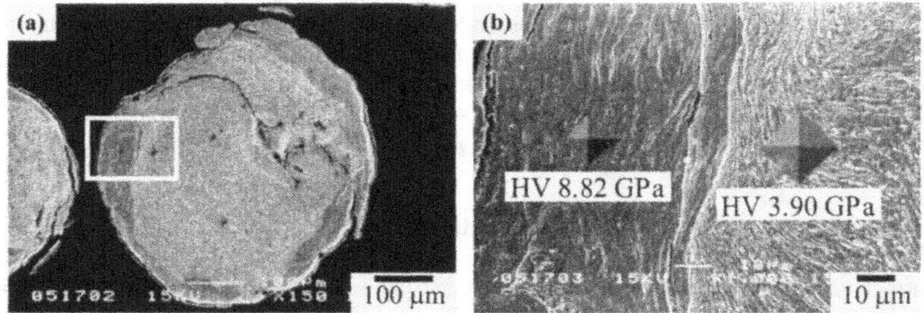

Figure 2: SEM micrographs of Fe-0.10C martensite ball milled for 360 ks. (a) Full view of a milled powder and (b) enlarged picture of (a) showing the hardness of nanocrystalline region (left hand side) and deformed structured region (right hand side) near powder surface.

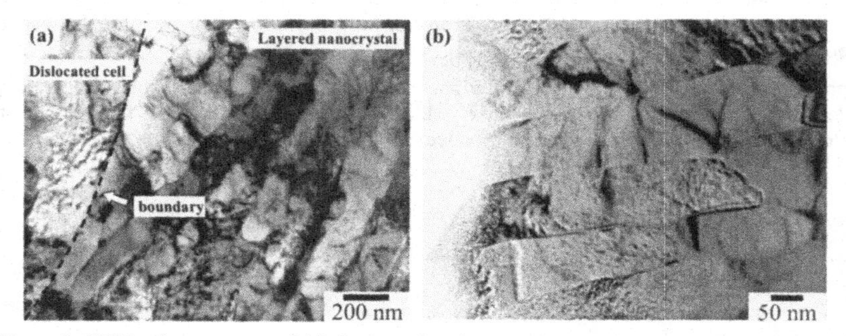

Figure 3: TEM microstructure of (a) the boundary between layered nanocrystalline region (left hand side) and dislocated cell structured region (right hand side) and (b) the layered nanocrystalline region observed in Fe-0.004C ball milled for 360 ks.

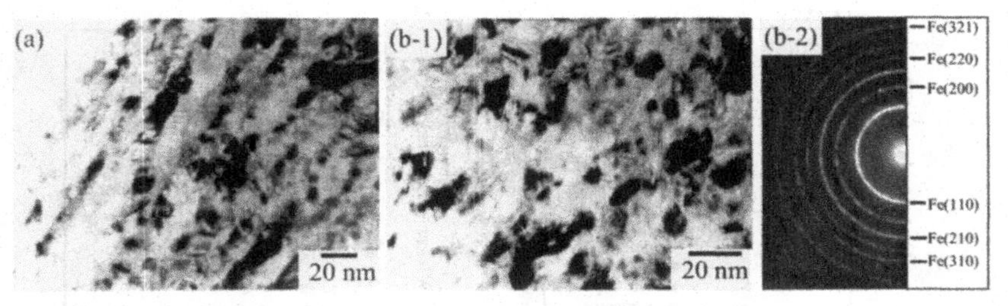

Figure 4: TEM micrographs of Fe-0.89C spheroidite ball milled (a) for 360 ks and (b-1, 2) for 1800 ks.

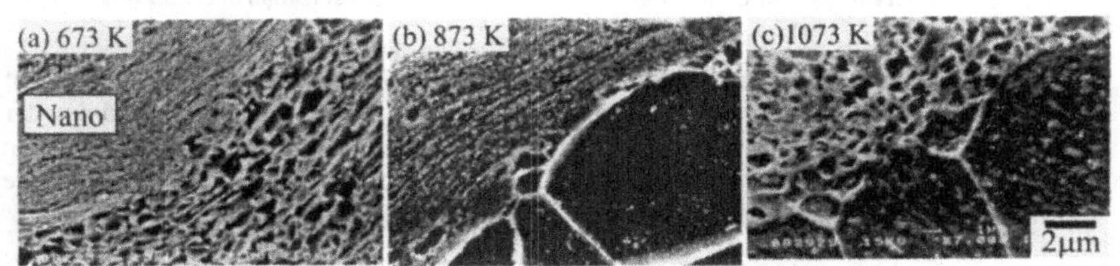

Fig. 5 SEM micrographs of pure iron (Fe-0.004C) after ball milled for 360 ks and annealed for 3.6 ks (a) at 673 K, (b) at 873 K and (c) at 1073 K.

The TEM micrographs of milled powder are shown in Fig. 3 and Fig, 4. Figure 3 (a) shows the area around the boundary of two regions in the Fe-0.004C alloy. The layered nanocrystalline structure with an average thickness of about 100 nm (right hand side) and dislocated cell structure (left hand side) are seen. These two regions are separated with sharp boundary and there is no transition region. Figure 3 (b) shows a high magnification of the layered nanostructure. It is seen that layered nanostructure consists of elongated grains with sharp grain boundaries. Figure 4 (a) [14] shows the layered nanostructure in ball milled Fe-0.89C alloy with spheroidite structure. The thickness of the layers is around 20 nm which is much smaller than that of pure iron (Fig. 4 (b-1)). The layers are subdivided into small grains as seen with different contrast. After milled for a long time uniformly distributed fine grains about 5-10 nm are produced (Fig. 4 (b-1)). These nanograins are randomly orientation as is seen from the continuous diffraction rings (Fig. 4 (b-2)). These observations suggest that randomly oriented equiaxed nanograins are produced by the subdivision of layers and rotation of divided regions. The diffuse contrast of equiaxed nanograin boundaries suggesting that they are highly disordered.

The annealing behaviors of nanograined region are quite different from those of the deformed structured region. Figure 5 shows the microstructural change with annealing temperature (annealed for 3.6 ks) in ball milled pure iron (Fe-0.004C). By annealed at 673 K (Fig. 5 (a)), the deformed structured region (lower right) becomes equiaxed grained structure with an average grain size of about 0.5 μm by recrystallization. On the contrary, nanograined structure region showed almost no detectable change (upper left). After annealing at 873 K (Fig. 5 (b)), the recrystallized grain size in the deformed structured region become 10 μm. While, the nanograined region remains almost unchanged. By annealing at 1073 K (Fig. 5 (c)) , detectable grain growth to about 0.7 μm occurred in the nanograined region.

The above difference in annealing behavior in the two regions can be explained as follows. In the deformed (dislocated cell) structured region, conventional recrystallization takes place due to high dislocation density. While, recrystallization does not occur in the nanograined region since almost no dislocations in the interior of grains. It seems that the grain growth of nanograined region is substantially

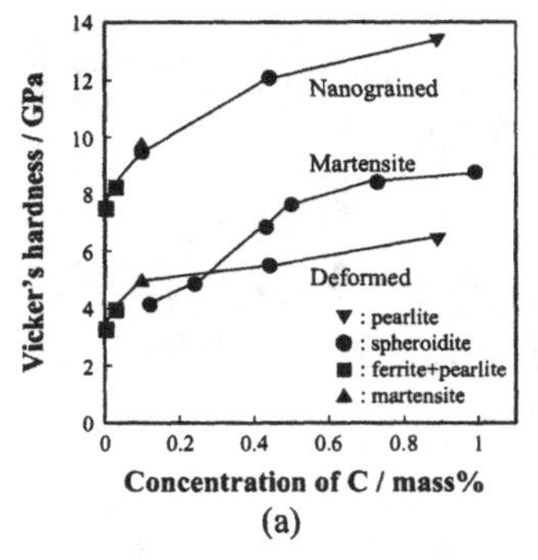

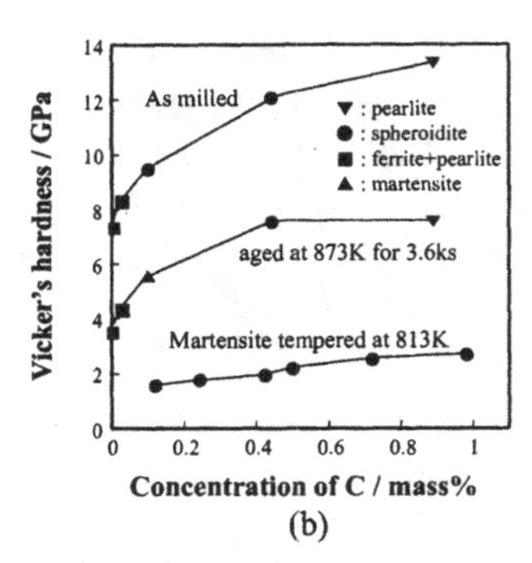

Fig. 6 (a) Hardness of nanocrystalline region produced by ball milling as a function of carbon content. The hardness of the dislocated cell structured region in ball milled powder and martensite [15] are shown for comparison. (b) Hardness of nanocrystalline region in ball milled powder after annealed at 873 K for 3.6 ks as a function of carbon content. The hardness before annealing and that of tempered martensite [15] are shown for comparison.

slower than that in a coarse grained counterpart. The slow grain growth rate has been observed in various nanograined materials and it has been suggested that the low mobility of triple junction is responsible.

Figure 6 (a) [16] shows the hardness of NS and dislocated cell structured regions in ball milled powders as a function of carbon content. The hardness of both regions increases with carbon content. In the nanocrystalline region, the grain size decreased with carbon content. The increase in hardness with carbon content is considered to be mainly due to the decrease in grain size and partly due to the solution hardening of carbon. In the figure the hardness of as quenched martensite [15] is shown for comparison. It should be noted that the hardness of nanograined region is about 4 GPa higher than its martensite counterpart.

Figure 6 (b) [16] shows the hardness of nanograined region produced by ball milling before and after annealing (at 873 K for 3.6 ks) as a function of carbon content. It is seen that the hardness of nanograined region decreases by annealing, but keeps high values of 4.5-8 Gpa depending on carbon content. In the figure the hardness of tempered martensite [15] (tempered at 813 K) is shown for comparison. It is clear that the hardness of annealed nanograined region is much higher than those of tempered martesite counterparts.

Nanocrystallization by a Ball Drop Test

NS can be produced near the surface of specimens after one or several times of a ball (weight attached) drops. Figure 7 [10] shows the cross section of the specimen (pearlite structure) deformed by 8 times of ball drops (a weight of 4 kg and a height of 1 m). A dark smooth contrast layer with a thickness of about 15 μm is seen near the surface of the indentation crater. The layer appears at top surface along the edge of crater and appears about 100 μm below the surface at the bottom of the crater. Figure 8 shows a typical dark contrast layer formed in Fe-0.80C specimen with pearlite (Fig. 8 (a)) and spheroidite (Fig. 8 (b)) structures, respectively. In the dark contrast layer the lamellar structure in pearlite or spherical cementite in spheroidite were disappeared. The microhardness and microstructure observed in the dark contrast layers produced by a ball drop test are similar to those observed in ball milled counterparts.

Figure 7: Nanocrystalline region (dark contrast layer about 15 μm thickness) formed near the surface of specimen (Fe-0.89%C with pearlite structure) by a ball drop test (1 m height, 4 kg weight, 8 times drops).

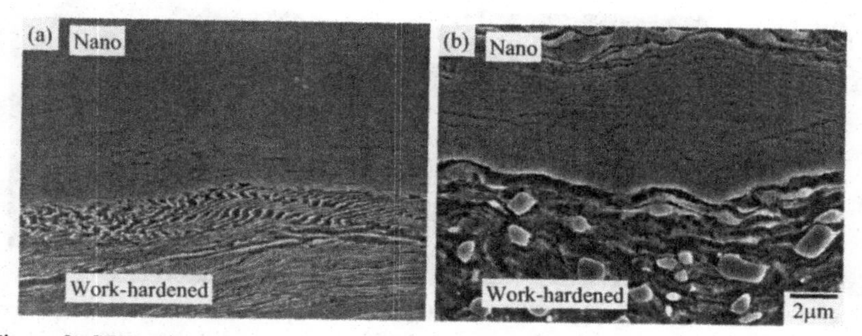

Figure 8: SEM micrographs of eutectoid (Fe-0.80C) specimen after a ball drop test. (a) Pearlite and (b) spheroidite.

Air Blast Shot Peening

Figure 9 (a) [13] shows nanocrystalline region formed in Fe-3.29Si specimen by 10 s of shot peening (coverage of 1000 %). The sharp boundaries are seen between the top surface nanocrystalline layer and subsurface deformed structured region. Figure 9 (b) is the TEM dark field (DF) image and selected area diffraction (SAD) pattern of the top surface region of Fe-3.29Si specimen shot peened for 60 s (coverage 6000 %). The DF image shows that the ferrite grain size is less than 20 nm, and the SAD pattern indicates that the ferrite grains are randomly orientated.

Figure 10 [13] shows cross section of near surface area of shot peened Fe-0.8C specimen (spheroidite structure pre-strained 84 % by cold rolling) after shot peening (a) and after shot peening and annealing (b). Nanocrystalline layer with 5 μm thickness is produced at the top surface. Before shot peening, the spherical cementite particles with diameter about 0.5 μm were uniformly distributed in the ferrite matrix. After shot peening, cementite particles are not visible in the nanocrystalline surface layer. This indicating that the dissolution of cementite takes place in the nanocrystalline layer similar to that observed in ball milled or ball dropped specimens. After annealed at 873 K for 3.6 ks, nanograined region did not show any detectable change while the deformed structured regions below the nanograined region showed recrystallized structure.

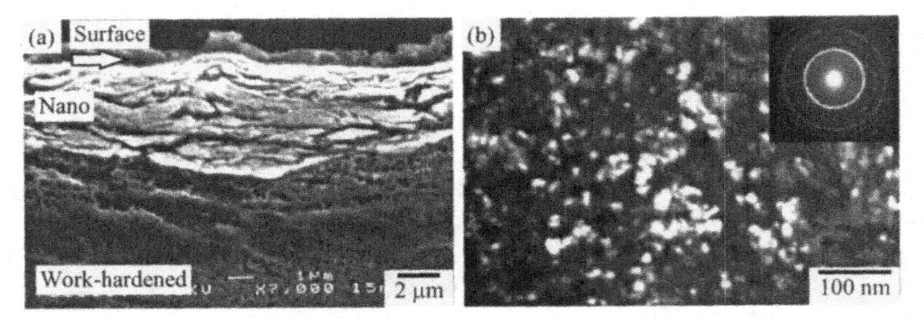

Figure 9: Micrographs of shot peened (< 50 μm in shot diameter, 190 m/s in shot speed) Fe-3.29Si steel. (a) SEM after peened 10 s (1000 % in coverage) and (b) TEM (dark field) after peened for 60 s (6000 % in coverage). The aperture size of diffraction was ϕ 1.2 μm.

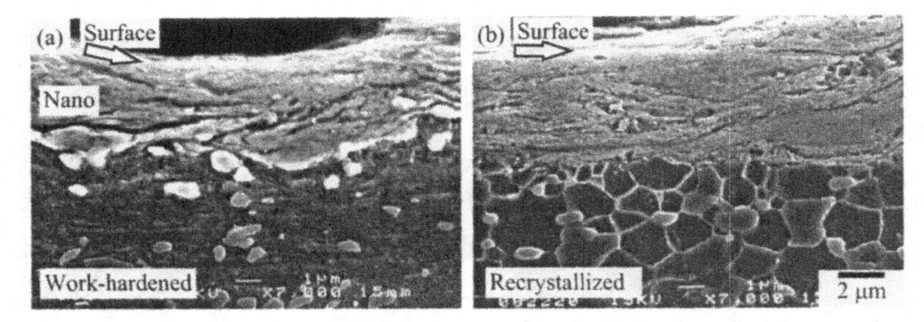

Figure 10: SEM micrographs of Fe-0.80%C steel with spheroidite structure (84 % cold rolled before shot peening) by shot peening (< 50 μm in shot diameter, 190 m/s in shot speed, 10 s in peening time and 1000 % in coverage). (a) As shot peened and (b) at 873 K for 3.6 ks after shot peening.

The production of nanocrystalline surface layer by air blast shot peening has considerable industrial importance since air blast shot peening is a popular process in industries and it can produce nanostructured surface layer with a high productivity. Industrial application of this new technology is expected especially to upgrade the traditional engineering materials.

Discussion

Impose a large strain seem to be the most important condition to produce nanocrystalline structure (NS). The reported amount of strain applied to obtain NS (grain size less than 100 nm) varies from 7 to 31 depending on the deformation techniques and materials employed.

The amount of strain can be measured well in drawing and torsion experiments. Langford and Cohen [17] studied the drawn Fe up to $\varepsilon = 6$ and reported that the flow stress increased linearly with strain up to 1.4 GPa. Extrapolating their flow stress vs strain data, the strain which gives the flow stress corresponding to d = 100 nm (2.1 GPa estimated from the Hall-Petch relationship for iron [18] (i.e., σ[GPa] = 0.12+20d-1/2, d in nm)) is expected to be attained at $\varepsilon = 12$. Tashiro [19] reported the flow stress of drown iron wire strained up to $\varepsilon = 11.5$. The flow stress corresponding to d = 100 nm (2.1 GPa) is achieved at $\varepsilon = 7$. The experiment of high pressure torsion straining in iron was carried out by Valiev *et al* [7]. After 3 turns ($\varepsilon = 108$) the hardness reaches 4.5 GPa and tends to reach a steady state. The grain size about 100 nm was obtained after 5 turns ($\varepsilon = 180$) at which the hardness was 4.6 GPa. Similar torsion experiment has been done by Kaibyshev *et al* [8] in Fe-3%Si. After straining to $\varepsilon = 31$ (0.5 turn), grains are reduced to 120-200 nm and the microhardness increased to 7.2 GPa. From the obtained microhardness it can be considered that the grain size less than 100 nm was achieved. From the hardness measurement d = 100 nm (6.3 GPa) is expected to be accomplished at a strain $4.3 < \varepsilon < 31$.

The amount of strain to produce NS was also estimated in a ball drop test and sliding wear test, although with less accuracy. In a ball drop test, Umemoto *et al* [10] estimated $\varepsilon = 7.3$ at strain rate of around 1.3×10^4 /s. Hughes and Hansen [20] studied the nanostructures in Cu produced by sliding wear. The amount of strain to reach d = 100 nm is estimated to be 8. Summarizing the above mentioned studies, the minimum amount of strain necessary to produce NS is considered to be around 7-8, although it depends on materials, microstructure, deformation techniques and deformation conditions employed.

The effect of strain rate on the strength of deformed steels was investigated using pure Fe (Fe-0.03%C) and Fe-0.10%C martensite. Specimens were deformed by rolling or by a weight drop technique. In rolling, specimens were rolled at strain rate of 0.4 /s to 60 % reduction with 10 % reduction per pass. In a weight drop technique, 5 kg weight was dropped on a cylindrical specimen (5 mm in diameter and 5 mm in height) from 2 m height either at R.T or at LN_2. The strain of 20 % height reduction was given in each weight drop experiment to totally 60 % height reduction. The strain rate was around 5000 /s. The Vicker's hardness measured after these deformation were listed in Table 1. It is seen that the difference in strain rate between 0.4 /s and 5000 /s does not make any detectable difference in the hardness of deformed specimens when the total strain is same.

This indicates that strain rate does not change the work-hardening rate in the strain rate range studied. Thus it is considered that the amount of strain necessary to produce NS dose not depend on strain rate appreciatedy.

Table I. Hardness of steels after rolling and weight drop deformation.

		Hardness [GPa]		
Alloy	Structure	Rolling	Weight drop at R.T.	Weight drop at LN_2
Fe-0.03%C	ferrite	1.67	1.48	1.58
Fe-0.10%C	martensite	4.10	4.19	4.08

Summary and Conclusions

In the present study, the formation of nanograined structure by various severe plastic deformation were demonstrated. The conditions to produce nanograined structure were discussed. The main results are summarized as follows.

(1) In ball milling, the nanograined regions appear near the surface of powder and are separated from the interior deformed structured region with sharp boundaries. The hardness of nanograined region is extremely high and increases with carbon content from 7 to 13 GPa. The cementite (either lamellar or spherical) dissolves completely by ball milling when the nanocrystallization of ferrite matrix starts. By annealing, nanograined structured region showed substantially slow grain growth without recrystallization.

(2) In a ball drop test, similar nanograined structure was obtained in the surface layer after several times of ball drops.

(3) In air blast shot peening, nanograined structure is produced when larger coverage (>1000 %) and higher shot speed (>100 m/s) than conventional operation are applied.

(4) To produce nanograined structure by deformation, the most important condition is to impose a large strain (larger than about 7).

Acknowledgements

This work is partly supported by the Grant-in-Aid by the Japan Society for the Promotion of Science.

References

1. H. Gleiter, *2nd Riso Int. Symp. Metall. and Mat. Sci.*, eds. N. Hansen, A. Horsewell and H. Lilholt (Riso National Laboratory, Denmark, 1981), 15.
2. L. Lu, M. L. Sui and K. Lu, Science 287 (2000), 1463.
3. G. Plumbo, S. J. Thorpe and K. T. Aust, Scr. Metall. Mater. 24 (1990), 1347.
4. J. S. C. Jang and C. C. Koch, Scr. Metal. 24 (1990), 1599.
5. H. J. Fecht, E. Hellstern, Z. Fu and W. L. Johnson, Met. Trans. 21A (1990), 2333.
6. J. Eckert, J. C. Holzer, C. E. Krill, III and W. L. Johnson, J. Mater. Res. 7 (1992), 1751.
7. R. Z. Valiev, Y. V. Ivanisenko, E. F. Rauch and B. Baudelet, Acta Mater. 44 (1996), 4705.
8. R. Kaibyshev, I. Kazakulov, T. Sakai and A. Belyakov, *Proc. of Int. Symp. on Ultrafine Grained Steels* (The Iron and Steel Inst. of Japan, 2001), 152.
9. P. Heilmann, W. A. T. Clark and D. A. Rigney, Acta Metall 31 (1983), 1293.
10. M. Umemoto, B. Haung, K. Tsuchiya and N. Suzuki: Scr. Mater. 46 (2002), 383.
11. N. R. Tao, M. L. Sui, J. Lu and K. Lu: NanoStructured Mater. 11 (1999), 433.
12. I. Altenberger, B. Scholtes, U. Martin and H. Oettel, Mater. Sci. Eng. A 264 (1999), 1.
13. M. Umemoto, Y. Todaka and K. Tsuchiya, Submitted to Mater. Trans.
14. Y. Xu, Z. G. Liu, M. Umemoto and K. Tsuchiya, Met. Mater. Trans. 33A (2002), 2195.
15. G. Krauss, *Steel: Heat Treatment and Processing Principles*, (ASM Int., 1990).
16. M. Umemoto, Z. G. Liu, K. Masuyama, X. J. Hao and K. Tsuchiya, Scr. Mater. 44 (2001), 1741.
17. G. Langford and M. Cohen, Trans. of the ASM 62 (1969), 623.
18. J. Yin, M. Umemoto, Z. G. Liu and K. Tsuchiya, ISIJ Int. 41 (2001), 1391.
19. H. Tashiro, Doctor thesis at Tohoku Univ. (1992).
20. D. A. Hughes and N. Hansen, Acta Mater, 48 (2000), 2985.

Processing and Properties of Structural Nanomaterials
Edited by Leon L. Shaw, C. Suryanarayana and Rajiv S. Mishra
TMS (The Minerals, Metals & Materials Society), 2003

Mechanical Properties and Fracture Mechanism of Nanostructured Al-20 wt% Si Alloy

Soon-Jik Hong and **C. Suryanarayana**

Department of Mechanical, Materials and Aerospace Engineering, University of Central Florida, Orlando, FL. 32816-2450, USA

Keywords: Nanostructure, Rapid solidification, Al-20Si alloy, Extrusion, Wear.

Abstract

The effect of powder size on the microstructure, mechanical properties and fracture behavior was studied in gas atomized Al-20 wt% Si alloy powders and their extruded bars using SEM, XRD, TEM, tensile testing and wear testing. The microstructure of the extruded bars showed a homogeneous distribution of eutectic Si and primary Si particles embedded in the Al matrix. The grain size of α-Al varied from 150 to 600 nm and the size of the eutectic Si and primary Si in the extruded bars was about 100 - 200 nm. The tensile strength of the Al-20Si (powder size < 26 μm) alloys at room temperature was 322 MPa while for the Al-20Si (powder size: 45-106 μm) it was 230 MPa; this value is higher than that of Al-20Si-3Fe (powder size: 60-120 μm) alloys. With decreasing powder size from 45-106 μm to < 26 μm, the specific wear decreased significantly at all sliding speeds due to the fine nanostructute. The wear resistance was also better than that of the Al-20Si-3Fe alloys. The fracture mechanism of failure in tension testing and wear testing was also studied.

Introduction

Al-Si alloys are used in many application areas such as automotive, electronics and aerospace industries [1] due to their good wear resistance and low thermal expansion coefficient. These properties are attributed to the high volume fraction of hard Si particles embedded in the soft Al matrix. Al-Si alloys are currently produced by casting and powder metallurgy methods. A relatively slow cooling rate, associated with the casting process, produces coarse and segregated primary Si and/or eutectic Si in the Al-Si alloys. It is common practice to increase the Si content to improve wear resistance and mechanical strength. However, with increasing Si content, above the eutectic composition, primary Si crystals become coarse, resulting in poor mechanical properties of the Al-Si alloys.

Techniques such as modification [2, 3], ternary alloying [4] or rapid solidification processing [5, 6] have been applied to refine primary Si crystals. Due to recent developments in the techniques for consolidation, use of powder-processed alloys seems to be most promising route to improve the strength and wear resistance of Al-Si alloys. Melt atomization provides chemical and microstructural homogeneity together with refinement in grain size and second phases in the powder particles. Further, rapidly solidified nanocrystalline Al alloys have high strength [7-9]. Therefore, it is expected that rapidly solidified Al-Si powder alloys will also show high strength and consequently improved wear resistance.

The present study was carried out to investigate the effect of powder size on the microstructure, mechanical properties, and fracture behavior of Al-20 wt% Si alloys in the as-solidified powders and their extruded bars.

Experimental Procedure

Master alloys of Al-20wt%Si (referred to as Al-20Si from now onwards) were prepared by induction melting of commercially pure metals in graphite crucibles in air. Al-20Si alloy powders were obtained by remelting the master alloy in a graphite crucible 200 K above the melting temperature of the alloy and by gas atomization. The size distribution of the alloy powder particles was measured by conventional mechanical sieving and the sieved powder was used for extrusion. The as-solidified powder was subjected to cold compaction, canning, degassing for 1 h at 673 K down to 10^{-2} torr. Subsequently, it was hot extruded after holding for 1 h at 673 K with a reduction ratio of 25:1 into a bar with a diameter of 15mm. None of the extruded bars showed blistering or cracking on the surface.

Microstructural analysis of both the as-solidified powder and the extruded bars was conducted using a scanning electron microscope (SEM, JSM 5410) equipped with Energy Dispersive X-ray Spectrometer (EDS). The crystal structures of both the powder and the extruded bars were determined by conventional X-ray diffraction using monochromatic Cu K_α radiation. The microstructures of the specimens were also observed using a CM20 transmission electron microscope operating at 200 kV.

Tensile specimens of the alloy bars were machined according to the ASTM-A370 specification and tensile testing was performed with a crosshead speed of 0.2 mm/min using an Instron 4206 machine. For wear testing, plate type specimens were polished and mounted on an Ohgoshi-type wear testing machine. In this method, a ring-shaped steel material (AISI 1045) is pressed and rotated against the specimen's surface, forming a wear track. During the test, the sliding distance and load were kept constant at 100 m and 2.1 Kg, respectively. However, the sliding speed was changed from 0.62 to 3.53 m/s.

Results and Discussion

Microstructure of the Powders and Extruded Bars

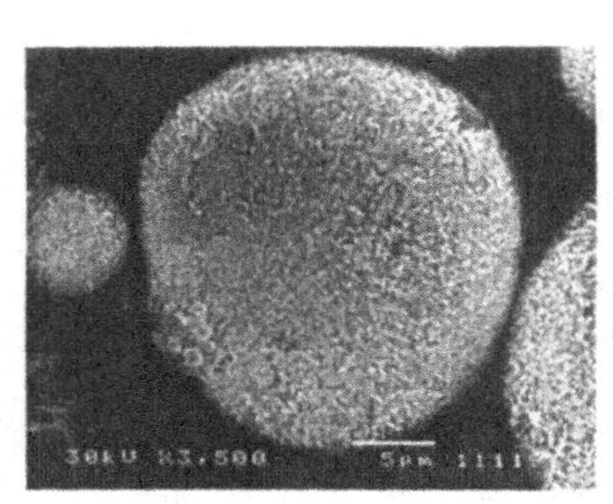

Figure 1: Cross sectional microstructure of gas atomized Al-20Si powders (< 26 μm).

The as-solidified powders were mostly spherical and had a smooth surface. There were no defects such as satellites or pores on the surface. The cross sectional micrograph (Figure 1) of the as-atomized powder showed an ultrafine microstructure due to the small powder size; an effect of high solidification rate. Therefore, it was impossible to distinguish between the primary Si and eutectic Si in the powder microstructure. However, it is possible to distinguish between the primary and eutectic Si particles in coarse Al-20Si alloy powders[10].

Figures 2(a) and (b) show SEM micrographs of the Al-20Si alloy extruded bars, with the particle sizes of < 26 μm and 45-106 μm, respectively. It may be noted that the microstructure in Figure 2(a) is very fine due to the combination of initial fine powders and the severe plastic deformation experienced by them during the hot extrusion process. It is also difficult to identify the extrusion direction in the micrograph because of the fine microstructure. Further, It is not easy to distinguish between the refined primary Si and eutectic Si due to the close similarity in their morphology and size.

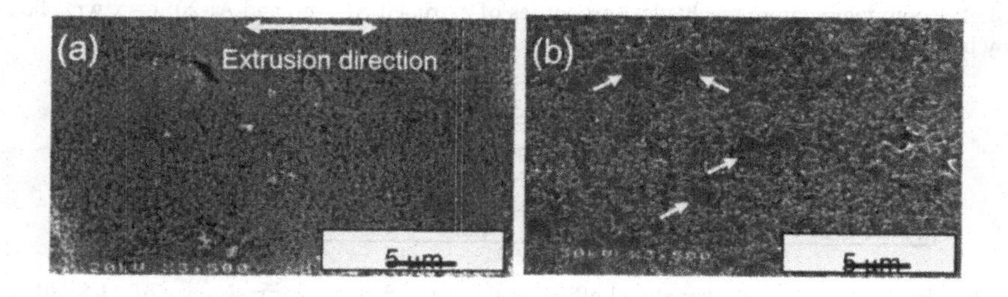

Figure 2: Scanning electron micrographs of longitudinal sections of as extruded Al-20 wt% Si alloy bars with different powder size: (a) < 26 µm and (b) 45 – 106 µm.

The size range of the primary and eutectic Si particles is between 100 and 200 nm and these are distributed homogeneously in the Al matrix. Although the phases present are the same, with increasing initial alloy powder size, Figure 2(b), one can easily distinguish between the primary Si (indicated by arrows) and the eutectic Si in the extruded bar. The extrusion direction can also be identified.

Figure 3: Electron micrograph of extruded Al-20Si alloys (< 26 µm) showing fcc-Al grains and Si particles.

The XRD patterns obtained from the rapidly solidified fine Al-20Si alloy powders (< 26 µm) and the perpendicular section of extruded bar show that all diffraction peaks could be indexed as a mixture of fcc-Al and Si phases. The relative intensities of the peaks from the two phases, however, are different in the atomized powder and the extruded bar.

TEM studies reveal significant inhomogeneity in the microstructure of the extruded bar even from the fine powders (< 26 µm) (Figure 3). The size of the α-Al grains (marked A) varied between 150 and 600 nm and the primary and eutectic Si particles, marked by arrows, are in the range of 100 to 200 nm. This variation appears to arise from the different sizes of the powder particles in the initial powder, which could vary from a few microns to 26 µm. Even at this level, it is difficult to distinguish between the primary and eutectic Si particles.

Mechanical Properties and Fracture Surface Morphology

The room temperature mechanical properties (ultimate tensile strength and elongation) of the extruded Al-20Si and Al-20Si-xFe (x = 3, 5wt%) alloy bars are presented in Table I. From the table it is clear that the Al-20Si alloy (< 26 µm) exhibits a higher UTS value than the coarse Al-20Si alloy (45-106 µm), but the elongation is not obviously deteriorated. This combination is directly related to the nano size of the α-Al grains and the Si particles as shown in Figure 3.

Table I: Room temperature mechanical properties of extruded Al-20Si and Al-20Si – x wt% Fe (x=3, 5) alloy bars.

Alloy (wt.%)	Powder Size (μm)	UTS (MPa)	Elongation (%)
Al-20Si	45 – 106	230	9
Al-20Si	< 26	322	8.5
Al-20Si-3Fe	60 – 120	301	8.4
Al-20Si-5Fe	60 –120	359	4.3

It is known that Fe has a beneficial effect on the strength and wear resistance of Al-Si alloys. Thus, with 5 wt% Fe addition to Al-20Si, the UTS is as high as 359 MPa for 60-120 μm powder particle size, much higher than that achieved in the Al-20Si with < 26 μm powder size. The present result indicates that a reasonably high tensile strength, without loss of elongation, can be achieved in Al-Si alloys by controlling the grain and primary Si size by rapid solidification processing, without the addition of a ternary element.

Figure 4 shows micrographs of the tensile fracture surfaces of extruded Al-20Si alloy bars with (a) < 26 μm, and (b) and (c) 45-106 μm powder size. Many large dimples with ridges are seen in the extruded bar from fine powders (Figure 4(a)) indicating better ductility. This is consistent with the tensile test results (Table I). Fractographs of the extruded bar from the coarse powder (45-106 μm) show few dimpled regions thus indicating a typical brittle fracture (Figure 4(b)). The microscopic fracture surface showed rough tear ridges comprising a large population of fine veins. Figure 4(c) shows cracking/fracture of primary Si particles on the tensile fracture surface. Typical fracture surface consists of distribution of large and small dimples associated with primary Si.

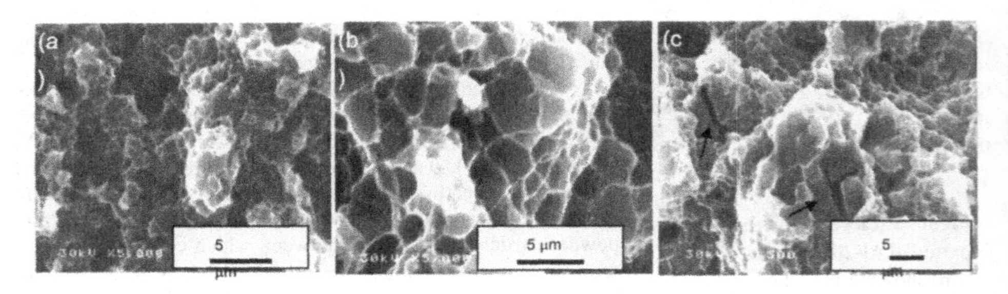

Figure 4: Scanning electron micrographs showing the tensile fracture surface of extruded Al-20Si alloy bars with powder size. (a) < 26 μm, (b) and (c) 45–106 μm.

It may be noted that fracture of Al-20Si alloys occurs through ductile fracture of the Al matrix and brittle fracture of primary Si particles. Thus, it may be concluded that fracture of the primary Si particles is the major factor responsible for the relatively low ductility of bars from the coarse powders. Most of the primary Si particles were found to contain fracture cracks and a fractured primary Si particle shows an obvious crack running through the center of the particles (Figure 4(c)). This indicates that the crack initiates in the primary Si particle, thus leading to typical brittle fracture. However, no such cracks were observed in the extruded bars of fine powders.

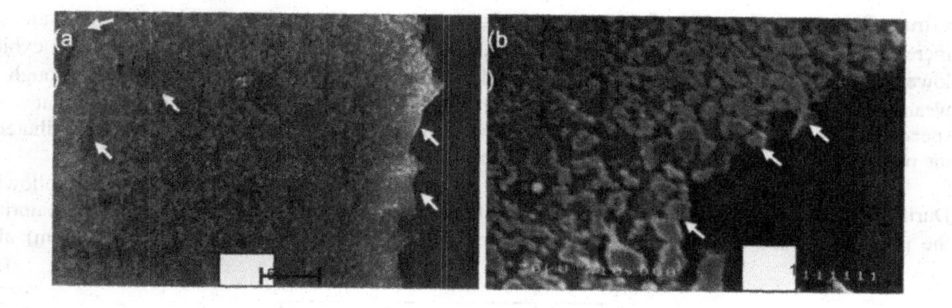

Figure 5: Cross sectional microstructure of tensile tested samples from (a) coarse powders (45-106 μm) and (b) fine powders (< 26 μm), respectively.

Damage and failure of Al-Si alloys is generally associated with cracking and growth of primary Si particles; these can be conveniently observed in cross-sectional microstructures of tensile-tested specimens as shown in Figure 5.

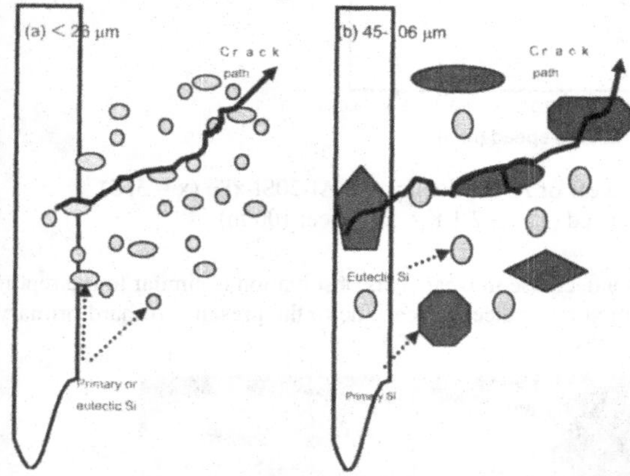

Figure 6: Schematic for the crack path during fracture of Al-20Si alloy with powder size.

Figure 5(a) shows the transverse structure of tensile-tested specimen showing initiation of cracks from primary Si particles at the edge (fractured area) of the specimen as indicated by arrows (right side). The matrix damage is associated with the brittle primary Si particles in the form of (i) cracking of primary Si and decohesion at the interface between the matrix and primary Si and (ii) fine microscopic voids, which have formed around the cracked primary Si particles. However, in the fine powder alloys (Figure 5(b)), the fracture surface observed was basically different. The transverse microstructure of the fracture surface shows particle boundary cracks at the fractured edge as indicated by arrows; no cracks were observed inside the matrix.

Figure 6 summarizes schematically the fracture mechanism of the Al-20Si alloy with different powder sizes. Microscopic analysis demonstrates that large particles were more prone to failure than smaller ones; the larger the particle size the greater is the probability to have cracks in it and consequently the lower the load it can withstand before fracture. Thus, it may be seen that in the alloy with fine powder size (Figure 6(a)), the cracks pass around the primary (or eutectic) Si particles, while in the coarse powder alloy (Figure 6(b)), the cracks pass through the large primary Si particles which have fractured

Specific Wear and Fracture Surface Morphology

Figure 7 shows the variation of specific wear of the Al-20Si and Al-20Si-xFe (x = 3, 5 wt%) alloy extruded bars as a function of sliding speed from 0.62 to 3.53 m/s. An increase in wear rate with increasing sliding speed is observed for all the alloys. However, the Al-20Si alloy (< 26 μm) exhibits lower wear rate than the Al-20Si (45-106 μm) and Al-20Si-3Fe alloys at all sliding speeds. Although, the wear rate of Al-20Si (< 26 μm) also increases with increasing sliding speed, it remains constant at speeds > 2.38 m/s. The specific wear was the lowest for Al-20Si-5Fe at all sliding speeds; attributed to the presence of fine primary Si and Al-Si-Fe intermetallic compounds.

Reasons for the improved wear resistance of the Al-20Si (< 26 μm) alloy may be the following. During the test, a larger number of primary Si particles will be in contact with the counterpart material if the primary Si size is small. The nanostructured particles formed in the Al-20Si (< 26 μm) alloy

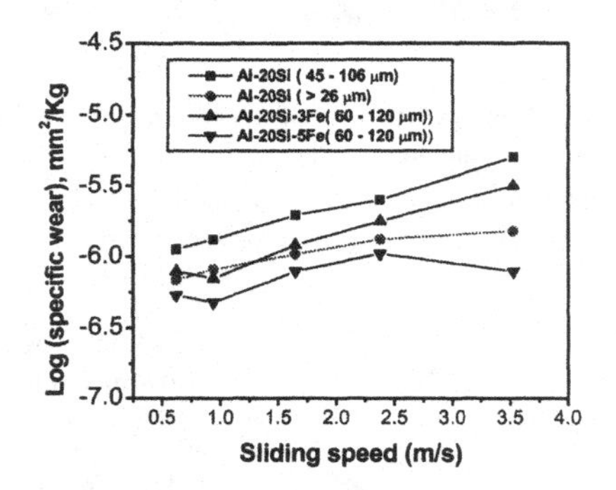

Figure 7: Variation of specific wear of Al-20Si alloy and Al-20Si-xFe (x = 3, 5) alloys as a function of sliding speed (Load: 2.1 Kg, Distance: 100 m)

contribute to an increase in the UTS and thus a decrease in wear. This observation is similar to the report that hypereutectic Al-Si binary alloys show high resistance to wear due to the presence of hard primary Si particles in the soft Al matrix [11].

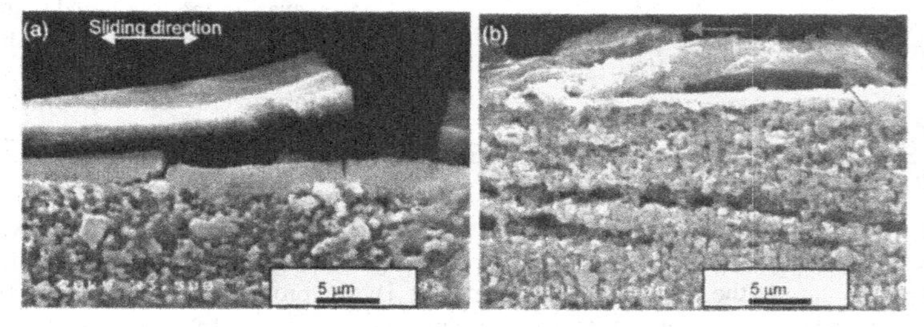

Figure 8: Scanning electron micrographs of transverse sections from wear tested Al-20Si specimens at 3.53 m/s sliding speed (Load: 2.1Kg, Distance: 100 m): (a) < 26 μm, (b) 45–106

SEM photographs were recorded from the longitudinal cross-sections (parallel to the worn surface) and the results were analyzed to investigate the wear mechanism. In the sample with < 26 μm powder particles, a deformed layer without any cracks has formed at 0.62 m/s sliding speed. With increasing speed, the thickness and area of the deformed layer increased further and cracks started forming between the deformed and undeformed layers due to the inherent brittle nature of the deformed layer. With continued increase in the speed to 3.53 m/s, these brittle surface layers cracked up, exposed fresh surfaces, and further wear took place. Figure 8(a) shows the micrograph for the bar extruded from the ≤ 26 µm particles after subjecting it to wear at 3.53 m/s. Accumulation of mechanical energy would promote crack nucleation first at the weak interface between the deformed and undeformed layers. With continued accumulation of energy, these cracks grow and also cracks start nucleating at the relatively stronger deformed layers due to work hardening. This process will eventually lead to continuous formation of a deformed layer and nucleation and growth of new cracks. Since this material cannot withstand much load, the material gets removed from the surface.

EDS patterns from the deformed layers of both the types of specimens showed only the presence of Al and Si and no oxygen; thus the brittle nature of the surface layer could not be due to the formation of oxides. It was reported that the hard primary Si particles in Al alloys help in forming a deformed layer on the worn surface [11]. A similar phenomenon was also observed in SiC-reinforced aluminum composites [12].

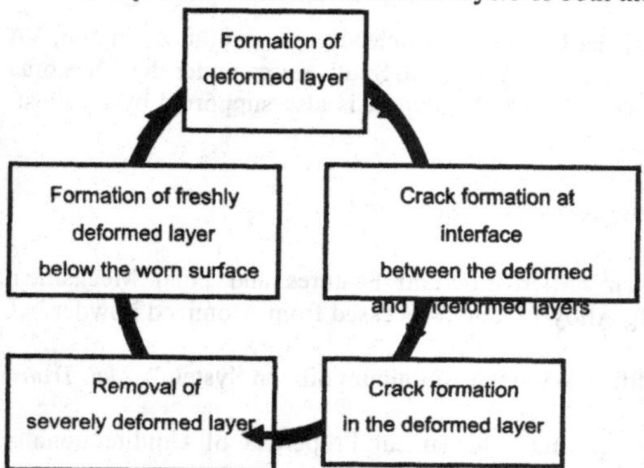

Figure 9: Schematic diagram of wear mechanism in Al-Si alloys with fine grain structure.

The situation is, however, different for the bar extruded from the coarse powder particles (45-106 µm). A deformed layer was also observed in this sample at all sliding speeds. But, the thickness of the deformed layer was larger in comparison to the bars from the fine powders; attributed to the lower hardness and strength of the bars with the coarse powders. Further, the deformed layer appears to be less brittle and therefore cracks were not easily observed. Figure 8(b) shows the micrograph of the bar extruded from the 45-106 µm particles after subjecting it to wear at 3.53 m/s.

From the above discussion, the wear of the Al-20Si alloy may be described to progress as shown schematically in Figure 9. The basic difference between the wear behavior of Al-Si bars with the two sizes of powders (< 26 µm and 45-106 µm) is as follows. Since the bar with the fine powders is very hard and brittle, the thickness of the deformed layer is small and cracks form more easily. But, the bar with the coarse powders is less hard and so the thickness of the deformed layer is larger and cracks also do not form easily. Thus, surface material removal in the softer material appears to be by a different mechanism. But, it may be noted that Al–20Si alloys (< 26 µm) have improved wear resistance due to the formation of fine nanostructured α-Al grains and increase of matrix strength due to a decrease in the primary Si size.

Conclusions

The microstructure of the extruded bar from < 26 µm powders showed a homogeneous distribution of primary Si and eutectic Si embedded in the α-Al matrix. The typical size of α-Al grains is about 150-600 nm and the primary (and eutectic) Si particles have a size of about 100-200 nm. The UTS and elongation of the fine Al-20Si alloy (< 26 µm) are 322 MPa and 8.5%, while for the Al-20Si-3Fe alloy these are 301 MPa and 8.4%, respectively. Fine Al-20Si alloy (< 26 μm) exhibits higher UTS values than the coarse Al-20Si alloy (45-106 μm), while the elongation is not very different. Transverse microstructures of the tensile-tested Al-20Si alloy (45-106 μm) showed cracks in the primary Si particles, while primary Si particle boundary cracks were observed in the Al-20Si alloy (< 26μm). The wear resistance of the Al-20Si alloy (< 26 μm) is higher at the all sliding speeds than the Al-20Si alloy (45-106 μm) and also the Al-20Si-3Fe alloy because of fine nanostructure. Al–20Si alloys (< 26 μm) have a significant effect on decreasing the wear rate due to a decrease in the thickness of the deformed layer than that of Al-20Si alloy (45-106 μm) owing to an increase in the matrix strength due to nanostructure formation.

Acknowledgments

The work reported in this paper is supported by the US National Science Foundation, Arlington, VA and the Korea Science and Engineering Foundation (KOSEF), Daejon, South Korea under the US-Korea Cooperative Science Project Grant Award No. INT-0196195. This work is also supported by the Post-doctoral Fellowship Program of KOSEF.

References

1. J. Zhou, J. Duszczyk, and B. M. Korevaar "Microstructural Features and Final Mechanical Properties of the Iron-Modified Al-20Si-3Cu-1Mg Alloy Product Processed from Atomized Powder", *J. Mater. Sci.*, 26 (1991), 3041.
2. M. D. Hanna. S. Lu, and A. Hellawell, "Modification in the Aluminum Silicon System", *Met. Trans. A*, 15A (1984), 459.
3. F. Yilmaz and R. Elliott, "The Microstructure and Mechanical Properties of Unidirectionally Solidified Al-Si Alloys", *J. Mater. Sci.*, 24 (1989), 2065.
4. A. Knuutinen, K. Nogita, S. D. McDonald, and A. K. Dahle, "Modification of Al-Si Alloys with Ba, Ca, Y and Yb", *J. Light Met.*, 1 (2001), 229.
5. J. Zhou, J. Duszczyk, and B. M. Korevaar, "A Study on an Atomized Al-Fe-Mo-Zr Powder to Be Processed for High-Temperature Application", *J. Mater. Sci.*, 26 (1991), 3292.
6. S. Das, A.H. Yegneswaran, and P.K. Rohatgi, "Characterization of Rapidly Solidified Aluminum-Silicon Alloy ", *J. Mater. Sci.*, 22 (1987), 3292.
7. Y. Kawamura, A. Inoue, K. Sasamori, and T. Massumoto, "High-Strength Powder Metallurgy Aluminum Alloys in Glass Forming Al-Ni-Ce-Ti (or Zr) Systems", *Scripta Met.*, 29 (1993), 275.
8. A. Inoue, K. Kakazato, Y. Kawamura, and T. Masumoto, "The Effect of Cu Addition on the Structure and Mechanical Properties of Al-Ni-M (M=Ce or Nd) Amorphous Alloys Containing Nano Scale fcc-Al Particles", *Mater. Sci. Eng.*, A 179-180 (1994), 654.
9. S.J. Hong, T.S. Kim, H.S. Kim, W.T. Kim, and B.S. Chun, "Microstructureal Behavior of Rapidly Solidified and Extruded Al-14wt%Ni-14wt%Mm (Mm: misch metal) Alloy Powders", *Mater. Sci. Eng.*, A271 (1999), 469.
10. S.J. Hong, C. Suryanarayana, and B.S. Chun, "Section-dependent Microstructure and Mechanical Properties of Rapidly Solidified and Extruded Al-20Si Alloy", *Mater. Res. Bull.*, (2003) in press.
11. V.K. Kanth, B.N.P. Bai, and S.K. Biswas, "Wear Mechanism in a Hypereutectic Aluminum Silicon Alloy Sliding Against Steel", *Scripta Metall. Mater.*, 24 (1990), 267.
12. B. Venkataraman and G. Sundararajan, "The Sliding Wear Behavior of Al-SiC Particulate Composites-I. Macrobehavior", *Acta Metall.*, 44 (1996), 451.

Processing and Properties of Structural Nanomaterials
Edited by Leon L. Shaw, C. Suryanarayana and Rajiv S. Mishra
TMS (The Minerals, Metals & Materials Society), 2003

High-Strength, High Conductivity Bulk Nanostructured Ag-Cu Alloys

R. B. Schwarz and T. D. Shen

Los Alamos National Laboratory, MST-8, MS G755, Los Alamos, NM 87545, USA
Keywords: Nanocrystalline, Nanostructured, Strength, Property, Pulsed Magnet

Abstract

Researchers are interested in developing high-strength and high conductivity alloys for use in high-field pulsed electromagnets. We have used a fluxing and quenching technique to prepare 5-mm diameter rods of Ag-Cu alloys with nanosized microstructure. Fluxing removes heterogeneous nucleation centers, enabling a higher degree of undercooling. The purified melt is quenched at rates of approximately 100 K/s. The as-quenched alloys are a mixture of metastable Ag-6 at.% Cu and Cu-3.5 at.% Ag solid-solutions and have a eutectic microstructure with approximately 50-nm thick Ag(Cu) and Cu(Ag) lamellae. The alloy's room-temperature compressive yield strength is 370 MPa and its electrical conductivity is 78% of the International Annealed Copper Standard (IACS). These results are evaluated in terms of current understanding on solidification and mechanical properties.

Introduction

The magnetic field delivered by a pulsed electromagnet is often limited by the tensile strength and electrical conductivity of the alloy used to make the outer and insert coils. The high-strength limitation is due to the Lorentz forces on the coils during its operation. The maximum hoop stress in the coils during the generation of a magnetic pulse of induction B_m is approximately equal to $B_m^2/(4\mu_0)$ where μ_0 is the magnetic permeability of free space [1]. For B_m = 50 Tesla, the maximum stress is approximately 500 MPa. A high conductivity, typically above 70% of IACS, is required to minimize the joule heating generated by the current pulse. Another important property that must to be considered in the design of pulsed magnets is fatigue life. Because of its high manufacturing cost, the lifetime of the electromagnet should exceed 10,000 cycles.

The high conductivity requirement necessitates use of Cu- or Ag-based alloys. Strength in these alloys is usually achieved through (a) precipitation hardening [2], (b) strain hardening [3-13], and (c) the manufacture of composites [3-13]. Copper reinforced by nanosized Al_2O_3 particles has both relatively high strength (above 330 MPa) and high conductivity (above 90% IACS) [2]. Strain hardening is useful to raise the yield strength of two-phase Cu-X composites (X = Ag, Nb, V, Cr, Ti, and stainless steel). To obtain the desired effect, the strain is usually several hundred percent. The "melt-and-deform" and "bundle-and-deform" methods have been used to achieve the required high plastic strains [3-13]. In the "melt-and-deform" method, the as-cast alloys are mechanically deformed to a strain of 4 to 5. This high strain limits the cross section of the product. In the "bundle-and-deform" method, several tens of wires are placed in a tube, which is heavily drawn to a strain of about 1. The drawn tube is then cut and the pieces are placed in a new tube. The drawing procedure is repeated four to five times to accumulate a strain of 4 to 5. These materials usually have an ultimate tensile strength of above 1 GPa. Solution hardening is, in general, not desirable because the strengthening effect is rather weak (compared to strain hardening) and solutes decrease the conductivity.

A common feature of all the strong wire synthesis methods is development of a fine microstructure, in the nanometer range. The present paper explores the synthesis of ***bulk*** nanocrystalline alloys by casting, without the need of an extensive mechanical deformation.

Experimental Procedures

A eutectic $Ag_{60}Cu_{40}$ alloy was selected for this study. Elemental Ag (99.999%) and Cu (99.999%) shots were mixed and placed in a fused silica tube together with dehydrated B_2O_3 (99.9995%). The silica tube was heated to 1350°C, forming a Ag-Cu melt covered by liquid B_2O_3. B_2O_3 acts as a flux, dissolving and reducing the oxide particles present in the molten alloy. During melt purification, O_2 bubbles form at the metal-flux interface and rise to the B_2O_3 surface. The purification is complete when oxygen bubbles no longer leave the molten B_2O_3, at which point the silica tube containing the melt was quenched in water. The as-quenched alloys were in the form of 3-to-5-mm diameter and 45-to-50-mm long rods.

The as-quenched $Ag_{60}Cu_{40}$ rods were investigated using DC conductivity tests, x-ray diffraction (XRD), optical microscopy (OM), transmission electron microscopy (TEM), and compression tests. For the conductivity measurements we used a DC current of 1.5 A. The XRD measurements were done in copper-anode powder x-ray diffractometer. The OM observations were done on a disk cut from the alloy rod, mechanically polished, and etched in a solution (5g $FeCl_3$ + 12 ml HCl + 40 ml H_2O) for five seconds. The TEM specimens were prepared by mechanically thinning, polishing, and ion milling a 3-mm diameter disk. Compression testing was done on cylindrical specimens, 2.9-mm diameter and 5.8-mm long, which were cut from the as-quenched rod by electrical discharge machining. The compression tests were performed in a screw-type machine at a strain rate of 2×10^{-4} sec^{-1}.

Results

The conductivity of the as-quenched $Ag_{60}Cu_{40}$ alloy was 4.52×10^5 S/cm, which is 78 % of the IACS (IACS = 5.80×10^5 S/cm).

Figure 1 shows a XRD pattern for as-quenched $Ag_{60}Cu_{40}$. The Bragg peaks indicate the presence of only fcc Ag(Cu) and fcc Cu(Ag) phases. The lattice parameters for Ag(Cu) and Cu(Ag) solid solutions, as determined by a Bradley and Jay's extrapolation method [14], are 0.4066 ± 0.0004 nm and 0.3635 ± 0.0002 nm, respectively. From these values we deduce that the Ag phase contains approximately 6 at.% Cu in solution, and the Cu phase contains approximately 3.5 at.% Ag in solution. Thus, the quenching is fast enough to develop supersaturated Ag(Cu) and Cu(Ag) solid solutions since in equilibrium Ag and Cu have negligible mutual solubilities. The diffraction peaks are rather wide, indicative of a small crystallite size and/or large residual strains. The Scherrer formula (neglecting the effect of strain) gives an average crystallite size of 38 nm for both Ag(Cu) and Cu(Ag).

Figure 2 shows the optical micrograph of the cross section of an $Ag_{60}Cu_{40}$ alloy rod. The alloy contains large grains, with sizes ranging from approximately 100 μm to 500 μm. It is apparent that these grains formed via a regular eutectic growth, starting from a low density of isolated nucleation centers. The shape of the grains indicates that the growth proceeds from the surface of the rods inwards, as expected from the .radial heat flow in the quenched rods

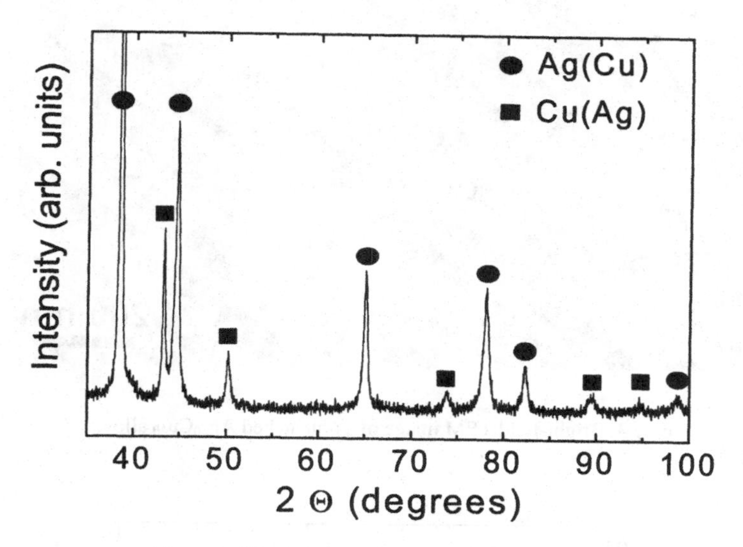

Figure 1: XRD pattern of as-quenched $Ag_{60}Cu_{40}$ alloy.

Figure 2: Optical micrograph of the cross section of an as-quenched $Ag_{60}Cu_{40}$ alloy rod.

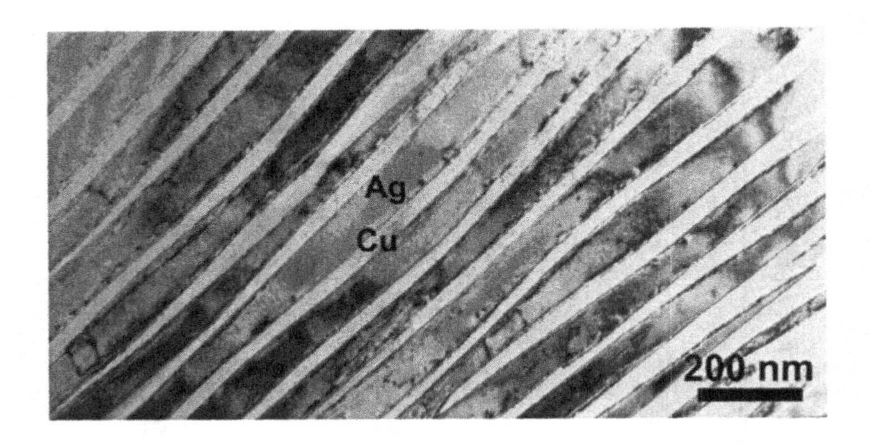

Figure 3: Bright-field TEM image of as-quenched $Ag_{60}Cu_{40}$ alloy.

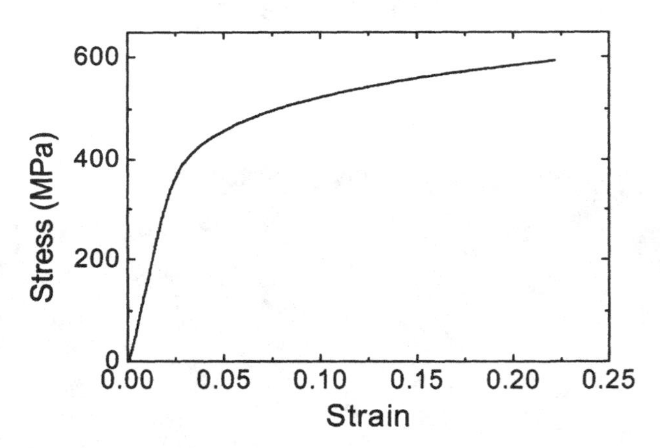

Figure 4: True stress vs. true strain curve for $Ag_{60}Cu_{40}$ alloy tested in compression.

Figure 3 is a TEM image showing the lamellar Ag(Cu) / Cu(Ag) structure of the as-cast alloy. The Ag and Cu phases are gray and white, respectively. The average eutectic spacing is approximately 50 nm.

Figure 4 shows the compressive stress-strain curve for $Ag_{60}Cu_{40}$ tested at room temperature. The yield strength ($\sigma_{0.2}$) is 370 MPa and the strength at 10% strain is higher than 500 MPa. The strain-hardening exponent, $m = d(ln\sigma)/d(ln\varepsilon)$, is 0.10±0.02.

Discussion

Depending on composition, the microstructures formed upon solidifying a binary eutectic system such as Ag-Cu can be either eutectic or dendritic, and these features can be present in a wide range of size scales. If the casting is done at near-eutectic compositions, the melt has a low solidification temperature and low viscosity. The resulting eutectic microstructure usually has relatively small grains and thus high strength. Typically, lamellae in eutectic alloys are about one tenth the size of dendrite trunks formed in the same alloy at off-eutectic compositions, both alloys prepared under similar cooling rates [15].

The microstructure of the present flux-purified and water-quenched Ag-Cu alloys is composed of large grains (100 to 500 μm) and very small crystallites (~ 50 nm). Clearly, the mechanical property of this alloy are determined by its crystallite size rather than grain size. Assuming both the grains and the crystallites are spherical, there are approximately 10^{10} crystallites in one grain. We believe the effective melt purification obtained by fluxing is necessary for obtaining such fine microstructure. The small number of nuclei suggests that most of heterogeneous nucleation centers have been removed by the flux-melting technique. Indeed, as shown in Fig. 2, the density of nuclei triggering the formation of grains in the undercooled melt is small. Crystal nucleation may have occurred at heterogeneous catalytic centers (e.g., oxide particles not removed by the flux) or by homogeneous nucleation. Although the B_2O_3 flux neutralizes most oxide particles in the melt (either reducing them altogether or encapsulating them in the flux, thus rending them harmless), it is unlikely that heterogeneous nucleation can be completely avoided.

The removal of most heterogeneous nucleation centers by fluxing enables us to highly undercool the Ag-Cu melt. Undercooling achieved in previous studies on binary alloys with a liquidus temperature above 1000 K is usually lower than 0.2 T_m (where T_m is the melting temperature) [16]. Crystallization at this degree of undercooling usually results in micron-size crystals, but not nano-size crystals. With our flux-melting technique and cooling rates of 10^2 to 10^3 K s^{-1} we have been able to undercool Pd- and Fe-based melts by as much as 0.4 T_m, resulting the formation of amorphous alloys [17-19]. Although we have not measured the undercooling in the present Ag-Cu alloy, which does not from an amorphous phase, we expect the undercooling to be at least 0.2 T_m. This undercooling has two effects: (1) it induces the formation of metastable, supersaturated Ag(Cu) and Cu(Ag) solid solutions, and (2) it forces the formation of nanosized grains of Ag(Cu) and Cu(Ag), since the width of the lamella forming during eutectic solidification is inversely proportional to the undercooling [15].

Other experiments in our Laboratory indicate that the cooling rate achieved in the present tests (water quenching 3 to 5 mm diameter melt cylinders inside 1-mm wall-thickness vitreous SiO_2 tubes) ranges from 10^2 to 10^3 K/s. In the absence of copious nucleation and at this relatively fast quenching rate, the eutectic lamellae are very thin.

Strengthening in our nanocrystalline alloys arises from crystal size refinement and from solid-solution hardening. The hardening from crystal size refinement can be estimated from the Hall-Petch relation, $\sigma = \sigma_o + K d^{-1/2}$, where σ is the yield stress and d is the crystallite size. With $\sigma_{o,Cu} = 0.06$ GPa, $K_{Cu} = 2.05$ GPa $nm^{-1/2}$ [20], $\sigma_{o,Ag} \approx 0.11$ GPa, and $K_{Ag} \approx 1.40$ GPa $nm^{-1/2}$ [21], the Hall-Petch relation averaged over the Ag and Cu volume fractions present in $Ag_{60}Cu_{40}$ predicts a yield strength of 337 MPa. Therefore, grain refinement seems to provide 90% of the strength, whereas solution hardening provides only 10% (~ 33 MPa). We found no data for solid-solution hardening in metastable Ag(Cu) and Cu(Ag) alloys. An estimate of solid solution hardening in Cu(Ag) can be obtained from the known hardening in Cu(Ge) alloys [22], which have a value of $d(\ell n\ a)/d(c)$ similar to that of our Ag-Cu alloys. Here, a is the lattice parameter and c the solute content [23]. This comparison predicts a solid-solution strengthening in Cu-5 at.% Ag of about 10 MPa. This result agrees with the previous discussion, which attributed most of the strengthening to grain refinement.

An additional contribution to the strength of the alloy may be provided by strain hardening, but that contribution is very small in the present as-cast Ag(Cu)/Cu(Ag) alloys. Our nanocrystalline Ag-Cu alloy has a strain-hardening exponent of about 0.10, significantly smaller than that of polycrystalline Cu, typically 0.30 to 0.35 [24]. This result agrees with previous observations in nanocrystalline materials [25]. Nanocrystalline alloys are assumed to have a small strain-hardening exponent because most of the deformation-generated defects such as dislocations recover dynamically at the grain boundaries.

Conclusions

Purifying $Ag_{60}Cu_{40}$ melt with B_2O_3 flux removes solid particles (e.g., oxides). This significantly decreases the heterogeneous nucleation rate in the undercooled melt, enabling a higher degree of undercooling before crystallization begins. The reduction in the density of nucleation centers results in the formation of large grains whereas the increase in the undercooling and the relatively fast cooling rate enables the formation of supersaturated solid solutions and ultra-fine lamellar eutectic microstructure. Bulk nanostructured Ag-Cu alloy has both high electrical conductivity (78% of IACS) and high compression strength (~ 370 MPa). The high strength is mainly attributed to the fine grain structure, with a weak strengthening contribution from solution hardening.

References

1. S. Askénazy, "Analytical Solution for Pulsed-Field Coils Placed in a Magnetic-Field", *Physica B*, 211 (1-4) (1995), 56-64.
2. T. J. Miller, S. J. Zinkle, and B. A. Chin, "Strength and Fatigue of Dispersion-Strengthened Copper", *Journal of Nuclear Materials*, 179-181 (1991), 263-266.
3. K. Han et al., "The Fabrication, Properties and Microstructure of Cu-Ag and Cu-Nb Composite Conductors", *Materials Science and Engineering A*, 267 (1999), 99-114.
4. F. Dupouy et al., "Microstructural Characterization of High Strength and High Conductivity Nanocomposite Wires", *Scripta Materialia*, 34 (7) (1996), 1067-1073.
5. A. Shikov et al., "High Strength, High Conductivity Cu-Nb Based Conductors with Nanoscaled Microstructure", *Physica C*, 354 (2001) 410-414.
6. W. Grünberger, M. Heilmaier, and L. Schultz, "Development of High-Strength and High-Conductivity Conductor Materials for Pulsed High-Field Magnets at Dresden", *Physica B*, 294-295 (2001), 643-647.
7. V. Pantsyrnyi et al., "High-Strength, High-Conductivity Macro- and Microcomposite Winding Wires for Pulsed Magnets", *Physica B*, 294-295 (2001), 669-673.
8. K. Han et al., "Material Issues in the 100T Non-Destructive Magnet", *IEEE Transactions on Applied Superconductivity*, 10 (1) (2000), 1277-1280.
9. Y. Sakai and H. –J. Schneider-Muntau, "Ultra-High Strength, High Conductivity Cu-Ag Alloy Wires", *Acta Mater.*, 45 (3) (1997), 1017-1023.
10. L. Thilly et al., "Recent Progress in the Development of Ultra High Strength "Continuous" Cu/Nb and Cu/Ta Conductors for Non-Destructive Pulsed Fields Higher than 80 T", *IEEE Transactions on Applied Superconductivity*, 21 (1) (2002), 1181-1184.
11. F. Dupouy et al., "High Strength and High Conductivity Composite Wires for High Pulsed Magnet", *Mechanics of Composite Materials*, 31 (4) (1995), 377-383.
12. C. Renaud, "A High Strength, High Conductivity, Low Radiation Activation Nanocomposite", *IEEE Transactions on Applied Superconductivity*, 10 (1) (2000) 1273-1276.

13. D. Raabe, K. Miyake, and H. Takahara, "Processing, Microstructure, and Properties of Ternary High-Strength Cu-Cr-Ag in situ Composites", *Materials Science and Engineering A*, 291 (2000), 186-197.
14. Harold P. Klug and Leroy E. Alexander, eds., *X-ray Diffraction Procedures* (New York, NY: John Wiley & Sons, Inc. 1974), pp. 591.
15. H. Biloni and W. J. Boettinger, "Solidification," *Physical Metallurgy*, ed. R. W. Cahn and P. Haasen (Amsterdam: Elsevier Science B. V., 1996), 667-842.
16. Merton C. Flemings and Yuh Shiohara, "Solidification of Undercooled Metals", *Materials Science and Engineering*, 65 (1984), 157-170.
17. Y. He, T. D. Shen, and R. B. Schwarz, "Bulk Amorphous Metallic Alloys: Synthesis by Fluxing Techniques and Properties", *Metallurgical and Materials Transactions A*, 29 (1998), 1795-1804.
18. T. D. Shen and R. B. Schwarz, "Bulk Ferromagnetic Glasses Prepared by Flux Melting and Water Quenching", *Applied physics Letters*, 75 (1) (1999), 49-51.
19. T. D. Shen and R. B. Schwarz, "Bulk Ferromagnetic Glasses in the Fe-Ni-P-B System", *Acta Mater.*, 49 (2001) 837-847.
20. R. S. Iyer, C. A. Frey, S. M. L. Sastry, B. E. Waller, and W. E. Buhro, "Plastic Deformation of Nanocrystalline Cu and Cu-0.2wt.% Cu", *Materials Science and Engineering A*, 264 (1999), 210-214.
21. X. Y. Qin, X. J. Wu, and L. D. Zhang, "The Microhardness of Nanocrystalline Silver", *NanoStructured Materials*, 5 (1) (1995), 101-110.
22. P. Haasen, "Mechanical Properties of Solid Solutions", *Physical Metallurgy*, ed. R. W. Cahn and P. Haasen (Amsterdam: Elsevier Science B. V., 1996), pp. 2009.
23. T. B. Massalski, "Structure and Stability of Alloys", *Physical Metallurgy*, ed. R. W. Cahn and P. Haasen (Amsterdam: Elsevier Science B. V., 1996), pp. 182.
24. R. W. Hertzberg, *Deformation and Fracture Mechanics of Engineering Materials* (New York, NY: John Wiley & Sons, Inc. 1983), pp. 17.
25. D. Jia et al., "Deformation Behavior and Plastic Instabilities of Ultrafine-Grained Titanium," *Applied Physics Letters*, 79 (5) (2002), 611-613.

Processing and Properties of Structural Nanomaterials
Edited by Leon L. Shaw, C. Suryanarayana and Rajiv S. Mishra
TMS (The Minerals, Metals & Materials Society), 2003

Optical and Nanomechanical Chararacterization of a Tin Sulfide-Silica Multilayer System

M. Deopura, Y. Fink and C.A. Schuh

Department of Material Science and Engineering
Massachusetts Institute of Technology
Cambridge, MA, 02139

Abstract

A tin sulfide-silica multilayer system has been developed and characterized for optical as well as nano-mechanical properties. This multilayer system acts as a 1-D photonic band-gap material and exhibits omnidirectional reflectivity for visible wavelengths. A refractive index contrast of 2.6/1.46 is achieved, one of the highest values demonstrated until now in photonic band-gap systems for the visible frequency range. Instrumented nanoindentation procedures have also been developed to assess Young's modulus and hardness for these very fine laminates, and reveal that the mechanical properties are in line with expectations based on a rule-of-mixtures composite model.

Keywords: Omnidirectional mirrors, dielectric mirrors, visible spectroscopy, nanoindentation, multilayers

Introduction

Development of new materials systems for optical applications has largely focused upon the identification of materials exhibiting low intrinsic absorption and scattering losses at the wavelengths of interest. While optical transparency is obviously a necessary requirement for these applications, it is not sufficient in itself. To develop practical applications for such optical materials, they need to have many other properties including: amenability to inexpensive processing methods, structural stability, excellent mechanical properties and long term environmental robustness. The difficulty associated with identifying a single material exhibiting all of the necessary properties has led to successes being the exception rather than the rule in the development of new materials systems for optical applications.

In the present study a new tin sulfide-silica multilayer system has been developed for potential use in optical applications at visible wavelengths (600–800 nm). The tin sulfide-silica multilayer system has been studied because of the feasibility of the system to behave as a one dimensional photonic crystal [1,2]. 1-D photonic crystals provide a low-cost alternative to the more expensive 3-D photonic crystals, while maintaining optical properties by using geometrical confinement methods. An example of such a structure is the omnidirectionally reflecting one-dimensional photonic crystal. [3,4]

The potential applications for 1-D tin sulfide-silica multilayer systems would be several: efficient reflectors, calibration standards for experimental optics, high-frequency waveguides for communications and power delivery, and high Q cavities. All these applications require that the multilayer films be mechanically stable with reasonable strength; high resistance to abrasion and contact damage would also be desirable to maintain good optical properties in service. Unfortunately, traditional methods of mechanical property characterization like tensile tests and even microindentation would be impossible to carry out on these multilayer specimens, given their total thickness on the order of just a few microns. Accordingly, it is important to develop nanoindentation procedures to obtain the Young's modulus and hardness from very fine-scale structures.

Experimental

Materials Design and Synthesis

High refractive index contrast is essential to the formation of large photonic band gaps. However, at the visible frequency regime, the lower values of electronic and ionic polarizability lead to an inherently low refractive index value for most materials. Tin sulfide has been chosen as the high refractive index material ($n_b \sim 2.6$ at 600 nm for a thin film, as measured by standard techniques [5]), as it seems to have the highest refractive index in the visible frequency range while at the same time being optically low-loss. It exhibits very low absorption below its characteristic 2.2 eV (590 nm) band edge [6]. Also, tin sulfide is easily amenable to thin film deposition [7]. Silica ($n \sim 1.46$) has been chosen as the low refractive index material as it has been extensively studied and used in the past to make thin films [8,9].

Based upon prior theoretical work [5,10], the target structure for reflection in the 600-800 nm range was identified as a 19 layer structure with layer thicknesses of h_1= 60 nm for tin sulfide and h_2 = 110 nm for silica. To fabricate this multilayer structure, the two materials were deposited sequentially using separate vacuum deposition chambers, on a square glass substrate 22 mm on each side. Tin sulfide layers were deposited using a thermal evaporator (CVC) at 10^{-6} torr and 10 A; the layer thickness and deposition rate were monitored in situ with a crystal thickness monitor (Sycon STM100). Silica layers were vacuum deposited using an electron beam evaporator (NRC 119) at 5×10^{-7} torr, operating at 10 kV and 50 A; again the layer thickness and deposition rate were monitored in situ by a built-in crystal monitor (Filtech). For all deposition procedures the substrate was maintained at room temperature.

The layer thickness of the nineteen-layer structure was characterized by performing cross-sectional scanning electron microscopy (SEM) using a field emission instrument (JOEL). The micrograph in Figure 1 depicts the final nineteen layer structure. Here the darker layers are silica while the bright layers correspond to tin sulfide. The RMS surface roughness of the nineteen layer structure was measured at approximately 30 nm, about 1.5 % of the total ~2 μm thickness of the nineteen layer structure. Additional details about the structure and composition of the multilayer specimens are available in Ref. [5].

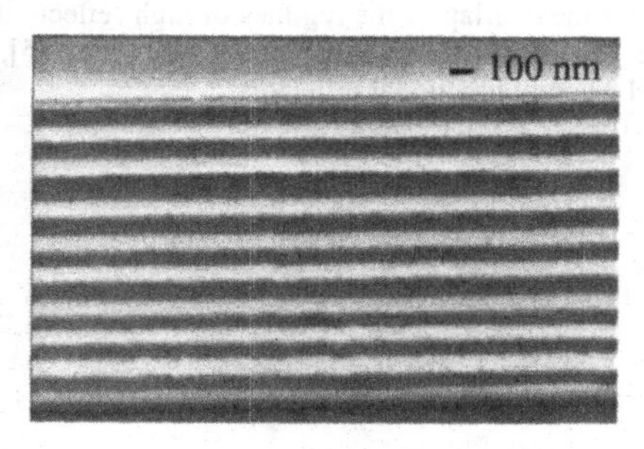

Figure 1 Cross-sectional SEM micrograph of 19-layer tin sulfide-silica multilayer. Bright regions correspond to tin sulfide. Dark regions correspond to silica.

In addition to the multilayer structure described above, one thin-film specimen of each tin sulfide and silica were prepared for comparison with the multilayer. These films were ~3 μm and prepared under the same conditions described above.

Optical Characterization

The optical response of the multilayer structure was measured using a UV-VIS-NIR spectrophotometer (Cary 500; Varian) fitted with a polarizer. Normal incidence reflectivity measurements were performed using a spectral reflectance accessory, and oblique angle measurements up to 70° inclination were carried out using a variable angle reflectivity stage. A freshly evaporated aluminum mirror was used as background for the reflectance measurements. The measured spectra from the spectrometer (for all the different angles and both polarizations) were corrected for absolute reflectivity values using a single frequency He-Ne Laser (632 nm) measurement.

Nano-mechanical Characterization

Indentation tests were performed using a Hysitron Triboindenter with a calibrated Berkovich tip. The applied indentation force in these experiments was between 1 and 3 mN; these conditions were carefully chosen to yield the most accurate results, as described in more detail later. The loading and unloading cycle times for the experiments were in the 1-10 second range, with no obvious rate-dependencies observed in these materials. The standard Oliver-Pharr analysis was used to extract values of Young's modulus and hardness. All of the data presented in what follows represent the average of at least ten independent experiments.

Optical Results

The tin sulfide-silica multilayer system was designed to have a high reflectivity for any angle of incidence in the 600-800 nm regime; in the experiment, we measured at angles from 0 to 70°. The thin film mirror had a golden appearance when observed in white light both at normal incidence as well as at oblique angles of incidence. When held against white light, the mirror allowed blue-green light to transmit through.

Figure 2 illustrates the measured reflectance spectra for normal (0°), 45° and 70° angles of incidence, for both the transverse electric (TE) and transverse magnetic (TM) modes of light polarization. As expected, the reflectivity is 100% near the target optical wavelength range around 700 nm. The shaded band shows the overlap of the regimes of high reflectivity, which denotes the range of the omnidirectional band. As discussed in more detail elsewhere [5], these results agree with the expectations of theory for 1-D omnidirectional reflectors.

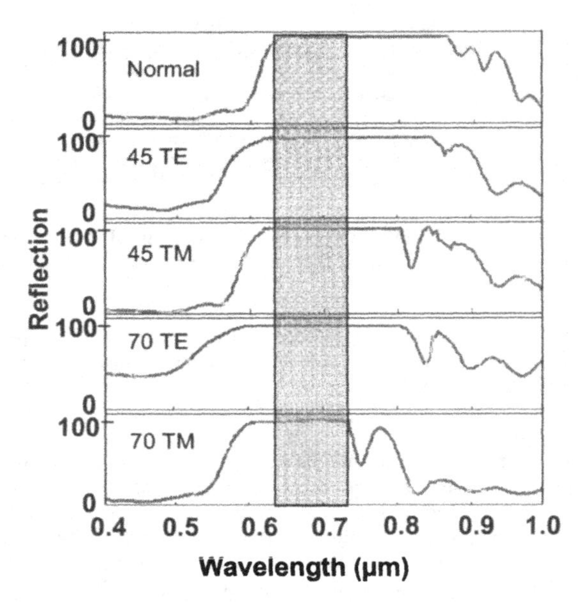

Figure 2 Measured reflectance spectra as a function of wavelength for TM and TE modes at normal, 45°, and 70° angles of incidence showing an omnidirectional reflectivity band. The semi-transparent region shows the experimentally observed omnidirectional band-gap.

Nano-mechanical Results

Test Restrictions for Multilayer Structures

In order to use nanoindentation tests for fine-scale systems, certain guidelines and procedures need to be established to avoid artifacts. For example, the well-known 10% rule [11] dictates that for indentation of a film atop a substrate, the maximum indenter penetration should not exceed 10% of the film thickness. In the present case, we are indenting a thin film system of 2 μm thickness on a glass substrate, giving an upper bound indentation depth of ~200 nm. At the same time, the system being indented is a multilayer with a characteristic structural length scale on the order of ~100 nm. Accordingly, very shallow indentations will sample the properties of just a single phase, and a lower bound for the indentation depth is required to ensure that the test samples the mechanical properties of both the phases in their true proportion.

Following the 10% rule, we assume that the indenter sees a volume of material that is roughly spherical, with a radius about an order of magnitude greater than the instantaneous indentation depth. With this approximation, it is straightforward to calculate the 'effective volume fraction' of the multilayer phases that are sensed by the indenter as it penetrates the specimen. Figure 3 presents the results of such simple calculations for the 19 layer tin sulfide-silica multilayer structure. Here the effective volume fraction of tin sulfide is plotted against the depth of the indent. For shallow indentation depths the volume fraction of tin sulfide sensed is equal to unity, as none of the silica comes within the plastic deformation zone. For large values of the indentation depth, the effective volume fraction of tin sulfide converges to the true volume fraction of tin sulfide in the multilayer (~40%). Although these calculations are quite approximate, they emphasize the important point that in thin-film multilayer systems, there is a minimum as well as the usual maximum indetation depth. For the tin sulfide-silica system we approximate that indentations should be at least 125 nm and less than about 200 nm deep; this very narrow range emphasizes the great difficulty in accurately assessing mechanical properties of these nano-scale laminates.

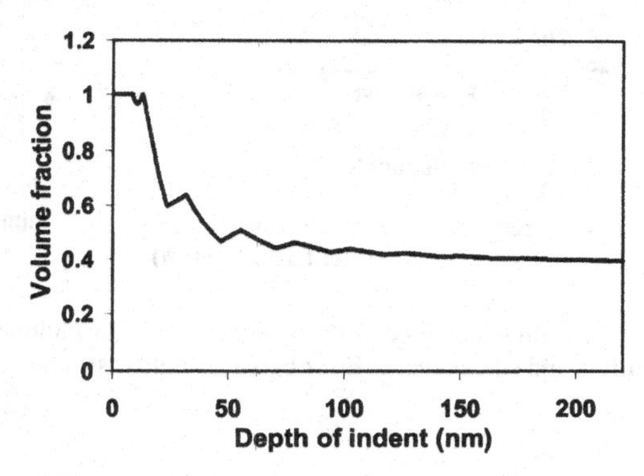

Figure 3 Volume fraction of tin sulfide sensed by the indenter for a tin sulfide-silica multilayer system

Nanoindentation results

The modulus and the hardness values obtained for both the multilayer and the individual tin sulfide and silica films has been presented in Figure 4a (modulus) and 4b (hardness). These data are shown as a function of the indentation depth; the pure silica and tin sulfide materials have properties that are effectively independent of depth, as expected. The variation of multilayer hardness with depth may be associated with sampling different fractions of the two phases, as described above. Average values of modulus and hardness are presented for all three materials in Table I. For the silica films, both the modulus and hardness values are lower than the bulk values reported in literature; as discussed elsewhere [5] this is most likely because thin film silica obtained using e-beam evaporation has a low density due to nano-scale porosity, to the point where the mechanical properties are significantly reduced [12].

Table I. Modulus and Hardness values for silica, tin sulfide and multilayer specimen

Sample	Modulus(GPa)	Hardness(GPa)
Silica	13.3 (±1.2)	2.87 (±0.19)
Tin Sulfide	39.6 (±2.3)	2.14 (±0.12)
Multilayer	31.6 (±0.9)	2.68 (±0.24)

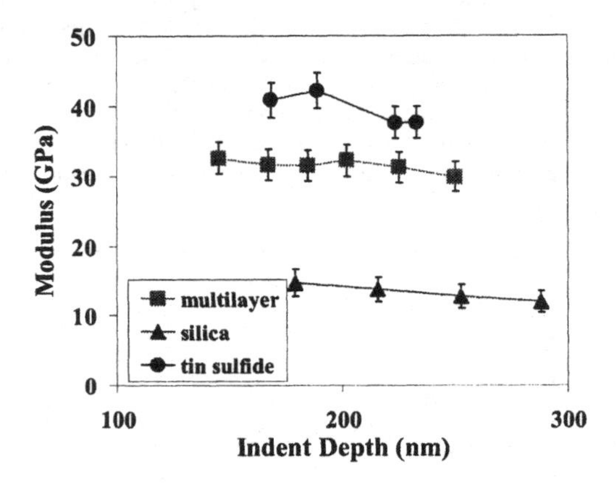

Figure 4a Modulus values obtained from indentation of the multilayer as well as the individual tin sulfide and silica layers. Error bars represent instrumentation error

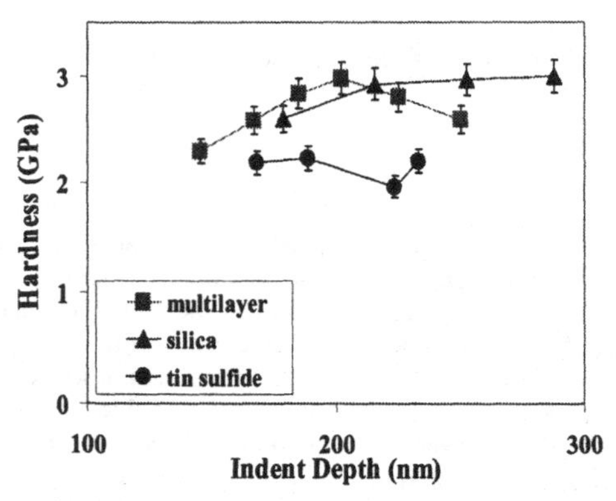

Figure 4b Hardness values obtained from indentation of the multilayer as well as the individual tin sulfide and silica layers. Error bars represent instrumentation error.

Examining the modulus values of the multilayer system in Figure 4a, no immediate anomaly is observed in the data; the modulus of the multilayer lies between the moduli of silica and tin sulfide, roughly in line with the rule of mixtures. Examining the hardness curves of the multilayer in Figure 4b it is seen that the hardness values fall within a range of 2.2 GPa to 2.7 GPa. Comparing these values to the those for tin sulfide and silica (~2.2 GPa and ~2.9 GPa respectively), the results again seem reasonable in light of a simple rule-of-mixtures interpretation. These results indicate that the mechanical properties of these dielectric multilayer systems are primarily dictated by the intrinsic properties of their constituents, rather than by any special aspects of their engineered structure. Accordingly, improvements in the contact damage resistance of these materials can best be achieved by improving the character of the deposited films of tin sulfide and silica. For example, careful adjustment of the deposition parameters may produce silica layers closer to the theoretical density. While this would have little impact on the optical properties of the mirrors, the hardness and modulus would be expected to increase by about a factor of two to four.

As a final note, we have observed some interesting discrete plastic events during nanoindentation of tin sulfide. Specifically, during nanoindentation of monolithic tin sulfide films (~2.5 μm thick) it was observed that the indenter would occassionally "pop-in" in a burst of rapid displacement. Figure 5 shows a typical indentation load-depth curve for the thick tin sulfide film, illustrating this effect. This kind of discrete displacement burst is normally associated with nucleation of strain-accommodating defects like dislocations, cracks, or shear bands [13]. The present films of tin sulfide are crystalline, but with a very fine grain size (< 200 nm), so the pop-in phenomenon may be related to dislocation activity or perhaps rapid grain boundary shearing events; no evidence for cracking was observed using atomic force microscopy around the indentations.

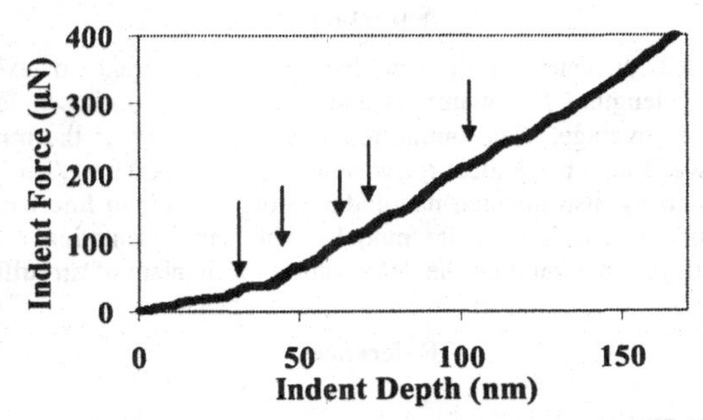

Figure 5 Typical loading during indentation of tin sulfide. Several pop-ins are observed as identified by the arrows.

In stark contrast to the above results on monolithic tin sulfide, no pop-ins have been observed in any of the multilayer specimens, even though tin sulfide was the outermost material on the multilayers and therefore the first material indented. Based on this result, we speculate that pop-ins may be suppressed in the multilayers due to the small thickness of the individual tin sulfide layers and the constraint from the silica layers. To test this hypothesis, we conducted an experiment in which an initially 3 μm thick tin sulfide film was sequentially polished to successively thin the specimen, with nanoindentation carried out on the film at each stage. The average number of pop-in events observed during several identical indentations is plotted as a function of the film thickness in Figure 6. It is found that the number of pop-ins reduce significantly when the film thickness is below the 500 nm range. This result corroborates with the earlier observation that pop-in is suppressed in the multilayer films, where the tin sulfide layers are ~60 nm thick. These results suggest that a fundamental change of deformation mechanisms comes into play at the nano-scale.

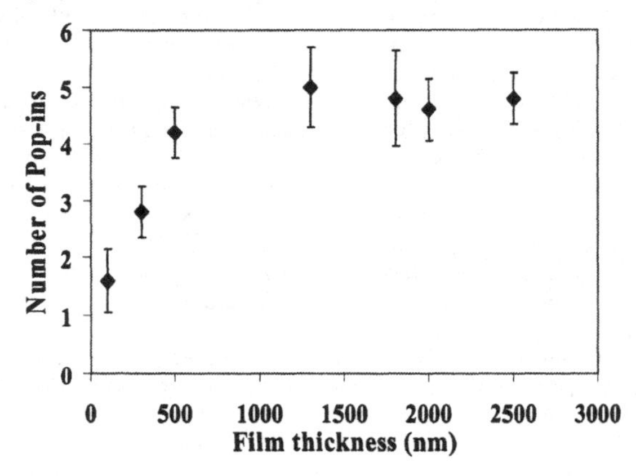

Figure 6 Number of pop-ins versus films thickness of tin sulfide. Number of pop-ins reduce dramatically as film thickness is reduced below 500 nm.

Summary

A tin sulfide-silica dielectric multilayer has been developed to exhibit omnidirectional reflectivity at visible wavelengths. The materials and processes were chosen for their low cost and applicability to large area coverage. The omnidirectional reflectivity of the mirrors was verified by spectrophotometry, and was found to be ideal for wavelengths near 700 nm. The mechanical properties of the mirror were assessed by instrumented nanoindentation, and are in line with expectations on the basis of a simple rule-of-mixtures composite model. The nanoindentation results also suggest an interesting effect of multilayer constraint on the deformation mechanism of tin sulfide.

References

1 J. N. Winn et al., *Opt. Lett.*, 23, 1573 (1998).
2 Y. Fink et al., Science 282, 1679 (1998).
3 D.N. Chigrin et al., J. Lightwave Technol. 17, 11, 2018 (1999).
4 M. Deopura et al., Optics Letters, 26, 17, 1370 (2001).
5 M Deopura, SM thesis, Massachusetts Institute of Technology (2003).
6 P. A. Lee et. al., J. Phys. Chem. Solids 30, 2719 (1989).
7 K. Kawano, R. Nakata, and M. Sumita, J. Phys. D: Appl. Phys 22, 136 (1989).
8 S. A. Khodier and H. M. Sidki, J. Mater. Sci: Materials in Electronics, 12, 107 (2001).
9 A. Brunet-Bruneau et al., J. Appl. Phys., 82 (3), 1330 (1997).
10 F. Abeles, *Ann. Phys.* 5, 706 (1950).
11 B. Bhushan, Handbook of Micro/NanoTribiology, Chapter 10 (CRC Press, 1999).
12 K. N. Rao, L. Shivlingappa and S. Mohan, Mater. Sci. & Engg., B8, 38 (2003).
13 D.F. Bahr, D.E. Kramer and W. W. Gerberich, Acta Mater., Vol. 46 no. 10, 3605 (1998).

Processing and Properties of Structural Nanomaterials
Edited by Leon L. Shaw, C. Suryanarayana and Rajiv S. Mishra
TMS (The Minerals, Metals & Materials Society), 2003

Processing and Mechanical Properties of Bulk Nanocrystalline Ni-Fe Alloys

J.S. Lee[1], X.Y. Qin[2], S.K. Kwon[1], and Y.S. Kang[1]

[1]Dept. of Metallurgy and Materials Science, Hanyang Univeristy, Ansan, 426-791 KOREA
[2]Laboratory of Internal Friction and Defects in Solids, Institute of Solid State Physics, Academia Sinica, 230031 Hefei, China

Keywords: Nanocrystalline, γ-Ni-Fe, Mechanical property

Abstract

The nano-processing has been known to be a potential and promising way to improve and change drastically the mechanical properties of the nanostructured (ns) materials. In this paper we overview a new processing route for fabricating ns Ni-Fe alloy and on its related mechanical properties. The bulk ns Ni-Fe alloys were fabricated by hot-isostatic pressing of mechano-chemically processed Ni-Fe nanoalloy powder. The temperature dependence of mechanical properties was investigated by compression and deformation morphology observations at temperatures from 162℃ to 600℃. At the temperature T<100℃ or T>400℃ yield strength decreased gradually with increasing temperature, while it dropped catastrophically in the range 100℃<T<400℃. Low-temperature deformation produced two sets of macroscopic bandlike traces which consisted of ultrafine straight lines; in contrast, high-temperature deformation led to formation of microcavities accompanying which some nanograins arranged locally in regular form. The results of both compression behavior and deformation morphologies suggest that low-temperature deformation mechanism was different from that at high temperatures.

Introduction

Nc materials have considerable scientific and technological interests due to their properties affected by reduced dimensions of the grains (less than 100 nm) as well as the presence of a large fraction of atoms at the grain boundaries [1, 2]. Especially mechanical properties of nc materials have aroused much attention such as high strength in nc metals [3] and high ductility in nc ceramics [4]. A promising route for fabricating bulk nc metals such as nc-Ni-Fe is provided by powder technology in which the synthesis of nanosized metal powder less than 100 nm is the most important task. In recent years Lee and his colleagues [5, 6] have attempted to find a new processing route for fabricating bulk nc γ-Ni-Fe material by conventional (pressureless) sintering of mechano-chemical processed γ-Ni-Fe nanoalloy powder from metal oxides. In order to achieve full density of nc Ni-Fe sintered alloy, the full densification process during a pressureless sintering of nc Ni-Fe powder should be understood in view of powder characteristics and microstructure. Especially sintering kinetics of nc Ni-Fe powder should be dealt with based on densification mechanism in terms of diffusion process in nc Ni-Fe.

Since fully densified nc Ni-Fe via pressureless sintering is limited the size and shape of specimen for the mechanical testing, bulk nc Ni-Fe produced by HIP process was used for the investigation of mechanical behavior. We report our investigations on microstructures and mechanical behavior of bulk nc γ-Ni-Fe, synthesized by hot isostatic pressing (HIP). Microstructures and deformation features were characterized using XRD, atomic force microscope (AFM) and optical microscopy (OM). Both conventional Vickers hardness and compression tests were utilized to characterize its mechanical behavior.

Experimental

Pressureless Sintering

The powders of nc Ni-Fe powders consisting of γ-Ni-Fe particles of 30 nm in size were prepared by the mechano-chemical process described in detail elsewhere [7, 8]. The pressureless sintering experiment of nanoalloy powder compact was aimed at explaining the relation of densification process to initial structure of green compact. For this purpose the powders were compacted with various compacting pressures of 125~1250 MPa to have 45~73 % relative green density. A sintering experiment was performed using laser-photo dilatometry in the course of heat-up to 950℃ at different heating rates of 5~20℃/min. The densification process during heat-up was examined based upon microstructure development especially with respect to temperature dependence of inter- as well as intra-agglomerate porosity.

HIP Process

The γ-Ni-19wt%Fe nc alloy powder was compacted, under uni-axial pressure of 900 MPa, into rectangular bars with dimensions of 24×8×3 mm^3. They have the green density of 60-65%. These raw bars were then pre-sintered at 650℃ for 1.5h in H_2 so as to remove any oxides present. High-density bars were prepared by using HIP [9]. To do this, the pre-sintered bars were embedded in high purity Al_2O_3 powder that was sealed in an evacuated stainless can and then pressed at 750℃ for 1h under the pressure of 190 MPa in Ar atmosphere.

Mechanical Tests

Compression tests were conducted in the temperature range from –162 to 600℃ using an Instron type testing machine operating at constant rate of anvils displacement [9]. To avoid indenting anvil heads and increase the accuracy in reading yield strength, two well-machined hard-metal plates were placed in between specimen ends and the anvil heads. MoS_2 powder was sprayed in the plates for lubrication. The initial strain rate was $\sim 1.4\times10^{-4}\ s^{-1}$. In order to investigate the effect of grain size of the material on its mechanical behavior (yield strength and hardness) a series of specimens cut from a same bulk bar were isothermally annealed at 850℃ for different time in sealed quartz tubes filled with argon atmosphere.

Results and Discussion

Consolidation Process under Pressureless Sintering

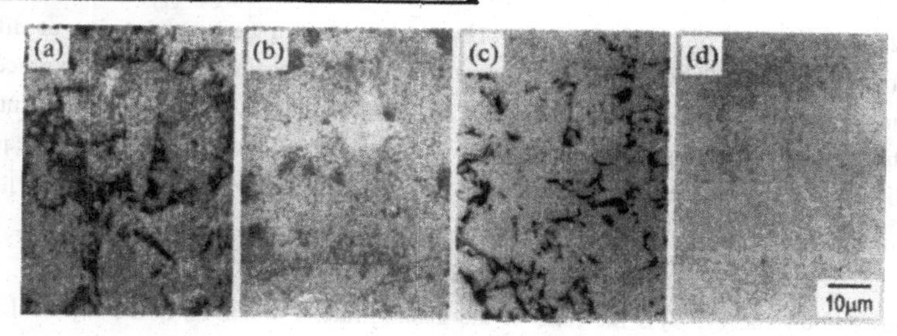

Figure 1: Microstructures of (a, b) green compact and (c, d) sintered compact of γ-Ni-Fe powder, compacted with (a, c) 120 MPa and (b, d) 1250 MPa and sintered during heat-up until 950℃.

Figure 1 shows the microstructures of nc γ-Ni-Fe powder compacts observed before and after heat-up sintering to 950℃ with different compacting pressure. In the green compact with 125 MPa compacting pressure (Fig. 1(a)), a number of micron sized inter-agglomerate pores, which are between agglomerates, are observed. However high pressure compact (1250 MPa) is composed of only nano sized intra-agglomerate pores in agglomerates (Fig. 1(b)). This indicates that the weakly bonding agglomerate powder was broken and rearranged into a dense packing structure by increasing compacting pressure. A difficult-to-sinter bimodal pore distribution, which is due to the coexistence of intra- and inter-agglomerate pores, leads to incomplete densification of the low pressure compact, as seen in Fig. 1(c). Thus it is easily expected that high-pressure compaction, which can minimize formation of inter-agglomerate pores, is the efficient way to reach full density through a homogeneous sintering process. This argument is confirmed by the full-densified microstructure of Fig. 1(d). From the observation of a TEM micrograph of full-densified nc Ni-Fe alloy shown in Fig. 1(d), the microstructure is composed of grains of about 100 nm in size [6].

Sintering behavior can suffer dramatically when the nc powders are formed agglomerate and large inter agglomerate pores are in the green compact, which have the dominant role to reducing driving force for shrinkage. The difficulties, imposed by the problem of agglomeration, can be overcome by increasing the density of the green compacts, i.e. by increasing the compacting pressure on the powder from the beginning.

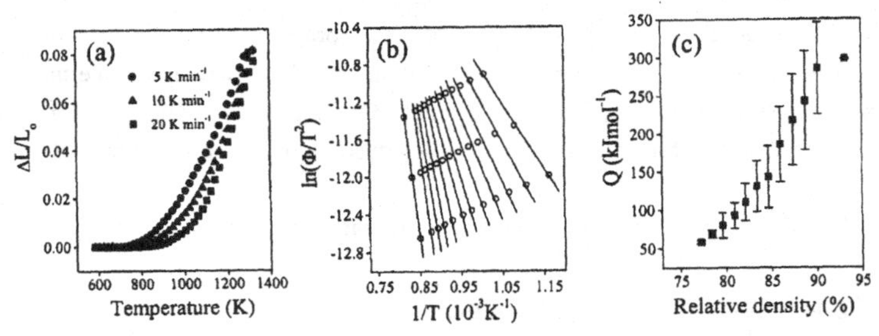

Figure 2: Densification process of the 73%TD compact during heat-up sintering at different heating rates. (a) shrinkage behavior, (b,c) apparent activation energy for densification process.

Figure 2(a) shows the densification process of 73%T.D. compact during heat-up sintering at various heating rates. From this result, depending on the heating rate, the parameters required to calculate activation energy on the basis of Equation 1 were obtained and plotted in Fig. 2(b) [5];

$$\ln\left(\frac{\Phi_i}{T_i^2}\right) = \ln\left(\frac{CR}{Y^n Q}\right) - \frac{Q}{RT_i} \tag{1}$$

where Q is activation energy, Φ_i is the ratio of the heating rate, $Y \equiv \Delta L/L_0$ is the identical value of shrinkage, C is a pre-exponential constant, R is gas constant, and T_i is temperature, respectively. Fig. 2(c) represents apparent activation energy derived from the slopes of the plots. By increasing the shrinkage value, in other words, as sintering proceeds, the activation energy gradually increases from ~50 kJ/mol to ~300 kJ/mol. It is quite reasonable that the steady state densification process appears to be controlled by high diffusion process such as grain boundary diffusion. Very recently we reported that Fe and Ni self diffusion in the same nc Ni-Fe alloy as this study, measured by the radiotracer technique, takes place along the nano-sized intra-agglomerate boundaries and micron sized inter-agglomerate boundaries [10-12]. Comparing the diffusion data with that of the densification process, activation energies for intra-agglomerate boundary- (177 kJ/mol for Ni and 187 kJ/mol for Fe) and inter-agglomerate diffusion (134 kJ/mol for Ni and 148 kJ/mol for Fe) approximately correspond to that for the intermediate sintering stage of 85~90% (120~200 kJ/mol). This implies that densification process of nc Ni-Fe powder is initiated by the diffusion process along high diffusion paths of intra-agglomerate- and inter-agglomerate boundaries.

Mechanical Properties and Deformation Mechanisms

Grain size and morphology of nc bulk Ni-Fe alloy, which was synthesized with HIP process, was observed using AFM (Fig. 3) before deformation. It can be seen from this figure that the grains are arranged randomly with fairly homogeneous sizes, and largely display spherical shape. Most of the grains had a size of about 40 nm, which agreed with XRD results. The densities of the HIPed bars were determined to be ~8.5 g/cm^3 (~99%T.D.).

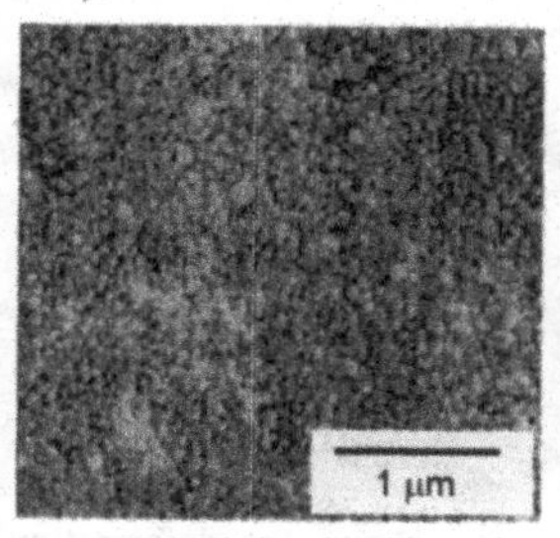

Figure 3: An AFM image for a specimen surface before compression.

Figure 4(a) gives yield strength ($\sigma_{0.2}$) and hardness at room temperature as a function of annealing time (at 850℃), which shows that yield strength or hardness decreased monotonously with increasing annealing time.

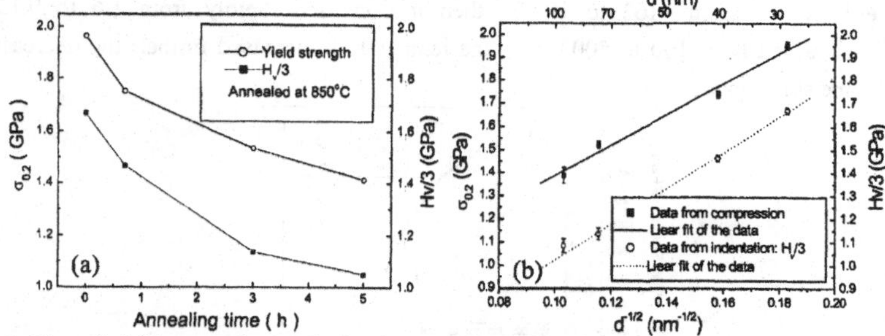

Figure 4: Plot of yield strength and the hardness (divided by 3), (a) versus time of isothermal annealing at 850℃ and (b) versus $d^{-1/2}$.

It can be seen that basically a straight line between yield stress and $d^{-1/2}$ can be drawn. There was a similar linear relation between hardness and $d^{-1/2}$ (Fig. 4(b)). At room temperature, the yield strength of nc Ni-Fe is over 1.5 GPa, which is about one order greater than that for conventional Ni-Fe (150 MPa) [13]. According to dislocation theory and Taylor theory [14], the critical normal stress for dislocation generation is $\sigma \approx m\tau \approx 3Gb/L$ (here τ is the critical shear stress, G is shear modulus, b is the Burgers vector and L is the length of dislocation segment). In a nano-grain, the length of a dislocation should be constrained by grain size; i.e., there is a relation $L \leq d$ (here d is grain size). As a qualitative evaluation, substituting b = 0.25 nm for Ni, G = 78 GPa [14] and d = 35 nm into the formula, one obtain σ = 1.7 GPa. This value coincides well with the yield stress of nc Ni-Fe. This suggests that multiplication of dislocations with the length of grain size could occur.

Figure 5 is a typical AFM image for the surface morphology of the specimen after deformation at room temperature. By comparing Fig. 5 with Fig. 3 one can conclude that these regular markings correspond to bandlike traces (BLTs) and were produced by deformation. AFM imaging showed that most of the nano-channels have the characteristics of running through in between the grains, implying deformation mainly occurred in grain boundary regions.

Figure 5: AFM image that shows the deformation morphology compressed at room temperature (a) scan size: 40×40 μm and (b) scan size: 3×3 μm.

To explore the temperature effect on the mechanical behavior compression tests were conducted at a constant strain rate of 1.39×10^{-4} s^{-1}. The yield strength is given in Fig. 6 as a function of temperature. It can be seen from this figure that yield strength decreased gradually from 1.8 to about 1.6 GPa as temperature increased from –162 to 100℃; then it decreased rapidly from 1.6 to 0.15 GPa as temperature increased from 100 to 500℃. As the temperature increased further, the decrease in yield strength became slow again.

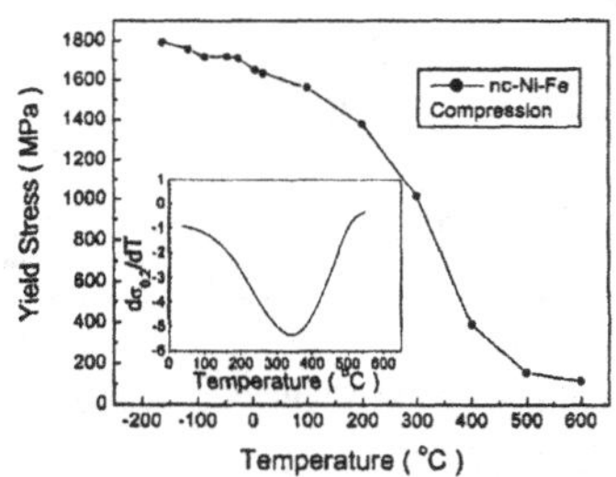

Figure 6: Yield stress ($\sigma_{0.2}$) as a function of temperature (strain rate $1.39\times10^{-4}s^{-1}$). The inset shows the changing rate of the yield stress ($d\sigma_{0.2}/dT$) with temperature.

Figure 7 gives an AFM image of the specimen surface compressed (to strain of 2%) relatively high temperature of 600℃. In contrast to the bandlike feature shown in Fig. 5, the specimen surface appeared very smooth and even, no substantial relief structure being resolved. However, the surface displayed porous structure. This feature turns out to be microcavities formed during deformation. Furthermore, accompanying formation of these microcavities a great number of nanograins aligned in a sub-micro-meter scale (Fig. 7). From the AFM morphology of nc bulk Ni-Fe, it was revealed that after deformation the nano-grains aligned in ordered ways. This phenomenon indicates adjustment of grain positions during deformation, which could be considered as an evidence of conventional grain boundary sliding (CGBS). Furthermore, it is generally believed [15] that CGBS is necessary to develop stress concentrations of magnitude necessary for cavity nucleation. Hence, the formation of micro cavities curing high-temperature deformation (Fig. 7) could be an indirect evidence of CGBS. The characteristic temperature T_c = ~340℃ could be considered as a symbol of deformation mechanism transition.

Figure 7: An AFM image that shows the deformation morphology compressed at 600℃. Arrows indicate microcavities during deformation.

Summaries

1. The densification process of γ-Ni-60wt%Fe nanoalloy powder was found to depend on the degree of agglomeration, which resulted in sintering homogeneity due to pore size distribution. The limited densification of nanoalloy powder originated from a high degree of particle agglomeration. Therefore, full density nc γ-Ni-Fe bulk material was successfully fabricated by high pressure compaction and pressureless sintering of γ-Ni-Fe nano-agglomerate powders. The elimination of inter-agglomerate pores by high-pressure compaction was explained to be responsible for this full density process of which kinetic is mainly controlled by diffusion in grain boundaries (inter- and intra agglomerate boundaries).

2. The mechanical behavior of nc Ni-19Fe was investigated in the temperature range from −162 to 600℃. The results indicated that yield strength decreased catastrophically in range 200℃ $< T <$ 400℃, with the fastest change located at about 340℃. Correspondingly, deformation morphologies at low temperatures were substantially different from those at high temperatures. The results of both compression behavior and deformation morphologies suggest that the low temperature deformation mechanism was different from that at high temperatures.

Acknowledgement

The authors gratefully acknowledge the financial support from the Korean Ministry of Science and Technology through the "National R&D Project for Nano Science and Technology"

References

1. H. Gleiter, "Nanocrystalline materials," *Progress in Materials Science*, 33(4) (1989), 223-315.

2. C. Suryanarayana , "Nanocrystalline materials," *Int. Met. Rev.*, 40(6) (1995), 41-65.

3. V. Y. Gertsman et al., "On the structure and strength of ultrafine-grained copper produced by severe plastic deformation", *Scripta Metallurgica et Materialia*, 30(2) (1994), 229-234.

4. A.H. Cottrell, *The Mechanical Properties of Matter*, (New York, Wiley, 1964), 282.

5. P. Knorr, J.G. Nam, and J.S. Lee, "Sintering behavior of nanocrystalline γ-Ni-Fe powders," *Metallurgical and Materials Transaction* , 31A (2000), 503-510.

6. J.S. Lee and Y.S. Kang. "Processing of bulk nanostructured Ni-Fe materials" *Scripta Materialia*, 44 (2001), 1591-1594.

7. J.S. Lee et al., "In-Situ alloying on synthesis of nanosized Ni-Fe powder," *Nanostructured Materials*, 9 (1997), 153-156.

8. J.S. Lee, J.G. Nam, and P. Knorr, "Synthesis and consolidation of γ-Ni-Fe nanoalloy powder," *Metals and Materials*, 5 (1999), 115-120.

9. X.Y. Qin et al., "Compression behavior of bulk nanocrystalline Ni-Fe," *J. Phys.: Condens. Matter*, 14 (2002), 2605-2620.

10. S. V. Divinski et al., "Gain boundary diffusion in nanostructured γ-Fe-Ni Part I," *Z. Metallkd*, 93 (2002), 256-264.

11. S. V. Divinski et al., "Gain boundary diffusion in nanostructured γ-Fe-Ni Part II," *Z. Metallkd*, 93 (2002), 265-272.

12. S.V. Divinski et al., "Tracer diffusion of ^{63}Ni in nano γ-FeNi produced by powder metallurgical method," *Interface Sci*, 11 (2003), 67-80.

13. E. P. Wohlfarth, ed., *Ferromagnetic Materials* (North-Holland Pub. Co. Amsterdam, 1980), 123.

14. J. Friedel, *Dislocations*, Pergamon Press (1964), Chapter 8.

15. R.W. Cahn et al., ed., *Materials Science and Technology*, (Weinheim, 1993), 434.

PROCESSING AND MICROSTRUCTURE DEVELOPMENT

Processing and Properties of Structural Nanomaterials
Edited by Leon L. Shaw, C. Suryanarayana and Rajiv S. Mishra
TMS (The Minerals, Metals & Materials Society), 2003

Microstructural Investigation of Nanocrystalline Bulk Al-Mg Alloy Fabricated by Cryomilling and Extrusion

Young S. Park[1], Kyung H. Chung[1], Nack J. Kim[2], and Enrique J. Lavernia[1]

[1]Department of Chemical Engineering and Materials Science,
University of California, Davis, Davis, CA 95616
[2]Center for Advanced Aerospace Materials, Pohang University of Science & Technology,
POSTECH, Pohang, 790-784, South Korea

Keywords: Nanocrystalline grains, Cryomilling, Anisotropy

Abstract

The microstructure of a nanocrystalline bulk Al-7.5wt.%Mg alloy produced by cryomilling and consolidation was investigated using optical microscopy (OM), scanning electron microscopy (SEM), and transmission electron microscopy (TEM). Nanocrystalline Al-Mg powders were prepared using cryogenic mechanical alloying (cryomilling). The powders were subsequently degassed, hot isostatically pressed, and extruded at different temperatures. The extrusion temperature played a crucial role in the evolution of the microstructural characteristics, which include anisotropy and grain size. TEM investigation revealed that the microstructure of nanocrystalline Al-Mg materials extruded at the lower temperature is more anisotropic than that of the alloy extruded at the higher temperature. In the case of low extrusion temperature, nano-sized grains have a tendency to rotate toward <111>; the direction is aligned along the extrusion direction. The lower extrusion temperature results in smaller grain size for the Al-Mg alloy (the average grain size is 114 nm) than that of alloy extruded at the higher temperature (197 nm). Small dislocation-free grains (<50 nm grain size) were observed at the boundary of nano grains in the higher extrusion temperature sample.

Introduction

Scientific investigations by materials scientists have been directed towards improving the properties and performance of materials [1]. In order to achieve advanced and novel properties in materials, many methods have been used; these include rapid solidification [2], mechanical alloying (MA) [1], plasma processing [3], vapor deposition [4], torsion straining under high pressure [5], and equal channel angular pressing [6]. Among them, mechanical alloying has received much attention as a powerful tool for fabrication of several advanced materials, including equilibrium, non equilibrium (e.g., amorphous, quasicrystal, nanocrystalline, etc.), and composite materials [7]. Recently, a number of published studies have addressed nanostructured Al-Mg alloys fabricated by mechanical alloying in liquid nitrogen slurry in order to improve their mechanical properties, thermal stability, and ductility [8,9]. In one particular study, the microstructural evolution during thermal annealing of a cryogenically ball milled Al-Mg alloy was studied [9]. However, inspection of the available literature reveals that there is limited fundamental information on the microstructural evolution during the consolidation processes such as extrusion of cryomilled Al-Mg alloys. A fundamental understanding of microstructural evolution in bulk nanostructured Al-Mg alloys during fabrication processes will provide the opportunity for process optimization, thereby yielding enhanced properties. Moreover, from a fundamental standpoint, the non-equilibrium microstructures that are typically present in these materials are sensitive to process conditions. The purpose of the present article is to present our research on the influence of the extrusion parameters on microstructure of a nanocrystalline bulk Al-7.5wt.%Mg alloy.

Experimental

Nanocrystalline Al-7.5Mg powders were produced by ball milling of a slurry of alloy powders in liquid nitrogen. The milling was carried out in a modified Union Process 01-HD attritor with a stainless steel vessel and balls (6.4mm in diameter) at a rate of 180 rpm. The ball-to-powder weight ratio was 36:1. Prior to milling, 0.25% of stearic acid was added to the powders as a process control agent to inhibit a cold welding process. The powders were subsequently degassed, hot isostatically pressed, and extruded at two different temperatures around 0.5 Tm. Hot isostatically pressed (HIPed) and extruded microstructures were examined using optical microscopy (OM), scanning electron microscopy (SEM), and transmission electron microscopy (TEM) operating at 100kV. TEM specimens were prepared by twin-jet electropolishing in a solution of 67% methanol and 33% nitric acid at ~14V and –35°C.

Results and Discussion

A. Microstructure of as-HIPped alloy

In order to develop a systematic understanding of the effects of extrusion temperature on microstructures, the as-HIPed material was first investigated. The optical micrograph in Figure 1 (a) shows an overall view of the as-HIPed sample. The microstructure can be divided into three areas; that is, the white areas indicated as area A in Figure 1(a), the dark gray areas as B, and the lamellar structures of white and the dark gray area, C. The white areas denoted by an A in the optical micrograph, as shown in Figure 1(c), consist of coarse Al grains. Their morphology is essentially irregular or distorted polygonal. However, they appear as dark areas in the SEM micrograph in Figure 1(b).

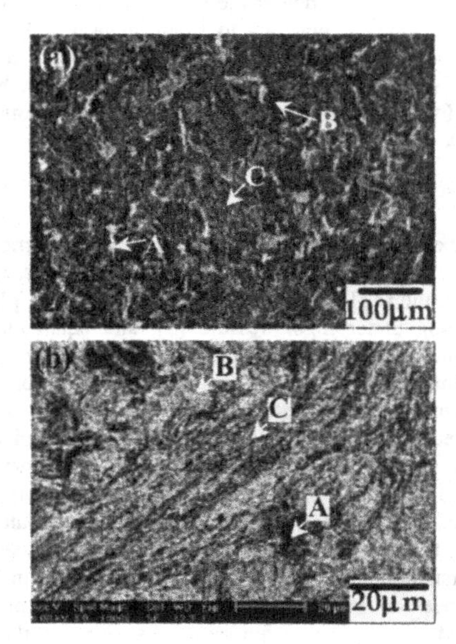

Figure 1: (a) Optical micrograph of as-HIPed sample, (b) SEM observation; dark areas are Al coarse grains, and (c) TEM observation showing a lamellar structure consisting of the coarse Al alloy grains and the fine Al grains.

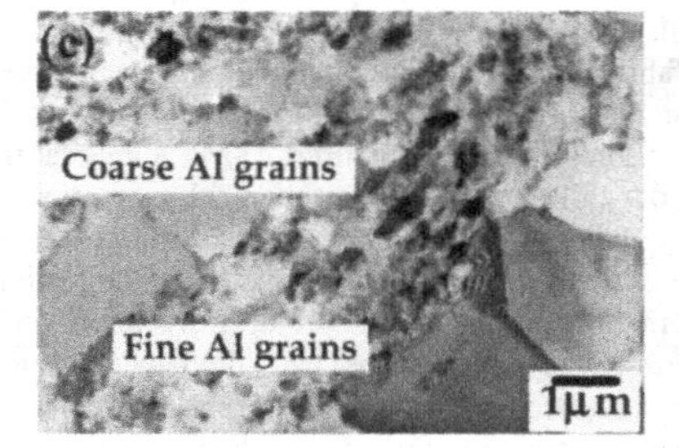

Although the origin of the coarse Al grains that are observed in the as-HIPed material (see Fig. 1) is not presently understood, some comments are in order. It is plausible that these coarse grains may evolve as a result of recrystallization and grain growth of nano-sized Al grains during HIPing. In related work on recrystallization and grain growth during HIPing, a mechanism was proposed by Jeon [10]. It was indicated that recrystallization in a duplex structure is due to the strain energy stored by the collapse of cast porosity during HIPing. This concept may be applicable to the case of cryomilled powders, because porosity is likely present as a result of incomplete inter-particle bonding, in combination with the stochastic nature of the milling process. Moreover, the nanostrucure that is formed during milling is likely to be locally unstable due to the presence of non-equilibrium boundaries, crystal lattice distortions, and a high level of elastic stress [11]. When pores that remain after consolidation are collapsed during HIPing, their surface will experience a large degree of plasticity. This deformation can affect the nanostructure and increase its strain energy; when the strain energy provided by the collapse event and simultaneously the thermal energy that is present during HIP are enough to exceed the activation energy for recovery and recrystallization, the formation of new grains as well as growth are likely to occur. There are, however, other mechanisms that will influence recovery and recrystallization such as: 1) subgrain boundary migration, 2) subgrain rotation and coalescence [12]. In the case of grain growth in nanocrystalline fcc metals, it has also been shown by a recent molecular-dynamics simulation that grain rotations play an equally important role as GB migration and that such rotations lead to the formation of large, elongated grains [13]. It is clear, however, that additional work is needed before the mechanisms that are responsible for the observed microstructure can be rationalized.

The dark gray areas, B as shown in Figure 1(a), are nanostructured Al grains areas. The average grain size of the area is 116 nm. SEM results [Figure 1(b)] show that area B contains sub-micron sized pores. SEM observation of an unetched HIPped sample did not reveal the presence of these pores, meaning that the pores were formed primarily during the etching. The average distance between pores is 148 nm, which is almost roughly the same as the average nanograin size. Consequently, the pores can be considered to be formed during the etching, because the high-energy sites such as nano-sized grain boundaries or their triple points are selectively attacked by the etchant.

Area C is a lamellar structure consisting of the coarse Al alloy grains and the fine Al grains, as shown in Figure 1(a), which is a magnified TEM micrograph of the mixture area. The average distance between layers is 726 nm. The microstructure in this study consists of coarse Al grains (<10μm), nano-sized Al grains (~100nm), and lamellar structure (nano-sized Al grains and coarse Al grains (<3μm). In recent studies it has been argued that a microstructure containing multiple length scales may be beneficial in terms of mechanical behavior [14,15,16]. For example, if a crack moving along the nanostructured Al grains meets a large, ductile Al grain, the crack tip is blunted, and the crack must now propagate through the ductile aluminum, thereby creating ductility on a macroscopic level [8].

B. Microstructures of extruded Al-7.5Mg alloy

The HIPed samples were extruded at two different temperatures around 0.5 Tm. The cross-sectional microstructures in both samples (Figures 2 (a) and (c)) have irregular shapes, similar as previously observed as-HIPped sample. However, the longitudinal-sectional microstructures in both samples (Figures 2 (b) and (d)) are anisotropic along the extrusion direction. The size of each coarse Al grain is 1~4 μm. Interesting features to note are that the microstructures of the lower temperature extruded sample (Figures 2 (a) and (c)) are finer and more anisotropic than those of the higher temperature extrusion sample (Figures 2 (b) and (d)).

Figure 3 shows area fractions about the coarse Al grains of HIPped, higher, and lower extrusion temperature samples. Several digital images of optical pictures (HIPped sample and extruded samples in longitudinal sections) were used to calculate the area fractions by analySIS™ (Soft Imaging System, Denver). The coarse Al grains were distinguished from the nano-sized grains by threshold intensity. The result shows that the area fractions of coarse Al grains increase after the extrusion process, especially at the higher extrusion temperature. This result shows that the extrusion temperature contributes to recrystallization and additional grain growth in the cryomilled Al-Mg alloy. In this study, as will be discussed later, it was found that nano-sized grains have a tendency to rotate toward <111>, which is parallel to the extrusion direction, especially at low extrusion temperature. The rotated grains are thought to accelerate the formation of large, elongated grains via a mechanism of subgrain rotation and coalescence [12,13]. Work in this area is continuing.

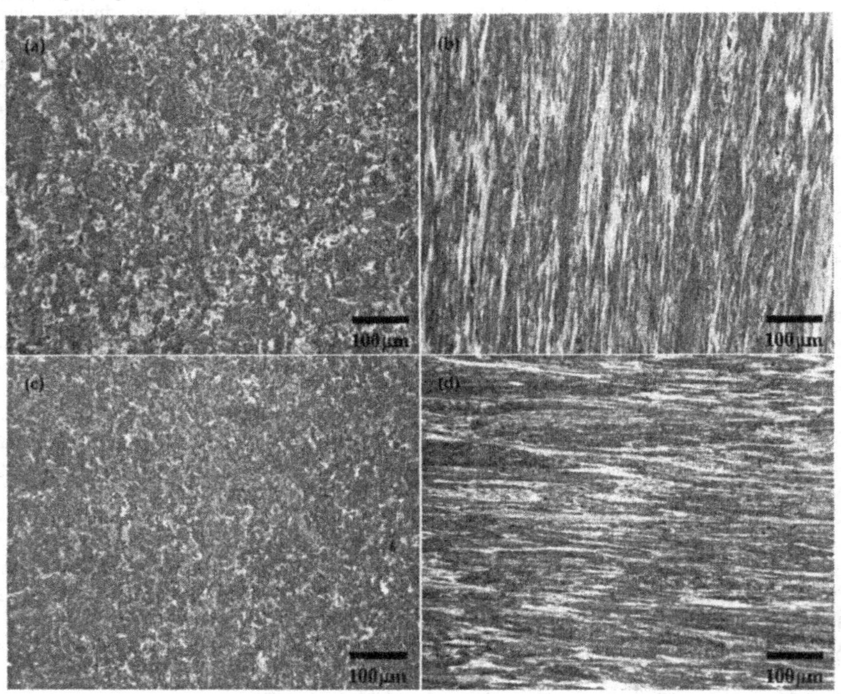

Figure 2: Microstructures of extruded Al-7.5Mg alloy showing (a) the cross section of the extruded one in the higher temperature (~0.5Tm), (b) the longitudinal section, (c) the cross section of the extruded one in the lower temperature, and (d) the longitudinal section; the microstructures of the lower temperature extruded sample are finer and more anisotropic than that of the extruded one in the higher temperature.

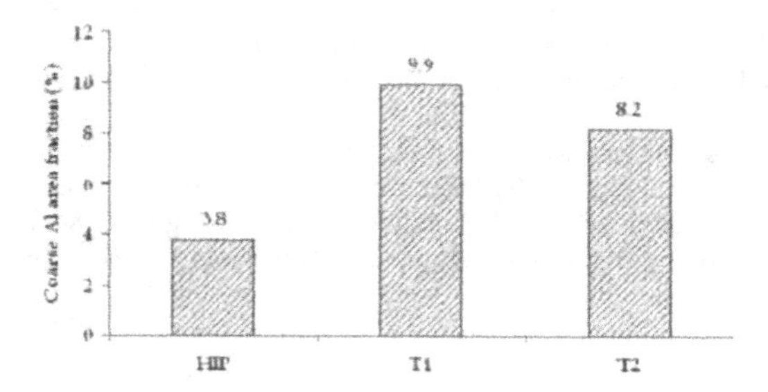

Figure 3: Area fraction of coarse Al grains in the HIPed material, as well as the as-extruded: both high (T1) and low temperatures (T2).

It is observed that new grains nucleate at the nano grain boundaries, as indicated by arrows in Figure 4(a). This is likely to occur because nano grain boundaries are high-energy regions having non-equilibrium characteristics and high elastic strain [11]. Strain fields are observed at the boundaries of the newly nucleated grains, as shown in Figure 4(a). They are due to the lattice distortion between deformed and strain-free volumes. In the lower temperature sample, nucleation of new grains was seldom observed.

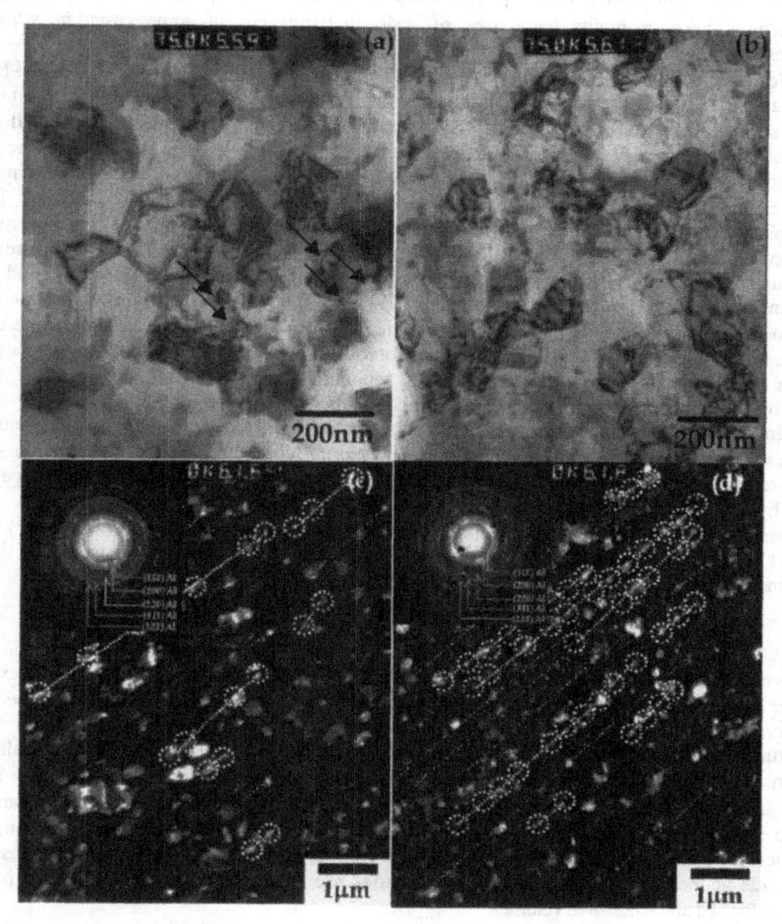

Figure 4: TEM observations showing the difference of grain sizes and an anisotropy between the higher and lower samples: (a) higher extrusion temperature, (b) lower extrusion temperature, (c) dark field micrographs in the longitudinal section extruded at higher temperatures and its SADP (selected area diffraction pattern), and (d) at lower temperatures and its SADP.

TEM observations show that the grain sizes of the higher and lower extrusion temperature samples (Figures 4(a) and (b)) are different. The average grain size of the higher extrusion temperature sample in the nanocrystalline area is 197 nm. However, the average grain size of the lower extrusion temperature sample in the nanocrystalline area is 114 nm. The average values were measured from several TEM photographs taken from different areas. The measurement was conducted only in the nanostructured microstructual area. It was also reported that raising extrusion temperature causes grain growth in commercial 5052 aluminum alloy [17]. The coarse Al grain bands appear to be fully recrystallized microstructures because grains were free of dislocations except in some cases in which some dislocations were observed during rotating the TEM goniometer (+40°~-40°).
Figure 4(c) and (d) show TEM dark field micrographs of the longitudinal section of Al-Mg alloy extruded at different temperatures. The ring pattern inserts in Figures 4(c) and (d) were taken using the first and second diffracted rings of each SAD pattern, the diameter of which is 5.5 μm. It is worth noting that the bright grains in Figure 4 (d) are more aligned than are those in Figure 4(c). The grains in the same direction are circled and lined, as shown in Figures 4(c) and (d). The SAD pattern insert in Figure 4(c) indicates that the individual grains are separated by high-angle grain boundaries. The SAD patterns observed in the nanostructure of cryomilled Ni alloys also show complete ring patterns similar to this result [18,19]. The SAD insert in Figure 4(d) shows a randomly oriented fine grain structure and a strong anisotropic feature, simultaneously. The dark field images shown in Figures 4(c) and (d) were taken using the first and second diffracted rings, (111) and (200). The conclusion is that the direction normal to (111), which has the brightest spot of the first diffracted beam, is parallel to the extrusion direction, after the consideration of the rotation calibration. This means that in the case of relatively low extrusion temperature, nano-sized grains have a tendency to rotate towards a particular direction, dictated by the extrusion direction, although the orientations are partially distributed in a random manner. In fcc materials subjected to uniaxial elongation, such as wire drawing, extrusion, tension, and swaging, axisymmetric flow produces a double fiber texture with <111> and <100> directions parallel to the wire axis [20]. In high stacking fault metals such as aluminum, the <111> texture is particularly favored by easy cross slip and thus predominates [21]. In the case of a relatively low extrusion temperature, the results in the present study show that nano-sized grains have a tendency to rotate toward <111>, which is parallel to the extrusion direction with a tendency similar to that of the micro-sized grains.

Conclusions

TEM investigation reveals that the microstructure of the nanocrystalline Al-Mg materials extruded at the lower temperature is more anisotropic than that of the alloy extruded at the higher temperature and that recrystallized structures are also partially present in both samples. In case of low extrusion temperature, nano-sized grains have a tendency to rotate toward <111>, which is parallel to the extrusion direction. The lower extrusion temperature results in smaller grain size of Al-Mg alloy (the average grain size is 114 nm) than for alloy extruded at the higher temperature (197 nm). It was observed that new grains (<50 nm) nucleate from the nano grain boundaries, due to that nano grain boundaries are a high-energy region including non-equilibrium characteristics and high elastic strain. Strain fields at the newly nucleated grain boundaries are due to the variation of lattice parameters between deformed and strain-free volumes.

Acknowledgments

The work was supported by the Office of Naval Research under grant number N00014-03-C-0163 with Mr. Rodney Peterson as program officer. In addition, we would like to thank Drs. Clifford Bampton and Daniel Matejczyk at Boeing Company for providing the samples used in this study. We would also like to acknowledge Dr. Yizhang Zhou of UCD and Dr. David Witkin of UCI for valuable comments and advises.

References

1. C. Suryanarayana, "Mechanical alloying and milling", Progress in Materials Science, 46 (2001), 3-6.
2. S.H Kim, D.H. Kim, and N.J. Kim, "Structure and properties of rapidly solidified Mg-Al-Zn-Nd alloys", Materials Science and Engineering A, 226 (1997) 1030-1034.
3. F.A. Khonsari, J. Kurdi, M. Tatoulian, and J. Amouroux, "On plasma processing of polymers and the stability of the surface properties for enhanced adhesion to metals", Surface and Coatings Technology, 142 (2001) 437-448.
4. S. Keller, P. Waltereit, P. Cantu, U.K. Mishra, J.S. Speck, S.P. DenBaars, "Electrical and structural properties of AlGaN/AlGaN superlattice structures grown by metal-organic chemical vapor deposition", Optical Materials 23 (2003) 187–195.
5. I.V. Alexandrov, K. Zhang, A.R. Kilmametov, K. Lu, and R.Z. Valiev, "The X-ray characterization of the ultrafine-grained Cu processed by different methods of severe plastic deformation", Materials Science and Engineering A234 (1997) 331-334
6. Y.G. Ko, W.S. Jung, D.H. Shin, C.S. Lee, "Effects of temperature and initial microstructure on the equal channel angular pressing of Ti-6Al-4V alloy", Scripta Materialia 48 (2003) 197-202
7. M.S. El-Eskandarany, Mechanical Alloying (Norwich: Noyes Publications, 2001), 1-21.
8. V.L. Tellkamp, A. Melmed, and E.J. Lavernia, "Mechanical Behavior and Microstructure of a Thermally Stable Bulk Nanostructured Al Alloy", Metallurgical and Materials Transactions A, 32 (2001) 2335-2343.
9. F.Zhou, X.Z. Liao, Y.T. Zhu, S. Dallek, E.J. Lavernia, "Microstructural evolution during recovery and recrystallization of a nanocrystalline Al-Mg alloy prepared by cryogenic ball milling", Acta Materialia, 51 (2003) 2777–2791.
10. J.H. Jeon, A.B. Godfrey, P.A. Blenkinsop, W. Voice, Y.D. Hahn, "Recrystallization in cast 45-2-2 XD™ titanium aluminide during hot isostatic pressing", Materials Science and Engineering, A271 (1999) 128–133.
11. R.Z. Valiev, R.K. Islamgaliev, I.V. Alexandrov, "Bulk nanostructured materials from severe plastic deformation", Progress in Materials Science, 45 (2000) 103-189.
12. F.J. Humphreys, M. Hatherly, Recrystallization and related annealing phenomena (Oxford, U.K., Pergamom, 1995) 127-171.
13. A.J. Haslam, S.R. Phillpot, D. Wolf, D. Moldovan, H. Gleiter, "Mechanisms of grain growth in nanocrystalline fcc metals by molecular-dynamics simulation", Materials Science and Engineering, A318 (2001) 293–312.
14. Y. Wang, M. Chen, F. Zhou, E. Ma, "High tensile ductile in a nanostructured metal", Nature, 419 (2002) 912-915.
15. M. Legros, B.R. Elliott, M.N. Rittner, J.R. Weertman, K.J. Hemker, "Microsample tensile testing of nanocrystalline metals", Philosophical Magazine A, 80 (2000) 1017-1026.
16. D. Witkin, Z. Lee, R. Rodriguez, S. Nutt, E. Lavernia, "Al-Mg alloy engineered with bimodal grain size for high strength and increased ductility", Scripta Materialia, 49 (2003) 297-302.
17. Y.C. Chen, Y.Y. Huang, C.P. Chang, and P.W. Kao, "The effect of extrusion temperature on the development of deformation microstructures in 5052 aluminum alloy processed by equal channel angular extrusion", Acta Materialia, 51(2003) 2005-2015.
18. K.H. Chung, J. Lee, R. Rodriguez, and E. J. Lavernia, "Grain Growth Behavior of Cryomilled INCONEL 625 Powder during Isothermal Heat Treatment", Metallurgical and Materials Transactions A, 33 (2002) 125-134.
19. J. Lee, F. Zhou, K.H. Chung, N.J. Kim, and E.J. Lavernia, "Grain Growth of Nanocrystalline Ni

Powders Prepared by Cryomilling", Metallurgical and Materials Transactions A, 32 (2001) 3109-3115.
20. H. Mecking, Texture in metals, preferred orientation in deformed metals and rocks: an introduction to modern texture analysis. 1985.
21. G. E. Dieter, Mechanical metallurgy (New York, U.S.A., McGraw-Hill Book Company, 1986) 237-240.

Processing and Properties of Structural Nanomaterials
Edited by Leon L. Shaw, C. Suryanarayana and Rajiv S. Mishra
TMS (The Minerals, Metals & Materials Society), 2003

Nanocrystalline Microstructure and Defects in Al Solid Solution Subjected to Surface Mechanical Attrition Treatment

J. Hui [1], X. Wu [1,2], N. Tao [3], Y. Hong [2], J. Lu [4], K. Lu [3]

[1] Chemical Eng Dept, SH455, Cleveland State University, 2121 Euclid Avenue, Cleveland, OH44115
[2] State Key Lab of Nonlinear Mechanics, Institute of Mechanics, Chinese Academy of Sciences, Beijing 100080, PR China
[3] Shenyang National Laboratory for Materials Sciences, Institute of Metal Research, Chinese Academy of Sciences, Shenyang 110016, PR China
[4] LASMIS, University of Technical of Troyes, 10000, Troyes, France

Keywords: Nanocrystalline Microstructure, Dynamic Recrystallization, Surface Mechanical Attrition Treatment

Abstract

The nanocrystalline (nc) microstructure was studied in a surface layer of an Al alloy induced by the surface mechanical attrition treatment. nc grains with the larger-end of the nanometer size regime (~50-100 nm) formed through continuous refinement *via* grain subdivision mechanism, in a sequence of deformation bands with elongated subgrains, submicro-, and nc-grains leveling off ~46 nm. Meanwhile, nc grains with the lower-end of the nanometer size regime (~<10 nm) were observed to occur at a critical strain. A dynamic recrystallization mechanism was suggested to account for their formation.

1. Introduction

Nanocrystalline (nc) materials processed by severe plastic deformation are receiving ever-growing attention because of their unusual microstructure and physical and mechanical properties compared with conventional coarse-grained counterparts [1,2]. The surface mechanical attrition treatment (SMAT) is capable of synthesizing nc surface layer of the bulk material and thus producing unusual surface-related mechanical properties [3-6]. During the SAMT, microstructures of various grain size regimes, i.e. from nano-sized grains to submicro-sized and micro-sized crystallites, can be obtained within the treated surface layer along the depth from the treated surface to the strain-free matrix. This gradient structure results from a gradient change in applied strain and strain rate along the depth, from very large (top surface layer) to zero (strain-free matrix) [3]. Therefore, the SMAT offers the particular advantage of being able to study the microstructural evolution at different levels of strain and strain rate to reveal the underlying mechanism for grain refinement down to the nanomater regime. In the present study, we have investigated the microstructures and defects in the nc surface layer of an aluminum solid solution subjected to the SAMT.

2. Experimental Procedure

The material used in this investigation was an Al alloy with the chemical composition (in wt%) of 6.2 Cu, 0.5 Mn, balance Al. Prior to the SMAT of the sample, the solution heat treatment was conducted to obtain a single solid solution phase with homogeneous coarse grains of 30 μm. The samples were in the form of 100 mm diameter and 10 mm thickness.

The SMAT method [3] was used to produce the nc surface layer. The principle of the treatment is the generation of plastic deformation on the top surface layer of a bulk material by means of repeated multidirectional impacts of flying balls (with mirror-like surface and a diameter of 7.8 mm) on the surface layer. Because of the high vibration frequency of the system (45 Hz in the present investigation), the sample surface under treatment was struck repetitively by a large number of balls within a short period of time, resulting in the sample surface becoming severely plastically deformed. The sample was protected by a high purity argon atmosphere to avoid oxidation during the SAMT at the room temperature for 15 min.

Microstructural characterization was performed using JEOL-2000FX transmission electron microscope (TEM) and JEM-2010FEF high resolution TEM for general and lattice image observations respectively. Cross-sectional thin films were prepared by means of usual sandwich method.

3. Experimental Results

3.1 nc Grains via Continuous Refinement

Figs.1 (a)-(c) are a set of cross-sectional TEM images showing the successive microstructural evolution along depth with increasing strain (corresponding to ~69, 40 and 5 μm deep from the top surface respectively). Fig. 1(a) shows lamellar deformation bands characterized by elongated subgrains. The electron diffraction pattern (EDP) is typical for a subgrain structure with weak misorientations between neighboring volumes. The absorption of dislocations into the subboundary, which may lead to increase in subboundary misorientation, is frequently observed (bottom-left inset). With increasing strain, equiaxed, submicro-sized (0.1-0.4 μm) grains are present within the subsurface, as shown in Fig. 1(b). Inset is a ring-like EDP giving an evidence for creation of high boundary misorientations. A dominant feature is the formation of subgrains in plastically deformed grains of various size regimes, as shown in bottom-left inset. With further increasing strain, the grain sizes decrease down to the nanometer regime (<100 nm) within the outer surface of the layer (from ~27 μm deep to the top surface). Fig. 1(c) exhibits equiaxed nc grains with the average grain size ~52 nm. According to the EDP, nc grains have high angle grain boundaries and random crystallographic orientations. Some grain boundaries are poorly defined and the contrast within the grain is not uniform, but often changes in a complex fashion that indicates a high level of internal stresses and elastic distortions in the crystal lattice due to the presence of a high density of dislocations at the boundaries [1]. The grain size of nc grains is observed to level off around 46 nm from ~10 μm deep to the top surface. X-ray diffraction analysis indicates that no other phases but FCC Al phase are detected in the SMATed surface.

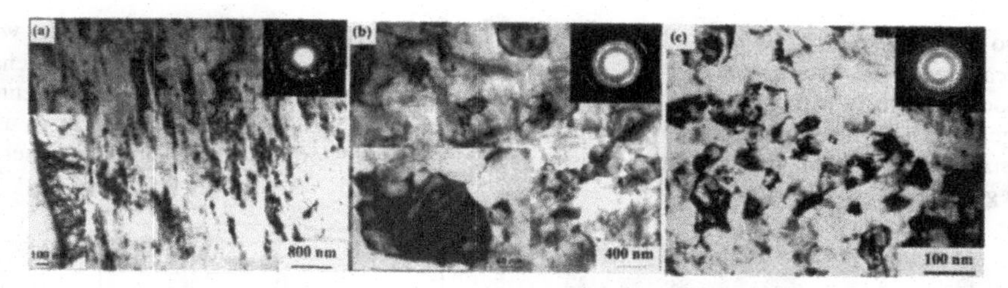

Fig. 1 Cross-sectional TEM images with EDP showing microstructural evolution with increasing strain. (a)-(c) correspond to ~69, 40, and 5 μm deep from the top treated surface respectively. (a) deformation bands of elongated subgrains. Bottom-left inset exhibits absorption of dislocations into the subboundary. (b) equiaxed, submicro-sized grains. Top-right inset shows subgrains in a deformed grain. and (c) nc grains.

3.2 nc Grains with the Lower End of The Nanoscale Regime

Fig. 2 (a) is a dark-field image of nc grains obtained from the partial (111) and (200) diffraction rings (circled) of the EDP (~52μm deep). Therefore, nc grains are Al crystals. They have the lower-end of the nanometer size regime (~<10 nm in size) as compared with those of large sizes in Fig. 1(c). Fig. 2(b) is an HREM image showing four nc grains. The fringe spacing is measured to be ~2 A, which matches well with the (111) interplanar spacing of the FCC Al crystal. Detailed HREM investigations indicate that the onset of nc grains of small sizes starts to occur at a critical strain (~57 μm deep), which is much low compared with that for nc grains of large sizes derived from grain refinement (~27μm deep). nc grains of small sizes are observed to form continuously and to increase their population with increasing strain.

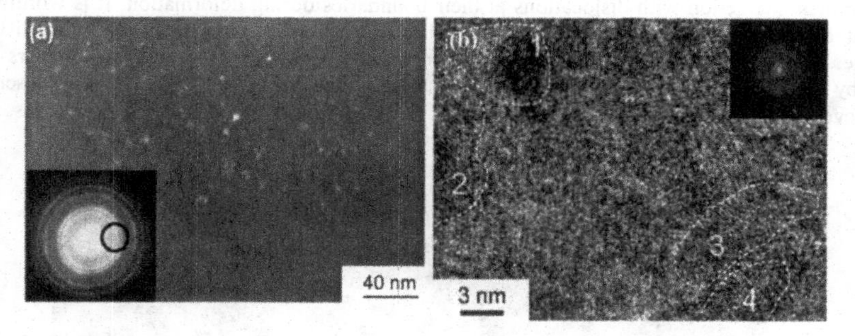

Fig. 2 (a) Dark-field TEM image showing nc grains. Inset is the EDP; (b) HREM image showing nc grains of high misorientation.

3.3 Dislocation Density

Fig. 3(a) is a Fourier reconstructed image showing one-dimensional [-11-1] lattice fringes taken near the grain boundary (GB) (~20 nm away from the GB) of an nc grain of ~100 nm in diameter. Inset is the EDP with [011] zone axis. Dislocations (a few are circled) are visible from the extra half planes of crystal lattices. The dislocation density is measured to be as high as 11×10^{12} cm^{-2}, using the number of dislocations divided by the area of the figure. The dislocation density on other (11-1) and (-200) planes are also measured to be $\sim 8.9\times10^{12}$ and $\sim 0.48\times10^{12}$ cm^{-2} respectively. The total dislocation density within this local area is thus around 20×10^{12} cm^{-2}. Extensive measurements are made to investigate the change in dislocation density with grain sizes and the result is shown in Fig. 3(b). Each data point represents the average from ~10 measurements in 3-5 grains. It is seen that the dislocation density decreases as the grain size reduces and that the dislocation density near the GBs is 1-2 order of the magnitude higher than that in grains.

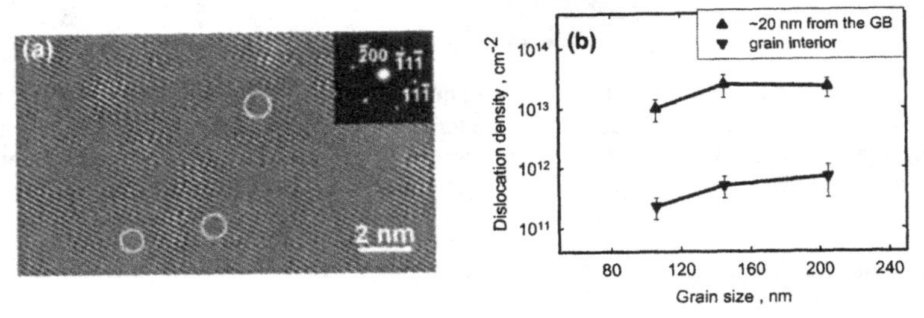

Fig. 3 (a) Fourier reconstructed image indicating high density dislocations (a few are circled). (b) dislocation density *vs* grain size.

Fig. 4(a) is a HREM image showing a group of nearly parallel areas (dashed line marks) where the lattice fringes are at best ill-defined or totally disappearing. When a large number of dislocations are present, they tend to form the more complicated configurations to reduce the elastic energy resulting from their mutual interactions. We therefore, prefer to call them dislocation complexes, probably due to severe strain and high strain rate during the SMAT. The disappearance of lattice fringes indicates the existence of extremely high dislocation density and intense strains. Their boundaries are rather waved, which implies interaction with dislocations at their boundaries during deformation. It is worthwhile to note that dislocation complexes subdivide the original grain into small subgrains with dislocation complexes being their subboundaries. Moreover, equiaxed subgrains of several nanometers in size appear by intersection of multi-directional dislocation complexes (Fig. 4(b)). HREM observations further reveal that the larger the strain, the more will be the population of dislocation complexes.

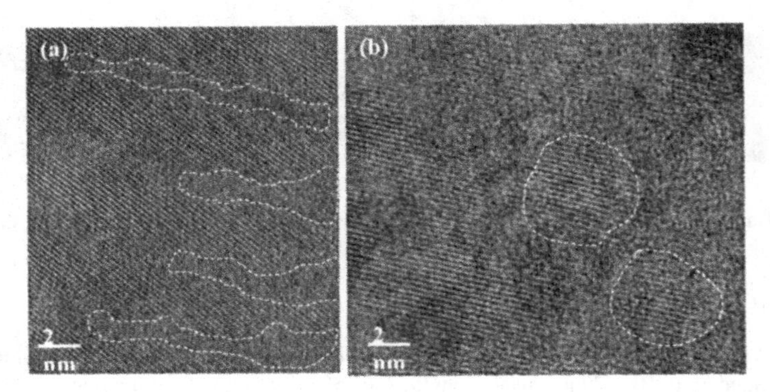

Fig. 4 HREM images showing dislocation complexes (marked dashed lines) (a) and resultant subgrains (b).

4. Discussion

TEM observations (Fig. 1) exhibit a microstructural evolution from well inside to the top surface, in a sequence of i) deformation bands with elongated subgrains, ii) equiaxed, submicro-sized grains, and iii) nc grains leveling off ~46 nm. The nc grains are observed to derive from continuous refinement of grains, due to the increment of strain over the whole deformed layer. The grain refinement is resulted from dislocation activity, which dominates still within the larger end of the nanometer size regime (~50-100 nm) [5-7]. Note that dislocations of high density are always present in deformed grains. As a result, the mechanism responsible for accommodating large amounts of plastic straining is to subdivide original grains into subgrains with dislocations being their boundaries [8,9], as indicated in Fig.1. With increasing strain, the microscopic grain subdivision will take place on a finer and finer scale. Simultaneously, with further increasing strain, the orientations of subgrains with respect to their neighbors become completely random and highly misoriented grain boundaries form. The increment of misorientations between neighboring subgrains may be realized by accumulating and annihilating more dislocations in subboundaries, or alternatively, by rotation of subgrains (or grain boundary sliding) with each other under certain strain [5,6]. The SAMT offers the multi-directional strain path and high strain rate, which are especially effective at promoting subgrain rotation. The present result is in good agreement with a lot of work devoted to large strain deformation [2,10].

nc grains of small sizes (Fig. 2) have the lower-end of the nanometer size regime and high misorientations. They begin to form at a critical strain, which is much low compared with that for nc grains of large sizes. They increase their population with increasing strain. Therefore, it is reasonable that nc grains are probably formed through dynamic recrystallization (DRX) triggered at a critical strain. The areas with high density dislocations will promote recrystallization nucleation as traditional materials do. It is envisaged that at these levels of dislocation density (Fig. 3), small energy fluctuations resulting from impacts during dynamic deformation process can trigger a recrystallization event. With the generation of high density dislocations and dislocation complexes, subgrains of the stored energy high enough provide favored sites for the recrystallization nucleation. The subboundary tends to be more misoriented due to accumulated rotation, which is the primary mechanism for further strain accommodation. Finally, recrystallized nc grains appear with high misorientation. The critical nucleus radius is only a few nanometers for recrystallized nc grains [11]. In the present investigation, DRX occurs at a critical strain and then operates simultaneously with grain subdivision during deformation. Such a case is very similar to the formation of nc grains *via* DRX during cryomilling Zn powders [11].

5. Summary

A nanostructured surface layer was synthesized on an Al solid solution by means of the SMAT. The crystal defects and microstructures were investigated to elucidate the mechanisms of the formation of nc grains. nc grains with the larger end of the nanometer size regime (~50-100 nm) were present *via* grain refinement operated by subdivision mechanism during deformation. nc grains with the lower end of the nanometer size regime (~<10 nm) were formed through dynamic recrystallization, which occurred at a critical strain and operated simultaneously with grain subdivision during deformation.

References

1. R.Z. Valiev, R.K. Islamgaliev, and I.V. Alexandrov, "Bulk Nanostructured Materials from Severe Plastic Deformation," *Prog. Mater. Sci.*, 45 (2000) 103-189.

2. Zhu YT et al., *Ultrafine grained materials* (Warrendale (PA): The Minerals, Metals and Materials Society, 2002).

3. K. Lu, J. Lu, J, Mater. "Nanostructured Surface Layer on Metallic Materials Induced by Surface Mechanical Attrition," *Sci. Eng.*, A (2003) inprint.

4. W.P. Tong, N.R. Tao, Z.B. Wang, K. Lu, J. Lu, J, "Nitriding Iron at Lower Temperatures," *Science*, 299 (2003) 686-88.

5. N.R. Tao, Z.B. Wang, W.P. Tong, M.L. Sui, K. Lu, J. Lu, J., "An Investigation of Surface Nnanocrystallization Mechanism in Fe Induced by Surface Mechanical Attrition Treatment," *Acta mater.*, 50 (2002) 4603-4616.

6. X. Wu, N. Tao, Y. Hong, B. Xu, J. Lu, K. Lu, "Microstructure and Evolution of Mechanically-Induced Ultrafine Grain in Surface Layer of AL-Alloy Subjected to USSP," *Acta mater.*, 50 (2002) 2075-2084.

7. C. Suryanarayana, C.C. Koch, "Nanocrystalline Materials - Current Research and Future Directions," *Hyperfine Interactions*, 130 (2000) 5-44.

8. D.A. Hughes, Q. Liu, D.C. Chrzan, N. Hansen, "Scaling of Microstructural Parameters: Misorientations of Deformation Induced Boundaries," *Acta mater.*, 45 (1997) 105-112.

9. D.A. Hughes, N. Hansen, "High angle boundaries formed by grain subdivision mechanisms," *Acta mater.*, 45 (1997) 3871.

10. Y. Iwahashi, Z. Horita, M. Nemoto and T. G. Langdon. "An investigation of Microstructural Evolution during Equal-Channel Angular Pressing," *Acta mater.*, 45 (1997) 4733-4741.

11. X. Zhang, H. Wang, J. Narayan, and C. C. Koch, "Evidence for the Formation Mechanism of Nanoscale Microstructures in Cryomilled Zn Powder," *Acta mater.*, 49 (2001), 1319-1326.

Processing and Properties of Structural Nanomaterials
Edited by Leon L. Shaw, C. Suryanarayana and Rajiv S. Mishra
TMS (The Minerals, Metals & Materials Society), 2003

Fabrication of Bulk Nanostructured Materials by Friction Stir Processing

Jian-Qing Su, Tracy W. Nelson and Colin J. Sterling

Department of Mechanical Engineering, Brigham Young University, 435 CTB, Provo, UT 84602, U.S.A.

Keywords: Friction stir processing (FSP), Nanocrystalline, microstructures, Dynamic recrystallization.

Abstract

Despite their interesting properties, nanostructured materials have found limited use as a result of the cost of preparation and difficulty in scaling up. Herein, the authors report a new technique, friction stir processing (FSP), to refine grain sizes to a nanoscale. Nanocrystalline 7075 Al with an average grain size of 100 nm was successfully obtained using FSP. The microstructure characteristics of the processed material were investigated, and furthermore the grain refinement mechanism has been revealed. It was found that the nanoscale grains were separated by high-angle boundaries. Dynamic nucleation followed by high-angle boundary migration is the mechanism of the nanostructure creation.

Introduction

Interest in nanostructured materials has been extremely high over the past two decades. Compared with their conventional coarse-grained polycrystalline counterparts, there are many potential advantages in the use of nanocrystalline metals and alloys for structural applications, such as high strength with ductility, high toughness, and excellent superplasticity at high strain rates and low temperature [1-4].

Many methods have been used to synthesize materials with nano-scale grain sizes, including inert gas condensation [5], high-energy ball milling [6], and sliding wear [7]. These techniques are attractive for producing materials with grain sizes below 100 nm, but there are disadvantages because some residual porosity remains after fabrication and it is difficult to use these techniques to make bulk samples. As a consequence of these difficulties, much attention has been paid to alternative procedures of introducing ultra-fine grains in materials by severe plastic deformation (SPD) [8-10]. The general experimental approach of SPD involves large-scale deformation using processes such as rolling, equal-channel angular pressing or extrusion (ECAP/E), or high-pressure torsional straining. Using this approach, nanostructured bulk materials have been produced from ductile metals and alloys of initial low to moderate strength [9]. However, high-strength metals and alloys are difficult to process by SPD methods. Furthermore, these processing techniques produce relatively small quantities of material, are very difficult to scale up, and are unlikely to be able to produce materials at low cost. Development of processing techniques which enable the production of bulk nanostructured materials large enough for many engineering applications is essential.

Recently, a new processing technique, friction stir processing (FSP), has been developed by adapting the concepts of friction stir welding (FSW) [11]. This technique involves plugging and traversing a cylindrical rotating tool through a material to produce intense plastic deformation. Localized heating is produced by severe plastic deformation of the material, and by friction between the rotating tool shoulder and the top surface of the base metal [12]. This new thermo-mechanical processing technique has been found to be an effective grain refinement technique for aluminum alloys [12-16]. Mishira et al [12] have reported that the grains of 3-4 μm were obtained in 7075 Al alloy by using FSW. The finest grain size obtained in a previous FSP study has been reported to be 0.5 μ [16]. Given the nature of the process and the mechanisms behind the microstructural evolution [17], the author's hypothesized that by combining this technique with rapid cooling, it should be possible to produce large scale nanocrystalline materials.

In the present paper, we will report the preliminary production of nanocrystalline 7075 Al alloy by using the FSP technique. In addition, the microstructure characteristics of the processed material will be investigated and the recrystallization mechanism during FSP will be discussed.

Experimental

The basic principle of FSP in the present research is schematically illustrated in Figure 1.

Figure 1: Schematic illustration of friction stir processing.

The tool shoulder diameter was 9 mm. The length and diameter of the headpin were 1.9 mm and 3 mm, respectively. Commercial 7075 Al sheet 2 mm thick was selected for this investigation. A single pass friction stir processed zone 30 cm long was produced at a travel speed of 10 cm/min and a rotational speed of 800 rpm. A mixture of water, methanol, and dry ice, was used to quench the plate immediately behind the FSP tool. The cooling rate was controlled by adjusting the volumetric flue rate of cooling liquid. In this investigation, two cooling rates were chosen to study the effect on microstructure.

The exact cooling rate was unknown. But a greater flow of cooling fluid was assumed to cause faster cooling.

Thin foil TEM samples were prepared by cutting the processed materials parallel to the processed surface at the mid-plane of the processed regions into discs of 3 mm in diameter using electrical-discharge machine (EDM). The discs were ground to a thickness of about 80 μm, then subject to twin-jet electropolishing. Microstructural investigations were carried out on a JEOL 2000FX instrument with a tungsten filament operated at 200 kV.

Results and discussions

The samples produced under different cooling rates will be denoted as sample A with faster cooling and sample B with slower cooling. Figure 2 shows TEM observation of the post-processed microstructures. Both specimen, A and B, were composed of very small recrystallized equiaxed grains. In particular, for specimen A, the grain size decreased to an average nano-scale level of about 100 nm (Figure 2a). Corresponding select area diffraction (SAD) pattern in Figure 2a, which was taken from a 1.7 μm diameter region, exhibits diffraction rings, indicating that there were many small grains with random misorientations in the region of analysis. The resulting microstructure clearly illustrates that a nanocrystalline structure was produced in a single pass during FSP. For specimen B, a larger grain size of roughly 300 nm was observed. This suggests that the obtained grain size can be controlled during FSP by changing cooling rate.

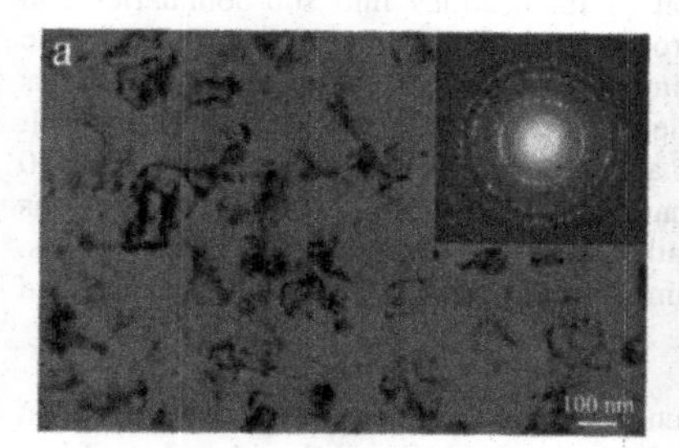

Figure 2: TEM micrographs of friction stir processed 7075 Al alloy (a) in specimen A with the corresponding SAD pattern and (b) in specimen B.

To reveal the microstructure evolution during FSP, the microstructure characteristics were further investigated. Grain sizes achieved in specimen A ranged from 30 nm to 180 nm. The extremely fine grains (< 50 nm) present in the FSP material were not sub-grains. As seen in Figure 3, the large difference in diffraction contrast of these extremely fine grains with the neighboring grains in the dark field image indicates that these are separated by high-angle boundaries. This microstructure feature indicates that the nanocrystalline grains were created by discontinuous dynamic recrystallization at elevated temperature during FSP.

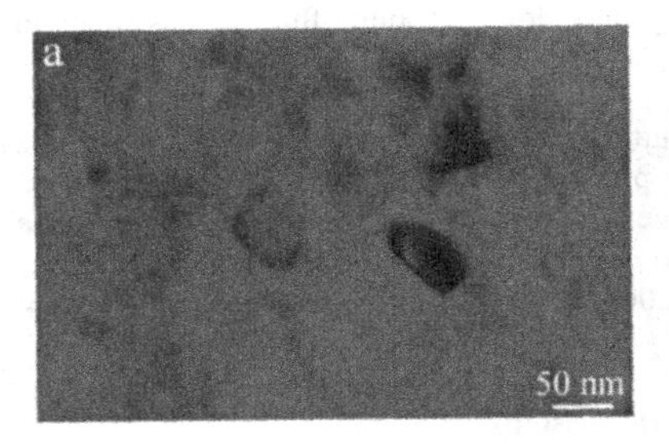

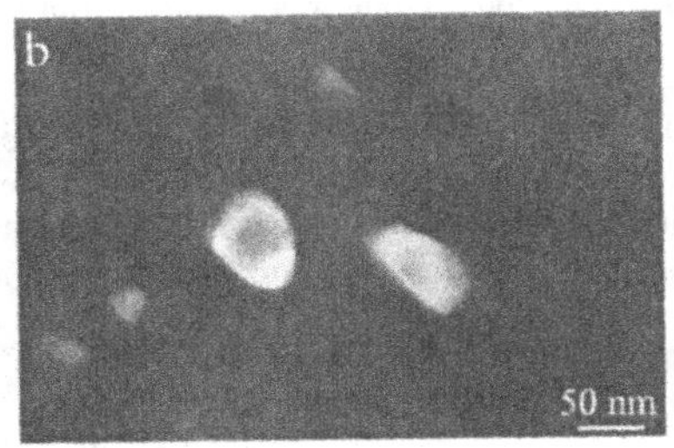

Figure 3: The extremely fine grains present in the specimen A. (a) bright-field image and (b) dark-field image.

Generally, two types of dynamic recrystallization (DRX) are discussed in the literature: 1) continuous DRX, and 2) discontinuous DRX [18, 19]. During continuous DRX new grains develop via a gradual increase in misorientation between subgrains. In contrast, during discontinuous DRX new grains exhibiting large-angle boundaries evolve, e.g. dynamic nucleation followed by grain growth from migration of large-angle boundaries. Although the exact formation mechanism of fine grain structures created during FSW or FSP are not well understood, to date these grain structures have been usually attributed to dynamic recrystallization [17, 20]. It has been proposed that the resultant fine grains are developed during the stirring operation by rotation of existing sub-grains within the parent microstructure [20], or by absorption of dislocations into sub-boundaries and growth of the subgrains developed during processing [17]. In other words, the dynamic recrystallization in FSW or FSP occurs continuously. However, the present study shows that the initial grains formed during FSP under the conditions given above are the result of nucleation and growth. Similar result was also reported by Rhodes et al [21] in 7050 Al alloy. It is likely that the evolution of dynamically recrystallized structures in FSW is affected by not only one mechanism but rather multi-mechanisms in different stages. These various mechanisms are currently under investigation and will be published elsewhere.

It is usually thought that grains developed under continuous DRX are much finer than those from discontinuous DRX during hot deformation [22]. Currently, a number of low temperature severe plastic deformation (SPD) processing routes are being developed to refine the microstructure of metals and alloys [9]. The ultrafine-grained structures developed by SPD at low temperature are the result of continuous DRX. However, it is worth mentioning that the development of SPD nanostructures with high angle grain boundaries, which result in qualitative changes in properties [9], is a rather difficult task. For example, equal channel angular extrusion (ECAE) can reduce the grain size to 0.5-1.0 μm in aluminum alloys, but requires a strain of >4.0 [9, 23]. Generally it takes >8 ECAE passes to achieve very fine grains exhibiting high grain boundary misorientations [24].

In contrast, microstructures represented by grains with high angle boundaries are produced directly as a result of dynamic nucleation during discontinuous DRX. For the formation of nanograined structures via this mechanism, the key ingredient is to achieve a high nucleation rate. It is believed that a complex stress state and strain components with very large strain gradients were produced in the processed material during FSP. Furthermore, large amounts of dislocations were introduced to accommodate the strain incompatibility. The complex stress state, complicated strain patterns and dislocation configurations, and high density of geometrically necessary dislocations are all beneficial in allowing copious nucleation during DRX.

Another feature of the obtained microstructure in specimen A is that for grains on the order of 50-150 nm, the grain interior and boundaries are free of dislocations (Figure 4). Specific "diffusive" diffraction contrast of inclined grain boundaries (GBs) was observed. In this orientation, an inclined grain boundary image contains only one central light band and two darker broader bands. This pattern style is a typical feature of; 1) non-equilibrium GBs with higher energy, and 2) long range stresses caused by an absorption of large number of lattice dislocations [25, 26]. These microstructure features indicate that the nanocrystalline grains created by discontinuous dynamic recrystallization at elevated temperature during FSP evolved by absorption of numerous dislocations.

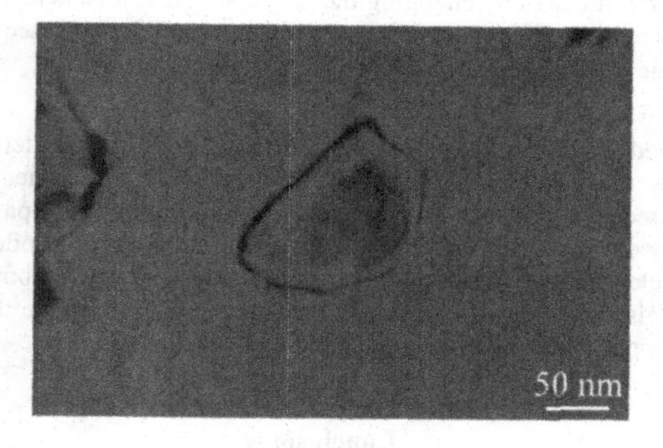

Figure 4: Dislocation free grain with non-equilibrium boundaries in specimen A.

For specimen B, the grains having same size as the grains in specimen A were found showing similar characteristics, e.g. the grains smaller than 200 nm are free of dislocations and have non-equilibrium boundaries. However, dislocation structures were clearly observed in the larger grains, as shown in Figure 5, suggesting that after dynamic recrystallization, additional dislocations were generated by subsequent plastic deformation within the larger (softer) grains, accommodating strains preferentially.

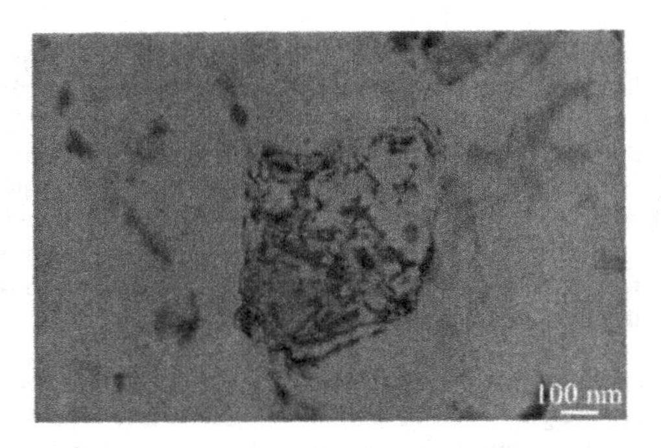

Figure 5: Dislocation structure in larger grain of specimen B.

Although process optimization has not been performed to study the minimum grain size possible by the FSP technique, these preliminary results illustrate that this technique is fairly effective in refining grain sizes to the nanometer level. Similarly, FSP has several advantages or potential advantages over other processing techniques: (i) It produces nanocrystalline with high angle boundaries in a single step. (ii) It is possible to control the resulting microstructures by changing the processing parameters and cooling rate. (iii) An entire sheet could be FS processed in a multi-pass overlapping sequence to obtain a desired microstructure over a large area.

So far, established techniques can not produce full-density nanomaterial in bulk forms large enough for most engineering applications. The FSP technique, in principle, can process any desired size thin sheet to nanostructure by running multi-passes. The authors believe that the work presented herein on nanostructured material produced via FSP will promote the development and investigation of bulk nano-materials at both the applied and fundamental levels. This processing technique may eventually pave the way for large-scale structural applications of nanostructured metals and alloys.

Conclusions

Combining FSP with rapid cooling, nanocrystalline 7075 Al with an average grain size of 100 nm was successfully obtained. The microstructures of processed material can be controlled by changing the processing parameters and cooling rate. It was found that the nanocrystalline grains were created by discontinuous dynamic recrystallization at elevated temperature during FSP whereas the grain boundaries evolved by absorption of numerous dislocations.

References

1. C.C. Koch, et al., "Ductility of Nanostructured Materials," *Mater. Res. Soc. Bull.*, 24 (1999), 54-58.
2. J.R. Weertman, et al., "Structure and Mechanical Behavior of Bulk Nanocrystalline Materials," *Mater. Res. Soc. Bull.*, 24 (1999), 44-50.
3. Y.M. Wang, et al., "High Tensile Ductility in a Nanostructured Metal," *Nature*, 419 (2002), 912-915.
4. S.X. McFadden, et al., "Low-temperature Superplasticity in Nanostructured Nickel and Metal Alloys," *Nature*, 398 (1999), 684-686.
5. H. Gleiter., "Nanocrystalline Materials," *Prog. Mater. Sci.*, 33 (1989), 223-315.
6. C.C. Koch and Y.S. Cho, "Nanocrystals by High Energy Ball Milling," *Nanostruct. Mater.*, 1 (1992), 207-212.
7. D.A. Rigney, "Sliding Wear of Metals," *Ann. Rev. Mater. Sci.*, 18 (1988), 141-163.
8. R.Z. Valiev, A.V. Korznikov and R.R. Mulyukov, "The Structure and Properties of Metallic Materials With a Submicron-Grained Structure," *Phys. Met. Metall.*, 73 (1992), 373-384.
9. R.Z. Valiev, R.K. Islamgaliev and I.V. Alexandrov, "Bulk Nanostructured Materials from Severe Plastic Deformation," *Prog. Mater. Sci.*, 45 (2000), 103-189.
10. H. Utsunomiya, et al., "Novel Ultra-high Straining Process for Bulk Materials--Development of the Accumulative Roll-bonding (ARMB) Process," *Acta Mater.*, 47 (1999), 579-583.
11. W.K. Thomas, et al., "Friction Stir Butt Welding", G.B. Patent Application No. 9125978.8, Dec. 1991; U.S. Patent No. 5460317, Oct. 1995.
12. R.S. Mishra, and M.W. Mahoney, "Friction stir Processing: a New Grain Refinement Technique to Achieve High Strain Rate Superplasticity in Commercial Alloys," *Materials Science Forum*, 357-359 (2001), 507-514.
13. R.S. Mishra, et al., "High Strain Rate Superplasticity in a Friction Stir Processed 7075 Al Alloy," *Scripta Mater.*, 42 (1999), 163-168.
14. S. Benavides, Y. Li, and L. E. Murr, "Ultrafine Grain Structure in the Friction-stir Welding of Aluminum Alloy 2024 at Low Temperatures," *Ultrafine Grained Materials*, ed. R.S. Mishra, et al., (The Minerals, Metals and Materials Society, 2000), 155-163.
15. N. Saito, et al., "Grain Refinement of 1050 Aluminum Alloy by Friction Stir Processing," *J. Mater. Sci.Lett.*, 20 (2001), 1913-1915.
16. Y. J. Kwon, N. Saito, and I. Shigematsu, "Friction Stir Process as a New Manufacturing Technique of Ultrafine Grained Aluminum Alloy," *J. Mater. Sci.Lett.*, 21 (2002), 1473-1476.
17. J.-Q. Su, et al., "Microstructural Investigation of Friction Stir Welded 7050-T651 Aluminium," *Acta Mater.* 51 (2003), 713-729.
18. F.J. Humphreys, and M. Hatherly, *Recrystallization and Related Annealing Phenomena* (Oxford Pergamon, 1995) 363.

19. T. Sakai, "Microstructural Development Under Dynamic Recrystallization of Polycrystalline Materials," *Thermomechanical Processing of Steels*, ed. S. Yue and E. Essadiqi (Montreal Metallurgical Society of the Canadian Institute of Metals, 2000), 47-62.
20. K.V. Jata and S.L. Semiatin, "Continuous Dynamic Recrystallization During Friction Stir Welding of High Strength aluminum Alloys," *Scripta Mater.*, 43 (2000), 743-749.
21. C.G. Rhodes, et al., "Fine-grain Evolution in Friction-stir Processed 7050 Aluminum," *Scripta Mater.*, 48 (2003), 1451-1455.
22. A. Belyakov, et al., "Grain Refinement in Copper Under Large Strain Deformation," *Phil. Mag. A*, 81 (2001), 2629-2643.
23. M. Furukawa, et al., "Factors Influencing the Development of Ultrafine Grained Materials Through Severe Plastic Deformation," *Ultrafine Grained Materials*, ed. R.S. Mishra, et al., (The Minerals, Metals and Materials Society, 2000), 125-134.
24. S.D. Terhune, Z. Horita, M. Nemoto, Y. Li, T.G. Langdon, and T.R. McNelley, in *proceeding of the Fourth International Conference on Recrystallization and Related Phenomena*, edited by T. Sakai and H.G. Suzuki, (Japan Inst. Of Metals, Sendai, Japan, 1999), p. 515.
25. R.Z. Valiev, V.Yu. Gertsman, and O.A. Kaibyshev, "Grain Boundary Structure and Properties Under External Influences," Phys. Stat. Sol. (a), 97 (1986), 11-56.
26. R.Sh. Musalimov and R.Z. Valiev, "Dilatometric Analysis of Aluminium Alloy With Submicrometre Grained Structure," Scripta Metall. Mater., 27 (1992), 1685-1690.

Processing and Properties of Structural Nanomaterials
Edited by Leon L. Shaw, C. Suryanarayana and Rajiv S. Mishra
TMS (The Minerals, Metals & Materials Society), 2003

Investigations of Glassy and Nanostructured Metal-Metal Type Alloys

D. V. Louzguine [1], S. Ranganathan [1,2] and A. Inoue [1]

[1] Institute for Materials Research, Tohoku University,
Katahira 2-1-1, Aoba-Ku, Sendai 980-8577, Japan
[2] Department of Metallurgy, Indian Institute of Science, Bangalore 560012,India

Keywords: Bulk Metallic Glasses; Nanocrystals; Nanoquasicrystals, Pettifor Maps

Abstract

Glassy, bulk glassy, nanostructured alloys and the composites obtained from them possess unique mechanical properties, high corrosion resistance and are promising candidates to be applied in tools, elastic, sportive and structural materials. In addition, the formation of an icosahedral quasicrystalline phase has been observed in many but not all of the bulk glass formers.

We present new results and also summarize a large number of earlier results obtained by our research group for various (Zr/Hf)-Ti-Ni-Cu with minor additions of noble metals. The emphasis is on bulk glass formers and the nanostructured alloys produced upon the devitrification of glassy alloys or directly upon solidification. The purpose is to describe and emphasize their most common features and properties as well as to show the future prospects of the field.

Attention is also given to the new class of Cu-based bulk metallic glasses derived by enrichment in copper of conventional Zr/Hf based bulk metallic glasses. These are found to posses high yield strength of above 2 GPa, high hardness >HV 600, high elastic energy and total (plastic and elastic) elongation of 2-2.5 %. New Cu-Zr-Ti-NM and Zr-Ti-Ni glassy alloys and bulk glass formers exhibiting formation of a nanoscale icosahedral phase on devitrification are also developed in the present work.

The icosahedral phase in Metal-Metal type glass formers, especially Zr and Hf-based alloys is found to be formed in composition ranges close to that of the cF96 cubic Ti_2Ni-type phase with a large unit cell size. Similarities in the structure of the icosahedral and cF96 phase as well as the reasons for the nanoicosahedral phase formation in Zr-, Ti-, Hf- and Cu-based alloys are discussed. Fine precipitates of the nanoicosahedral phase in a glassy matrix form a nanocomposite and improve the ductility of the glassy alloys.

Introduction

The advent of bulk metallic glasses in 1988 has signalled a renaissance in the science and technology of metallic glasses [1-2]. They offered for the first time sample dimensions in centimeters and made possible the determination of mechanical, magnetic and chemical properties with a precision previously not dreamt of. The availability of a large supercooled region permits the forming of components much in the same way as polymers can be formed. The devitrification of these glasses led to nanocrystals and nanoquasicrystals. Partial and complete devitrification led to the control of the resultant microstructure and hence properties. In addition to the in-situ composites, it is possible to add external reinforcements to produce composites. It is not surprising that the field has seen intense activity over the past decade and this excitement is captured in a number of recent reviews and conference proceedings [3-8].

The choice of systems leading to bulk metallic glasses has been largely guided by the semi-empirical rules formulated by Inoue [9] several years ago. These require multi-component alloying of elements with very different sizes, exhibiting highly negative enthalpies of mixing. These have in practice led to almost all the systems discovered to date. The structure of the metallic glasses continues to be studied. There are strong hints that many of them possess icosahedral order. The stability of the glasses is of utmost importance both for its intrinsic importance as well as the use of the glass as a precursor for creating useful microstructures. This paper addresses this aspect of the problem.

While a wide basis of elements exists for making bulk metallic glasses, intense attention has been focussed on the Group IV transition elements - Ti, Zr and Hf having large atomic radius alloyed with Ni and Cu [10]. In alloying with late transition elements such as Ni they display very high negative enthalpies of mixing. Cu is strictly not a transition metal but shows s-d hybridization. It is not surprising therefore that it forms alloys with Group IV elements with highly negative enthalpies of mixing. We have carried out comprehensive studies on (Zr/Hf)-Ti-(Cu, Ni) alloys ranging from ternary to quinary systems. Special mention must be made of the recent discovery of Cu-rich metallic glasses by Inoue and coworkers [11], which qualify as Cu based metallic glasses. The addition of elements such as Au, Pd and Pt to Zr-based bulk metallic glasses are shown to lead to a different devitrification behaviour and in particular the occurrence of icosahedral quasicrystals.

Experimental Details

Ingots of the alloys mentioned in subsequent sections (composition is given in atomic percentages) were prepared by arc-melting the mixture of the appropriate elements of Zr 99.7 mass% purity, Hf 99.7 mass% purity, Ti 99.5 mass% purity, Ni 99.9 mass% purity and Cu 99.99 % purity in an argon atmosphere. From theses alloys, ribbon samples of about 0.02 mm in thickness and 1 mm in width were prepared by rapid solidification of the melt on a single copper roller at the surface velocity of 42 m/s. In several cases bulk glasses were produced by copper mould casting. The structure of the samples was examined by X-ray diffraction with monochromatic CuK_{α} radiation. Transmission electron microscopy (TEM) was carried out using a JEOL JEM 2010 and 2000FX II microscopes operating at 200 kV and equipped with an energy-dispersive X-ray spectrometer (EDX). Transformation temperature and heat released during transformation were examined by differential scanning calorimetry (DSC) at heating rates ranged from 0.17 to 1.33 K/s. Isothermal annealing of the ribbon samples encapsulated in quartz tubes was carried out in vacuum of $1*10^{-3}$ Pa.

Results

Zirconium based Metallic Glasses

The devitrification of Zr-based glasses has attracted enormous attention due to two very intriguing features. These are the formation of nanocrystals of both metastable and stable phases as well as the appearance of nanoquasicrystals under special circumstances [5]. The occurrence of nanocrystals has been attributed to either quenched-in nuclei or phase separation in the glassy matrix [4-5].

The first observation of quasicrystals in zirconium alloys was correctly identified as arising from oxygen [12]. In turn it prompted two lines of investigations. The first is to vary the composition to see which element or combination of elements leads to quasicrystal formation. Extensive work by Saida and Inoue [13] has shown reproducible icosahedral phase formation in the devitrification of $Zr_{65}Al_{7.5}Ni_{10}Cu_{17.5}$ alloy by substituting noble metals such as Ag, Pd, Au or Pt for 5 or 10 at. % Cu. It was also shown that the addition of other elements such as V, Nb, Ta, and Mo also leads to quasicrystal formation [13]. The recipe given was that a slight deviation from the Inoue rules for glass formation led to quasicrystals. Effectively it meant that the minor alloying additions displayed lower enthalpy of mixing with one of the components already present. Though it was recognized as quasicrystal formation in Ti based alloys, it was known that Ti-Zr-Ni forms a stable quasicrystal and in fact is as much a Zr-based quasicrystal [14]. Thus this particular alloy combination in multinary alloys can also be expected to lead to quasicrystal formation. A second line of research focussed on binary and ternary alloys systems. It is clearly established that both Zr-rich Zr-Pd and Zr-Pt binary alloys are quasicrystal formers [15-17]. Thus the presence of Pd or Pt in multinary alloys leads to quasicrystal formation.

We briefly describe our work in Zr-Ti-Ni [18] and Zr-Ti-Ni-Cu [19] alloys leading to quasicrystals. A transmission electron micrograph and the nanobeam diffraction patterns from the melt spun $Zr_{41.5}Ti_{41.5}Ni_{17}$ alloy heat treated at 810 K are given in Fig. 1. The size of the precipitates is in the range of 30 to 50 nm. The electron diffraction patterns confirm that the precipitates are quasicrystals. Fig. 2 (a) shows the structure of the as-solidified $Zr_{55}Cu_{20}Ti_{15}Ni_{10}$ alloy to be amorphous. When heat treated at 652 K, which is slightly below the first exothermic peak revealed in a DSC run, it is seen that the pattern can be indexed as arising from an icosahedral quasicrystalline phase (Fig 2 b). The size of the particle is estimated to be between 3 and 7 nm. The concurrent presence of Zr, Ti and Ni in the amounts in the alloy has led to the phase selection. On further treatment equilibrium Zr_2Cu phase forms [19].

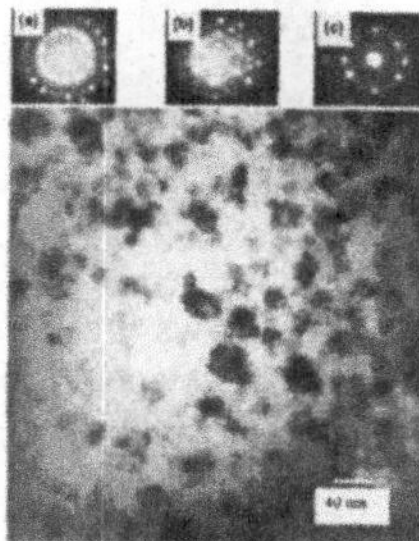

Fig. 1 Transmission electron micrograph of the Zr-Ti-Ni melt spun alloy annealed at 810 K for 600 s. (a) 5-fold (b) 3-fold (c) 2-fold electron diffraction patterns from the precipitates showing the presence of icosahedral quasicrystalline phase.

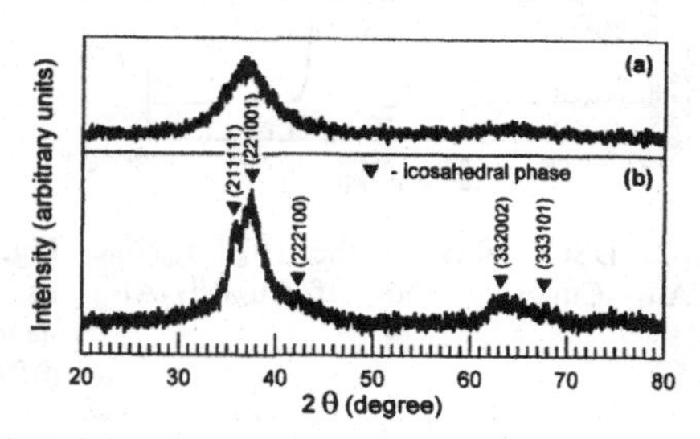

Fig. 2. X-ray diffraction patterns of the $Zr_{55}Cu_{20}Ti_{15}Ni_{10}$ alloy (a) as solidified and (b) heat treated for 1.2 ks at 652 K, indicating peaks corresponding to the icosahedral phase.

Hafnium based Metallic Glasses

In many ways hafnium alloying behaviour mimics that of zirconium, even though there are subtle variations. When Zr-based bulk metallic glasses were discovered, it was natural to enquire whether Hf-based bulk glasses can be formed in homologous alloys such as Hf-TM-Al and Hf-Ni-Cu [10]. Again the incidence of quasicrystals in Zr based alloys triggered similar investigations in Hf-based alloys. For example, Li et al. [20] showed that $Hf_{73}Pd_{27}$, $Hf_{70}Cu_{20}Pd_{10}$ and $Hf_{70}Cu_{20}Pt_{10}$ alloys led to the formation of quasicrystals on devitrification, but surprisingly $Hf_{73}Pt_{27}$ alloy did not form quasicrystals.

As in the case of Zr, in the homologous ternary Hf-Ti-Ni alloy both quasicrystals and rational approximants have been found [21]. Hf-Ti-Ni-Cu-Al with as little as 3% Ti produces quasicrystals. It is also established now that Hf-Ni-Pd-Al, Hf-Ni-Cu-Al-Ti and Hf-Ni-Cu-Al-Pd alloys produce quasicrystals on devitrification. The addition of ~5% Pd or the presence of Ti seems to be the essential ingredient in these alloys. Fig. 3 shows DSC traces from a few selected alloys from our extensive studies [22-24]. An amorphous single phase was formed in as-solidified ribbon sample of ternary $Hf_{60}Pd_{30}Ni_{10}$ and quaternary $Hf_{65}NM_{17.5}Cu_{10}Al_{7.5}$ and $Hf_{65}NM_{17.5}Ni_{10}Al_{7.5}$ (NM=noble metals, except for Ag) alloys. However, $Hf_{65}Cu_{20}Ni_{7.5}Al_{7.5}$ did not vitrify completely. No quasicrystalline phase was observed in $Hf_{60}Pd_{30}Ni_{10}$ alloy, a surprising observation as the binary Hf-Pd alloy shows the formation of quasicrystals [20], although our present studies show that nanoicosahedral clusters are formed together with nanocrystalline particles in the $Hf_{70}Pd_{20}Ni_{10}$ alloy heat treated at 828 K for 0.6 ks.

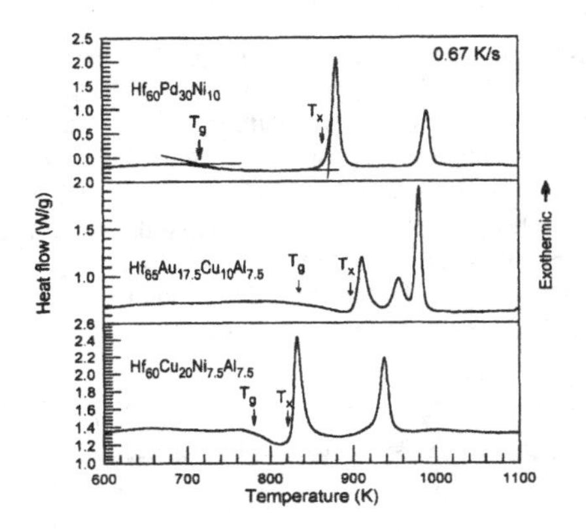

Fig. 3. DSC traces of the $Hf_{60}Pd_{30}Ni_{10}$, $Hf_{65}Au_{17.5}Cu_{10}Al_{7.5}$ and $Hf_{60}Cu_{20}Ni_{7.5}Al_{7.5}$ alloys.

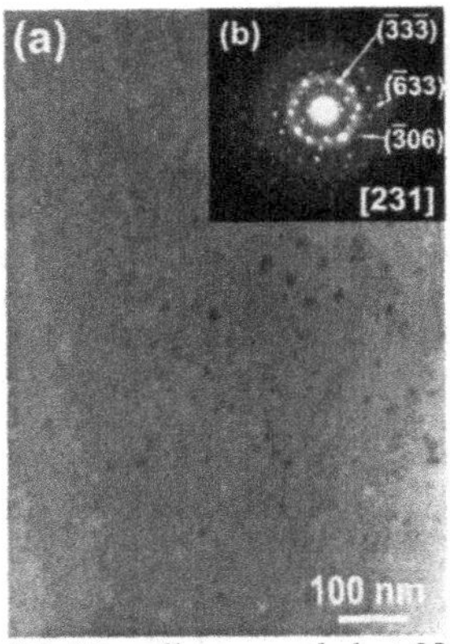

Fig. 4. $Hf_{60}Pd_{30}Ni_{10}$ alloy annealed at 923 K for 0.6 ks. (a) TEM micrograph shows nearly spherical nanocrystals indexed with (b) selected-area electron diffraction pattern.

The formation of the $Fd\bar{3}m$ Ti_2Ni-type phase with a large lattice parameter has been observed at the primary crystallization step in the $Hf_{60}Pd_{30}Ni_{10}$, $Hf_{65}(NM)_{17.5}Fe_{10}Al_{7.5}$ and $Hf_{65}(NM)_{17.5}Co_{10}Al_{7.5}$ alloys. At least two crystalline phases were formed by primary crystallization reaction in the $Hf_{60}Cu_{30}Ni_{10}$ alloy. The breadth of the X-ray diffraction peaks indicates the formation of nanoparticles that have been directly confirmed using TEM. Fig. 4 shows the structure of the $Hf_{60}Pd_{30}Ni_{10}$ alloy annealed up to the temperature of the initial crystallization step including selected-area electron diffraction patterns revealing the cubic $Fd\bar{3}m$ symmetry of the crystalline phase. The composition of the crystalline particles of the $Fd\bar{3}m$ Ti_2Ni-type phase was found to be qualitatively close to the composition of the glassy phase.

Cu-based bulk metallic glasses

It has been reported that a glass is formed in Cu-Zr and Cu-Hf systems in a wide composition range of 30 to 70 at% Zr or Hf. A significant increase in glass formability has been reported in Cu-Zr-Ti and Cu-Hf-Ti with 60 % of copper [11]. Earlier studied ternary and quaternary alloys had a Cu content less than

40 %. The effect of just 10 % Ti is remarkable. These alloys exhibit superior mechanical properties: Young's modulus 114-134 GPa, compressive yield strength 1785-2010 MPa , tensile yield strength 1780-1920 MPa, tensile fracture strength 200-2130 MPa and the elastic elongation of 1.5 to 2% [11].

While (Zr,Hf)-Ti-Ni quasicrystals are well studied, the replacement of Ni by Cu does not yield quasicrystals in ternary alloys. On devitrification $Cu_{60}Zr_{30}Ti_{10}$ and $Cu_{60}Hf_{25}Ti_{15}$ alloys produced crystalline phases [25]. The addition of Ti to the binary Cu-Zr or Cu-Hf alloys induces multistage crystallization. Nanoparticles of cubic phases were formed in Cu-Zr-Ti (Fig. 5) or Cu-Hf-Ti alloys. The Ni-Zr-Ti and Ni-Hf-Ti alloys formed quasicrystals. In the Cu-rich alloy the Early transition metal content was not adequate to promote the formation of quasicrystals.

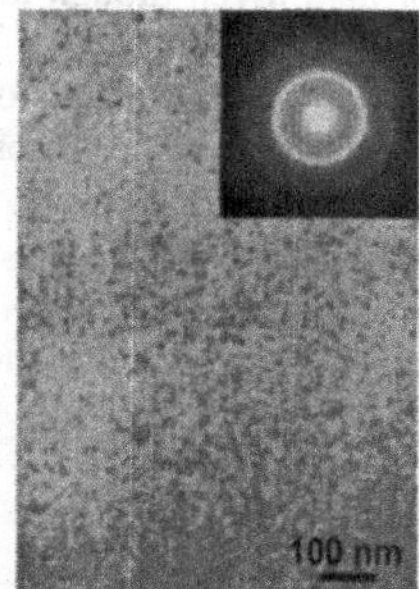

Fig. 5. $Cu_{60}Zr_{30}Ti_{10}$ heat treated in DSC run up to 780 K for completion of primary precipitation. TEM bright-field image. The insert – SAED pattern.

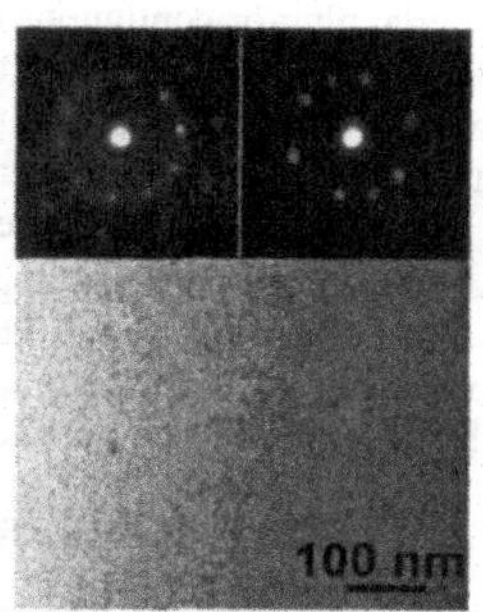

Fig. 6. $Cu_{55}Zr_{30}Ti_{10}Pd_5$ alloy annealed for 2 ks at 750 K showing icosahedral quasicrystals. Bright-field image. The insets are NBD representing five-, and two-fold symmetries of the quasicrystal.

Fig. 6 demonstrates that addition of 5% Pd replacing Cu in Cu-Zr-Ti alloy led to the formation of nanoicosahedral phase [26]. However a similar substitution in Cu-Hf-Ti alloy did not produce quasicrystals. Partial replacement of Cu by Ni led to primary formation of an equilibrium crystalline oC68 $Cu_{10}Zr_7$ phase, though one might have expected quasicrystals in the case of Ni substitution [27]. The replacement of Cu by 5 % Co or Ni stabilizes the supercooled liquid [27-28].

There is an interesting question about the true character of the Cu-based alloys. Their mechanical properties are remarkable and superior to that of Zr-based glasses. In a detailed TEM study of these alloys Jiang et al. [29] found that the material was actually a composite of nanocrystals embedded in the amorphous matrix. In $Cu_{60}Zr_{30}Ti_{10}$ alloy they found Cu-rich nanoparticles ranging from 5 to 10 nm in sizes. This unexpected result explains the remarkable properties. The volume fraction was estimated to be as high as 5 to 10 % and it is not entirely clear as to why such a relatively high volume fraction escaped detection by X-rays. They inferred that phase separation into Cu-rich and Cu-poor regions occurred prior to solidification. This idea was initially put forward in Ref. [25]. At the same time in Ref. [25] has been shown that in the supercooled liquid region the structure is essentially amorphous.

Discussion

The main objective of our study is to determine the conditions favoring the occurrence of quasicrystals in (Zr,Hf)-Ti-(Ni,Cu) systems. The effect of minor alloying additions also needs to be explained. Even though five components are involved, the system can be treated as a pseudo-quaternary system as Zr and

Hf play a very similar, if not, identical role. At another level, even though Ni and Cu have quite different characteristics, it seems fruitful to treat the alloys as a pseudo-binary system consisting of the early transition elements Ti, Zr, Hf with the late transition elements Ni and Cu. Even though all the associated binary systems are fully investigated, only partial information is available for the relevant ternary and quaternary systems. An additional complexity is introduced by the high chemical affinity of Ti, Zr and Hf for oxygen, leading again to stabilization of certain metastable phases.

Table I gives the stable and metastable phases formed at AB_2 stoichiometry in the systems under investigation. Nevitt and Koch [30] have discussed some important structures of fixed stoichiometry and gave special emphasis to the phases indicated in Table I. The three structure types Al_2Cu, $MoSi_2$ and Ti_2Ni occur in overlapping or contiguous domains in the AB_2 Pettifor map. Even though the structures and atomic environments are different, it is clear that these structures are very close in energy and compete for stability in some two component and higher order systems. This leads to coexisting polymorphic forms, whose stability depends on composition, temperature and other factors including interstitial (O, N) components. In these three structures, the structure displaying icosahedral coordination and serving as a rational approximant to quasicrystals is the cF96 Ti_2Ni phase. Hf forms this phase with Mn, Fe, Co, Rh, Ir and Pt. With oxygen impurity the phase occurs in Hf-Ni and Hf-Pd alloys. It is also possible to form this phase in ternary alloys Hf-Ni-(Ru, Re, Os), even though binary alloys of Hf with Ru, Re and Os do not show this phase.

Table I. Structure formation of Ti, Zr, Hf with Cu and Ni

Mendeleev Number	M_b	67	72
M_a	Elements	Ni	Cu
49	Zr	A, *C, G	B, G
50	Hf	A, *C, C-O, G	B, G
51	Ti	C, Q-O, G	B, C-O, G

A- (tI12) Al_2Cu; B- (tI6) $MoSi_2$; C- (cF96) Ti_2Ni
Q- Quasicrystals; G- Glass; O- Oxygen stabilised; *- Metastable

In Zr alloys it is now well established that binary Zr-Pt, Zr-Pd, ternary Zr-Pd-(Cu,Ag,Au) and quaternary Zr-Ni-Cu-Al and quinary alloys with Ag & Pd additions lead to quasicrystal formation. In some cases QC formation is promoted by oxygen. In contrast to Hf alloys, Zr alloys form quasicrystals more readily than the cubic phase. It is known that oxygen stabilises not only the cubic phase but also quasicrystals as well. Thus, the cubic phase forms at low oxygen content compared to the higher oxygen content that stabilizes quasicrystals.

Hafnium tends to favour the big cubic phase or with increasing number of electrons the $MoSi_2$ phase. Under metastable conditions, the formation of metastable big cubic or icosahedral quasicrystalline phase is observed. Their appearance is influenced by alloying additions as well as oxygen impurity content. The absence of a quasicrystalline phase in the $Hf_{60}Pd_{30}Ni_{10}$ alloy is surprising in view of the quasicrystal observed in binary Hf-Pd alloy The occurrence of quasicrystals during devitrification of $Hf_{65}Pd_{17.5}Ni_{10}Al_{7.5}$ and $Hf_{65}Au_{17.5}(Ni,Cu)_{10}Al_{7.5}$ alloys is to be noted. For example, the replacement of Ni in the $Hf_{65}Pd_{17.5}Ni_{10}Al_{7.5}$ and $Hf_{65}Au_{17.5}Ni_{10}Al_{7.5}$ alloys by Fe or Co causes a significant change in their crystallization behaviour and again no quasicrystalline phase is formed.

The Cu-based glasses are quite different from Zr and Hf - based glasses as they are based on the element having the smallest atomic diameter. The Cu-based alloys being rich in Ni tend to produce oC68

$(Cu,Ni)_{10}Zr_7$ phase on devitrification. This combination of elements does not promote the formation of quasicrystals. A higher content of Group IV transition element is required, as seen in the other cases leading to quasicrystal formation. For example, even when 10 % Ni is added, quasicrystals are not produced, even though Zr/Hf, Ti and Ni are all present.

As Nevitt and Koch [30] have remarked, these families of compounds occur at fixed characteristic A_mB_n stoichiometries. The quasicrystal in these systems must be added as a fourth member to this list. A or B is frequently a composite of several isovalent elements. This is a rephrasing of the observations of Saida and Inoue as deviation from the Inoue rules [13]. Thus Zr, Ti and Hf serve as A and Cu and Ni serve as B element. In A substitution with Ta, V, Nb or Mo is possible, while in B possible substitution is Ag, Au, Pt or Pd.

Conclusions

In the alloying of early transition elements Ti, Zr and Hf with the late transition elements Ni and Cu it is possible to demonstrate the link between the occurrence of bulk metallic glasses, icosahedal quasicrystals and the cF96 phase. All of them show icosahedral order. It is also shown that replacement by isovalent elements in arriving at the stoichiometry for these phases promotes the formation of quasicrystals.

Acknowledgements

The authors wish to express their gratitude to their colleagues (T. Sakurai, T. Zhang, A. Takeuchi, H. Kimura, J Saida, A. R. Yavari, K. Chattopadhyay, U. Ramamurty and J. Basu) for stimulating discussions.

References

1. A. Inoue, K. Ohtera, K. Kita and T. Masumoto, "New amorphous Mg-Ce-Ni alloys with high strength and good ductilty", *Jpn. J. Appl. Phys.*, 27 (1988), L2248.
2. A. Inoue, T. Zhang and T. Masumoto, "Al-La-Ni amorphous alloys with a wide supercooled liquid region", *Mater. Trans. JIM*, 30 (1989), 965.
3. A. Inoue, *Bulk Amorphous Alloys-Preparation and fundamental characteristics* (Trans. Tech. Publications, Switzerland) 1998.
4. W. L. Johnson, "Bulk glass forming metallic alloys: Science and Technology", *MRS Bull*, 24 (1999), 42.
5. A. Inoue, "Stabilization of metallic supercooled liquid and bulk amorphous alloys", *Acta Mater.*, 48 (2000), 279.
6. A. Inoue and A. Takeuchi, "Recent progress in bulk glassy alloys", Mater. Trans. JIM, 43 (2002), 1982.
7. T. Egami, A. L. Greer, A. Inoue and S. Ranganathan (Editors), *Supercooled liquids, the glass transition and bulk metallic glasses* (Warrendale, PA:Mater. Res. Soc) 2003.
8. J. Basu and S. Ranganathan, "Bulk metallic glasses: A new class of engineering materials", *Sadhana*, 28 (2003), 21.
9. A. Inoue, "High strength bulk amorphous alloys with low critical cooling rates" *Mater. Trans. JIM*, 36 (1995), 866.
10. T. Zhang, A. Inoue and T. Masumoto, "Amorphous (Ti,Zr,Hf)-Ni-Cu alloys with a wide supercooled liquid region", *Mater. Sci. Eng*, A181/182 (1994), 1423.
11. A. Inoue, W. Zhang, T. Zhang and K. Kurosaka, "High-strength Cu-based bulk glassy alloys in Cu-Zr-Ti and Cu-Hf-Ti systems", *Acta Mater.*, 49 (2001), 2645.

12. U. Koester, J. Meinhardt, S. Roos and H. Liebertz, "Formation of quasicrystals in bulk glass forming Zr-Cu-Ni-Al alloys", *Appl. Phys. Lett*, 69 (1996), 179.
13. J. Saida and A. Inoue, "Quasicrystals from glass devitrification", *J. Non-Crystalline Solids*, 317 (2003), 97.
14. V. V. Molokonov and V. N. Chebonitnikov, "Quasicrystals and amorphous alloys in Ti-Zr-Ni system: Glass forming ability, structure and propertiers", *J. Non-Crystalline Solids*, 117-118 (1990), 789.
15. J. Saida, M. Matsushita, C. Li and A. Inoue, Formation of icosahedral quasicrystalline phase in $Zr_{70}Pd_{30}$ binary glassy alloy, *Phil. Mag. Lett.*, 81 (2001), 39.
16. B. S. Murty, D. H. Ping and K. Hono, "Nanoquasicrystallization in binary Zr-Pd metallic glasses", Appl. Phys. Let., 77 (2000), 1102.
17. S. Ranganathan, J. Z. Jiang, J. Saida and A. Inoue, "Glass forming abilty and quasicrystal forming abilty: A comparative study in zirconium based ternary alloys", Phil. Mag. Lett. 2003, in press.
18. J. Basu, D. V. Louzguine, A. Inoue and S. Ranganathan, Nanocrystallisation and nanoquasicrystallisation in (Ti/Hf)-Zr-(Ni, Cu) ternary alloys, in Supercooled liquids, the glass transition and bulk metallic glasses (Warrendale, PA: Mater. Res. Soc) 2003.
19. D. V. Louzguine and A. Inoue, "Formation of a nanoquasicrystalline phase in Zr-Cu-Ti-Ni metallic glass", *Appl. Phys. Lett.*, 78 (2001), 1841.
20. C. Li, S. Ranganathan and A. Inoue, "Initial crystallization studies of Hf-Cu-M (M = Pd, Pt or Ag) amorphous alloys", *Acta Mater.*, 49 (2001), 1903.
21. V. T. Huett and K. F.Kelton, "Formation and hydrogen absorption properties of Ti-Hf-Ni quassicrystals and crystal approximants", *Phil. Mag. Lett*, 82 (2002), 191.
22. D. V. Louzguine, M. S. Ko and A. Inoue, "Nanoquasicrystalline phase produced by devitrification of Hf-Pd-Ni-Al metallic glass", *Appl. Phys. Lett.*, 76 (2000), 3424.
23. D. V. Louzguine, M. S. Ko, S. Ranganathan and A. Inoue, "Investigation of metallic glassy, nanocrystalline and nano quasicrystalline phases formed in Hf-based alloys", Proc of the Fourth Pacific RIM International Conference, PRICM-4, Japan Institute of Metals, 2001, p 67.
24. D. V. Louzguine, M. S. Ko, S. Ranganathan and A. Inoue, "Nanocrystallization of the Fd-3m Ti_2Ni type phases in Hf-based metallic glasses", *J. Nanosci. Nanotech.*, 1 (2001), 185.
25. D. V. Louzguine and A. Inoue, "Nanocrystallization of Cu–(Zr or Hf)–Ti metallic glasses" *J. Mater. Res.*, 17 (2002), 2112.
26. D. V. Louzguine and A. Inoue "Nanoparticles with icosahedral symmetry in Cu-based bulk glass former induced by Pd addition", *Scripta Mater.*, 48 (2003) 1325.
27. D. V. Louzguine and A. Inoue, "Influence of Ni and Co additions on supercooled liquid region, devitrification behaviour and mechanical properties of Cu-Zr-Ti bulk metallic glass", *J. Metastable and Nanocrystalline Materials*, 15-16 (2003), 31.
28. D. V. Louzguine and A. Inoue, "Structural and thermal investigations of a high strength Cu-Zr-Ti-Co alloy", *Phil. Mag. Lett.*, 83 (2003), 191.
29. J. Z. Jiang, J. Saida, H. Kasto, T. Ohsuna and A. Inoue, "Is $Cu_{60}Ti_{10}Zr_{30}$ a bulk glass forming alloy?" *Appl. Phys. Lett.*, 82 (2003), 4041.
30. M. V. Nevitt and C. C. Koch, *Chapter 6 Some important structures of fixed stoichiometry, in Crystal structures of intermetallic compounds*, Edited by J. H. Westbrook and R. L. Fleischer, John Wiley & Sons Ltd, New York ,1995.

Processing and Properties of Structural Nanomaterials
Edited by Leon L. Shaw, C. Suryanarayana and Rajiv S. Mishra
TMS (The Minerals, Metals & Materials Society), 2003

Nanocrystal Formation By Crystallization Of $Zr_{52}Ti_6Al_{10}Cu_{18}Ni_{14}$ Bulk Metallic Glass

G. K. Dey[1], R. T. Savalia[1], S. Neogy[1], D. Srivastava[1], R. Tewari [2], S. Banerjee[1]

[1]Bhabha Atomic Research Centre, Matls. Sci. Div., Trombay, Mumbai, Maharashtra, 400 085 India
[2]University of Cincinnati, Chem. & Matls. Engrg., Cincinnati, OH 45219 USA

Abstract

Bulk glass has been synthesized in $Zr_{52}Ti_6Al_{10}Cu_{18}Ni_{14}$ by the technique of copper mould casting. The as cast microstructure has been characterized by X-ray diffraction and transmission electron microscopy. High resolution electron microscopy (HREM) has revealed the presence of nanometer size quenched-in nuclei of crystalline phase in amorphous regions. Crystallization of the as cast microstructure was carried out in a furnace as well as in a differential scanning calorimeter (DSC). Crystallization at 923 K for 2 hours lead to a nanocrystalline microstructure with the nanocrystals lying in the size range of 15 to 50 nm. The nature of the interfaces present in this nanocrystalline microstructure like stacking faults, anti-phase domain boundaries and twins have been examined in detail using HREM. The structure of the nanograin boundary observed in this study has been compared with that of the grain boundary in large grained material. Coarsening of the nanocrystalline microstructure and its effect in the structure of various interfaces has been examined as a function of crystallization temperature and time.

Introduction

Nanocrystalline materials have been at the forefront of scientific research in recent times because of their many unusual and useful properties [1]. Several techniques have emerged by which nanocrystalline solids can be produced [1]. Transformation of an amorphous phase to nanocrystals is an important method of producing this class of materials [2]. Production of nanocrystals by crystallization of the amorphous phase is especially attractive in view of the fact that here the control of the crystal size is more accurate. The production of metallic glasses in bulk form and crystallization of the same to give a nanocrystalline microstructure has opened up the possibility of synthesizing nanocrystalline solids in bulk. Composites of bulk metallic glass and nanocrystalline phases have been found to have excellent mechanical properties [3]. In view of the technological importance of nanocrystalline materials, the study of their properties and the factors, which govern these properties, have acquired considerable significance. The various types of interfaces in nanocrystalline solids play an important role in deciding the properties of this material. The important interfaces in this type of material are not only the nanograin boundaries but interfaces like twins and stacking faults. Though the structure of the nanograin boundary has been reported in some studies [1, 4], information regarding the structure of interfaces like twin boundaries, stacking faults and antiphase domain boundaries is limited. The characterization of these defects is very important to understand the deformation behaviour of nanocrystalline solids. A recent study has shown that deformation of nanocrystals occurs by rotation of nanocrystals [5]. Grain boundary disclination dipoles have been found to be the primary carriers of rotational plastic deformation in this class of materials [5]. In this study bulk glass having the composition $Zr_{52}Ti_6Al_{10}Cu_{18}Ni_{14}$ was crystallized to obtain the nanocrystalline intermetallic phases. Detailed analysis of the interfaces present in these has been carried out using conventional and high-resolution transmission electron microscope (HREM). It was found from the HREM observations that the structure of the nanograin boundary is similar to that of the grain boundary in a large grained material in many ways. Besides nanograin boundary, the nature of other types of crystallographic defects were analyzed and compared with those observed in a large grained crystalline material. The effect of the heat treatments on the nature of interfaces was examined in this study. Some change in the nature of interfaces with heat treatment could be observed. The coarsening of the nanocrystals was examined as a function of heat treatment. Precipitation inside a nanograin was encountered in this study. The interface between the precipitate phase and the matrix phase was examined by HREM.

Experimental

Crystal bar zirconium and high purity elements were melted in the right proportion in a vacuum arc furnace for preparing the glass-forming alloy. Repeated melting of the arc-melted button was carried out for composition homogenization. The arc melted buttons were used to produce bulk metallic glass rods by induction melting these in quartz crucible and injecting the melt in a copper mould under protective argon atmosphere. Nanocrystallisation was carried out by heat-treating pieces of bulk metallic glass rod in a furnace at various temperatures for different durations and also in a Differential Scanning Calorimeter (DSC).

Specimens for TEM examination were prepared either by jet thinning in a twinjet electropolishing unit or by ion milling in a Gatan Duomill. Conventional transmission electron microscopy was carried out in a JEOL 2000 FX transmission electron microscope and HREM was carried out in JEOL 3010 microscope having a point to point resolution of 0.21 nm.

Results and Discussion

In the present study, the microstructure of the as cast bulk metallic glass was found to be fully amorphous under conventional microscopy, electron and x-ray diffraction. However, HREM observation revealed the presence of very few very small crystalline regions in the size range of 3 – 4 nm which can be categorized as quenched-in-nuclei of a crystalline phase. The presence of such regions has been noticed in rapidly solidified metallic glasses as well and they have been found to play an important role in the process of crystallization of the amorphous phase [2, 6]. The absence of interfaces such as twins or stacking faults inside these nuclei indicated that these interfaces do not develop during the solidification process. In the present study crystallites of small size could be obtained by treatment at various temperatures as well as by heating in the DSC beyond the crystallization peaks and then cooling these down to ambient temperature rapidly. In the following sections a description of these microstructures is presented.

Figure 1 shows the microstructure of the specimen heated in DSC at 20 K/min. upto 893 K, a temperature which is beyond the crystallization peaks. Nanogains in the size range of 10 – 50 nm could be seen in this microstructure. SAED and microdiffraction analysis confirmed that the phases forming on crystallization are isostructural with Zr_2Cu and Zr_2Ni. These observations are consistent with those made by other workers in a glass having composition close to this glass [7]. Crystallographic orientation between the nanograins of the Zr_2Cu and Zr_2Ni phases could not be established as was evident from the microdiffraction patterns and the HREM images. Figure 2 is a HREM image of a nanograin boundary in specimens crystallized in the DSC, which shows the extension of the lattice fringes right up to the grainboundary. No amorphous region could be seen at the boundary even after tilting the nanograins. In some of the grains the lattice fringes were distorted and curved which were manifested by a slight change in contrast of the lattice fringes near the nanograin boundary. It was possible to see a localized region of the order of 0. 5 nm along the grain boundary exhibiting a different contrast. Similar kind of features as mentioned above were also encountered in images of grain boundaries obtained from large grained material. It is known that to image the lattice in many grains at the same time the plane of the grain boundary must be parallel to low index planes on either side of the grain boundary plane [8]. In the present study this condition was satisfied for many grains simultaneously as the lattice was observed in many grains at the same time. Small grain size led to the appearance of a curved grain boundary in many areas. However, faceted grain boundary was also noticed in many areas since crystallization of metallic glasses involving the formation of intermetallic phases has been found to lead to the formation of faceted crystals in many instances [2, 6]. The grain boundary was devoid of disordered regions. Similar kind of observations were made by Thomas et al. [4] in case of nanocrystalline Pd where manifestation of grain boundary structures with random displacements of average magnitude greater than 12 % of nearest neighbour distance could not be observed. The nanograin boundary was mostly a large angle grain boundary without any coherency between the planes along any orientation. Voids were also absent at the nanograin boundaries. Dislocations could be seen in the specimen, however, their number density was quite small. Many of the nanograins were found to be divided into smaller grains by small angle boundaries. A regular array of edge dislocations with a periodic distance of about 3 nm and localized strain contrast around the dislocation cores could be seen. Presence of disclinations at the nanograin boundary has been found to contribute to the deformation behaviour of nanocrystals [5]. In this study the presence of disclinations at the nanograins could not be seen. The absence of disclinations may be due to the larger size of the nanocrystals or because the specimens had not been deformed. Figure 3 shows a HREM image inside a nanograin. A low angle grain boundary is seen in this image (shown by an arrow). The crystal orientation was found to change substantially across some of the interfaces (shown by a double arrow in Figure 3). In many of the nanograins planar defects resembling stacking faults could be seen (Figure 4). It was possible to see the presence of very small spherical particles in some of the nanograins.

Crystallization was carried out at 923 K and 1073 K for various lengths of time. Figure 5 shows a conventional TEM micrograph of the crystallized microstructure obtained by heat-treating the samples in a furnace at 923 K for 2 hours. Features similar to stacking faults and antiphase domain boundaries [8] were observed in some of the grains (Figure 6). The structure of the antiphase boundary and stacking faults observed in this study was found to be similar to those observed in large grains of similar phases [8]. The ordered intermetallic compound phases like the ones studied here provide an opportunity to study interfaces like these because these will not be readily encountered in pure elemental nanocrystalline materials. Figure 6 also shows many nanograin boundaries. These boundaries have features which have already been described in case of the specimens heat treated in the DSC.

HREM observation of DSC treated specimens as well as specimens treated at 923 K revealed the presence of twins in many of the nanograins examined in this study. Structurally the twins observed inside the nanograins was found to be identical to those seen in the case of large grained materials except for the fact that the propensity of these twins was found to be much larger. It was also possible to see twin-twin interaction in many instances. These observations are similar to those reported in studies carried out in many other nanocrystalline solids where extensive twinning has been found to occur [1]. The presence of twins in the nano-crystalline phases is most probably due to the fact that the stress in the small grains is too large because of their structure and the growth of the nanograins necessities the formation of twins. As the grain size increased due to coarsening the propensity of the twins came down because the need for accommodating the stresses generated reduced substantially.

Figure 7 shows the microstructure of a specimen, which had been treated at 1073 K for half an hour. It can be seen in the microstructure that the average grain size is quite large. Since this temperature is close to the melting point of the alloy, a rapid coarsening of the grains had taken place. It was possible to see the precipitation of the second phase in some of the larger nanogains. As in case of the DSC treated specimens this phase had a spherical morphology. The exact identity of this phase could not be established in this study because of the small size of the particles of this phase. No simple crystallographic orientation relationship seemed to exist between this phase and the matrix phase. The interface between the matrix phase and the precipitate phase was found to be coherent (Figure 8). The nanograins were found to have many types of interfaces of the type encountered earlier. However the number of small angle grain boundaries and the other types of interfaces seemed to have reduced in number.

The coarsening of the nanocrystals during the aforementioned heat-treatments was examined. Initially the crystallization process lead to the formation of isolated crystals. As crystallization progressed, formation of more crystals occurred. This resulted in impingement of the nanocrystals. The grain boundary features described above apply to such grain boundaries where impingement had already taken place. Once the impingement was over, the coarsening of the grains set in. The nanograins increased in size as a result of coarsening. HREM investigation of the nanograin boundary did not reveal a substantial difference in structure of the nanograin boundary as coarsening of the nanograins progressed.

Crystallization of rapidly solidified as well as bulk glass can lead to the formation of nanocrystalline structures. Dey et al. [2] have shown that in the Zr based ternary alloys having compositions $Zr_{76}Fe_{24-x}Ni_x$ where x <12, nanocrystal formation can occur under suitable conditions of crystallization. The formation of nanocrystals occurs only in glasses having certain compositions because the steady state nucleation rates of crystals are high in these glasses. This has been demonstrated by Dey et al. [2] in the Zr based ternary glasses. The requirement for the formation of the nanocrystalline structures is that a large number density of nuclei should form and these nuclei should grow to the stage of impingement without leaving any untransformed amorphous matrix. The formation of nanocrystals during crystallization in the case of rapidly solidified metallic glasses is expected in view of the fact that these are alloys where the steady state homogeneous nucleation rate is high. This is why these compositions require rapid solidification to prevent homogeneous nucleation and facilitate glass formation. In case of bulk metallic glass forming alloys, the homogeneous nucleation frequencies are low as compared to those compositions where glass formation requires rapid solidification. Threat to glass formation in such alloys is not from homogeneous nucleation and bulk metallic glasses can be obtained in these alloys at low cooling rates provided heterogeneous nucleation has been avoided. Though the homogeneous nucleation frequency is low in bulk glass forming alloys, formation of nanocrystalline phases have been noticed in many bulk glasses after crystallization. In this regard it is interesting to compare the nanocrystal formation behaviour of the bulk glass and the rapidly solidified Zr based glasses. It could be seen in this study and the study carried out earlier [2] in rapidly solidified Zr based glass (Figure 9) that the size of the nanocrystals can be made very fine by suitable heat treatment in the latter whereas in bulk glass the crystal size could not be made very small. This difference in size can be traced to the fact that in case of bulk glass the homogeneous nucleation frequency is low which leads to less copious nucleation of crystals as compared to rapidly solidified metallic glass.

Conclusions

The structure of the nanograin boundary had features, which were very similar to that seen in case of large grained material. The nanocrystalline microstructure had a very large amount of small angle boundaries and other type of interfaces. The number of such interfaces was found to come down as the nanocrystal size increased due to coarsening. Spherical particles of a phase were found to precipitate in many of the nanograins during crystallization

References

1. C. Suryanrayana, Nanocrystalline materials, *Int. Mat. Rev.*, 40 (1995), 41-64.

2. G. K. Dey, et al., Crystallization of Ternary Zr based glasses – Kinetics and microstructure, *J. Mater. Res.*, 13, (1998), 504 - 517.

3. A. Inoue, Stabilization of metallic supercooled liquid and bulk amorphous alloys, *Acta materilia*, 48(2000) 279.

4. G. J. Thomas, R. W. Siegel and J. A. Eastman, Grain boundaries in nanophase palladium: high resolution electron microscopy and image simulation, *Scripta Metal.*, 24 (1990), 201 - 206.

5. M. Murayama et al., Atomic-Level Observation of Disclination Dipoles in Mechanically Milled, Nanocrystalline Fe, *Science*, 295 (2002) 2433.

6. S. Banerjee, R. T. Savalia and G. K. Dey, Glass Formation and Crystallization in Rapidly Solidified Zirconium Alloys, *Mat. Sci and Eng.*, A 304 (2001), 26.

7. L. Q. Xing et al., Crystallization behaviour and nanocrystalline microstructure evolution of a $Zr_{57}Cu_{20}Al_{10}Ni_8Ti_5$ bulk amorphous alloy, *Philos. Mag.*, 79 A (1999), 1095 – 1108.

8. D. B. Williams and C. B. Carter, *Transmission Electron Microscopy* III (New York; Plenum Press, 1996), 459 - 482.

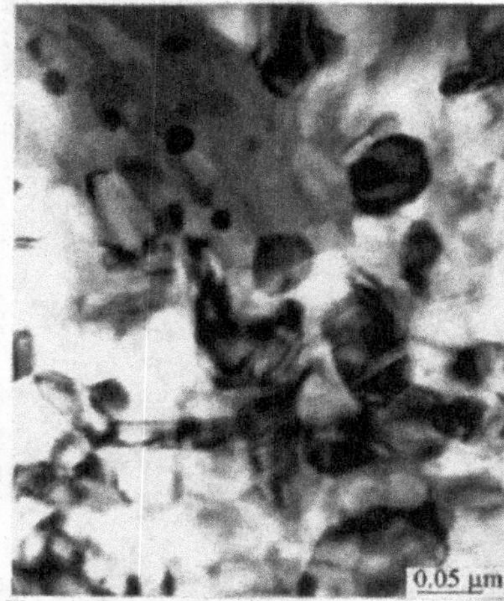

Figure 1 Bright field micrograph of a specimen crystallized in DSC.

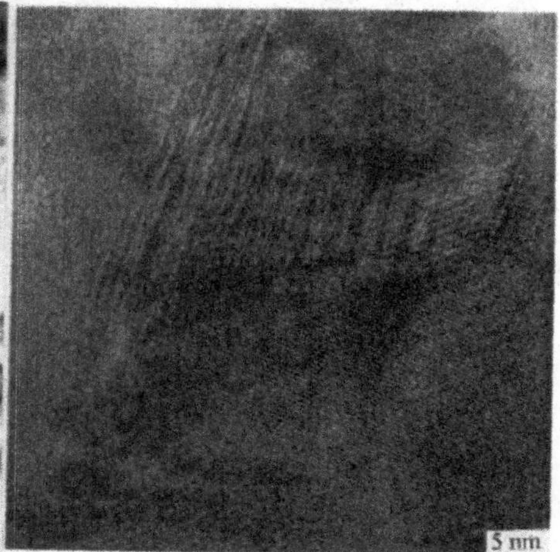

Figure 2 HREM micrograph showing the structure of a nanograin boundary in a specimen crystallized in DSC.

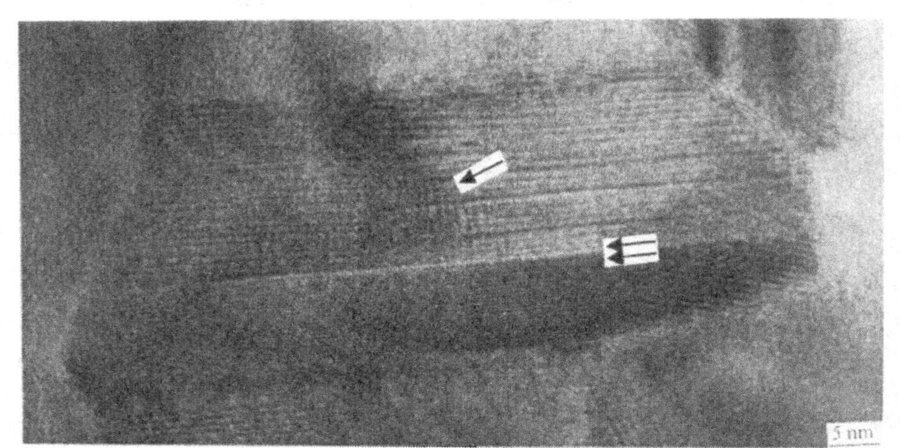

Figure 3 HREM micorgraph showing the structure inside a nanograin. A low angle boundary can be seen which has been marked by an arrow.

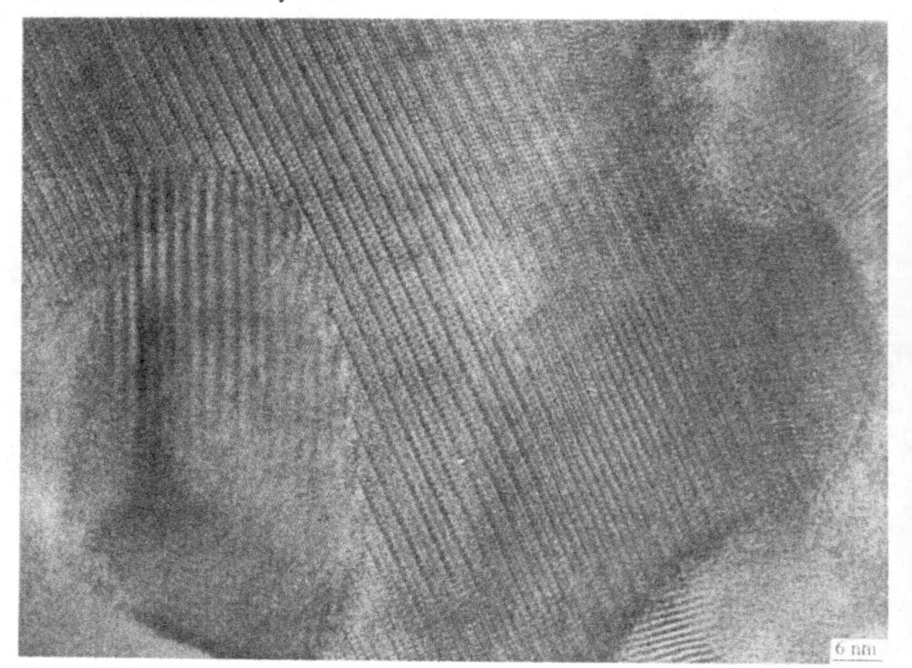

Figure 4 HREM micrograph of a DSC treated specimen showing stacking faults inside the nanograin.

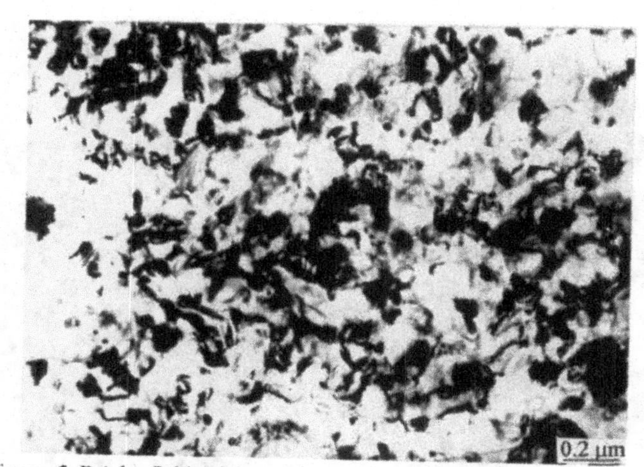

Figure 5 Bright field micrograph of the crystallized microstructure obtained by heat-treating the samples in a furnace at 923 K for 2 hours.

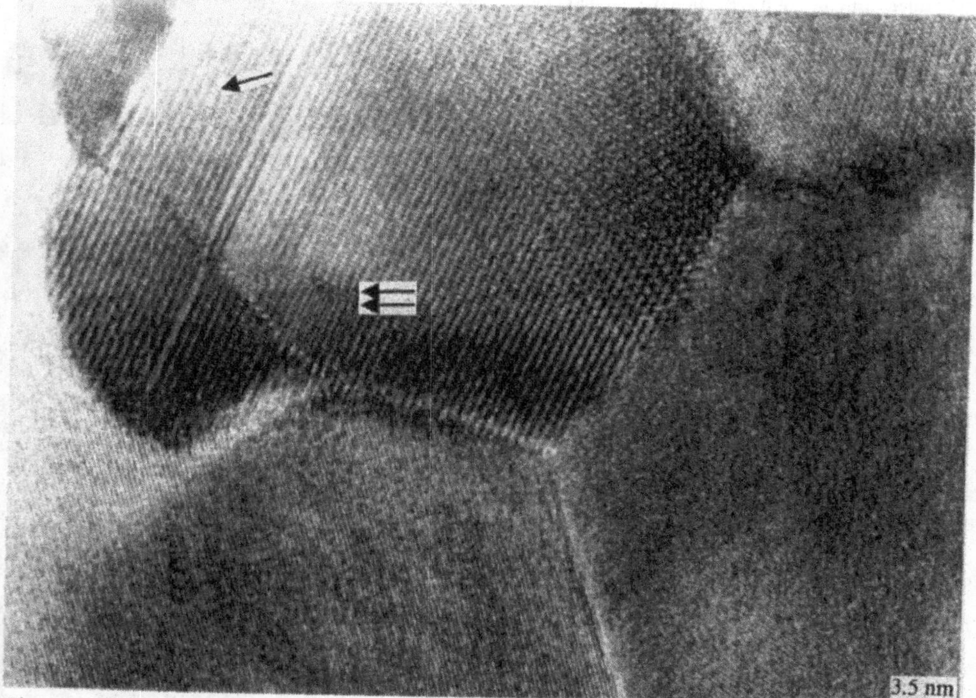

Figure 6. HREM micrograph showing many nanograin boundaries, stacking fault (indicated by single arrow) and antiphase domain boundary (indicated by double arrow).

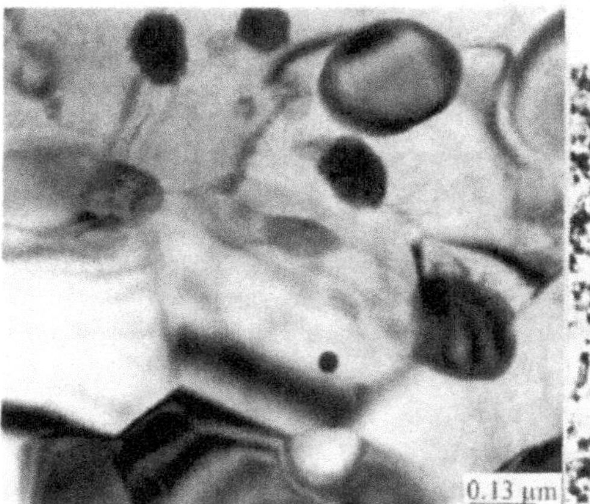

Figure 7 Bright field micrograph showing the microstructure of a specimen, which had been treated at 1073 K for half an hour.

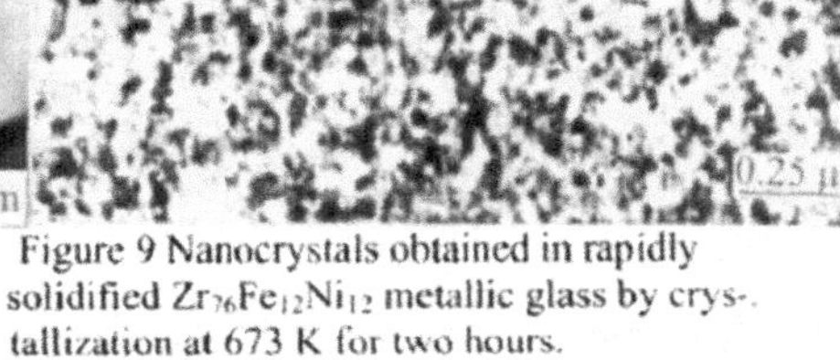

Figure 9 Nanocrystals obtained in rapidly solidified $Zr_{76}Fe_{12}Ni_{12}$ metallic glass by crys-. tallization at 673 K for two hours.

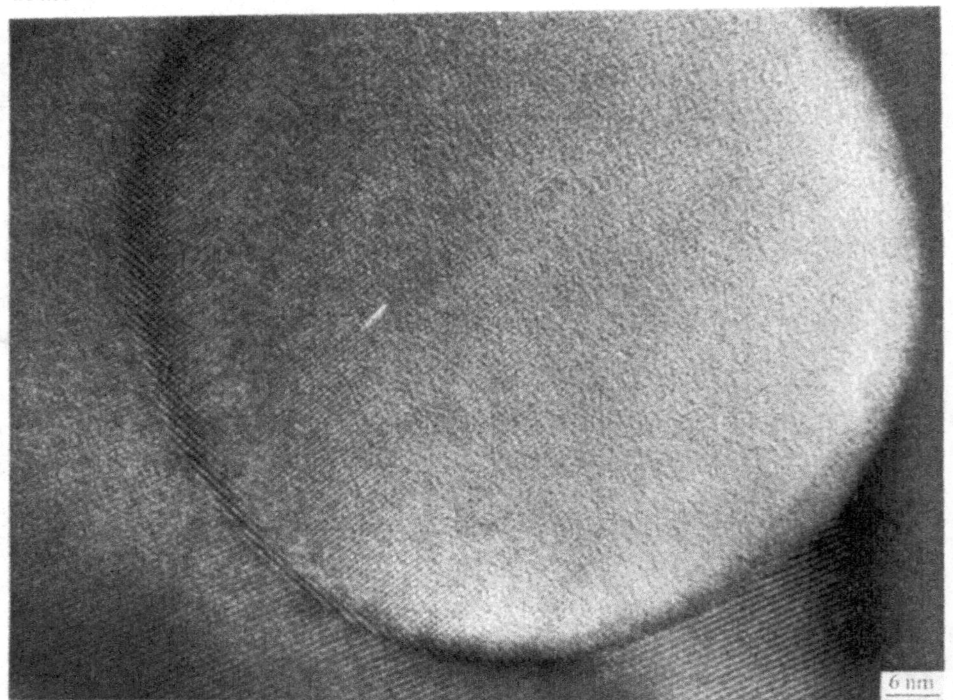

Figure 8. HREM micrograph showing the interface between a spherical particle and a nanograin in which this particle is located.

Processing and Properties of Structural Nanomaterials
Edited by Leon L. Shaw, C. Suryanarayana and Rajiv S. Mishra
TMS (The Minerals, Metals & Materials Society), 2003

Crystallization Behavior Of An Amorphous $Al_{85}Ni_{10}La_5$ Alloy

Zhihui Zhang[1], David Witkin[2], Enrique J. Lavernia[1]

[1] Dept of Chemical Engineering and Materials Science, University of California, Davis, CA 95616
[2] Dept of Chemical Engineering and Materials Science, University of California, Irvine, CA 92697

Keywords: Amorphous alloys, Crystallization, Al-Ni-La alloys

Abstract

$Al_{85}Ni_{10}La_5$ (at.%) alloy powders were fabricated using gas atomization. X-ray diffraction analysis revealed that powders in the size range of <500 mesh (<25μm) are amorphous. The crystallization behavior and kinetics of the amorphous $Al_{85}Ni_{10}La_5$ powders (<25μm) were investigated during continuous heating and isothermal annealing. The amorphous $Al_{85}Ni_{10}La_5$ alloy undergoes a three-step crystallization reaction in the temperature range of 250 °C ~ 390 °C. The activation energies for the first and the third exothermic reactions were determined as 344 kJ/mol and 198 kJ/mol, respectively, on the basis of the Kissinger analysis. However, the Kissinger analysis for the second exothermic reaction fails to yield a linear plot, indicating that the activation energy controlling the second reaction may vary under different heating rates. The mechanisms for such a crystallization behavior are discussed and may be attributed to the presence of the intermetallic compounds, along with precipitation of the Al phase during the first reaction. The isothermal annealing was carried out at temperatures of 235°C, 245°C and 250°C. Results from the isothermal annealing analysis further confirmed the observation that precipitation of the intermetallic compounds was concurrent with crystallization of the Al phase. The influence of isothermal annealing on the thermal stability of the $Al_{85}Ni_{10}La_5$ powders (<25μm) is also described and discussed.

1. Introduction

Interest in the study of high-strength and light-weight materials has intensified in recent years. For aluminum alloys, an upper tensile strength in the range of 550 to 600 MPa may be readily achieved via age hardening, and usually does not exceed 700 MPa even by other strengthening approaches, such as Hall-Petch strengthening, solid solution hardening, and dispersion strengthening [1]. However, since the Al-rich amorphous alloy family of Al-R-TM (R-rare earth metal, TM-transition metal) was discovered by two groups [2, 3], it has been suggested that this type of Al alloys may have a high strength, exceeding 1000MPa, in combination with good ductility [4]. The rare earth metal and transition metal provide this alloy with good glass forming ability and thermal stability. The Al-R-TM amorphous alloys have thus attracted interest as candidates to produce bulk forms that retain a nanocrystalline/amorphous microstructure by compaction and consolidation of the amorphous precursors (gas-atomized powders and melt-spun ribbons, etc.) [5]. Consolidation of these materials requires a clear understanding of their thermal stability and the crystallization behavior, especially the precipitation of the nanocrystalline aluminum crystallites, which is regarded as a likely source of the high strength of the alloy [4]. In this paper, the crystallization behavior and thermal stability of an amorphous $Al_{85}Ni_{10}La_5$ alloy was examined using differential scanning calorimetry (DSC), X-ray diffractometry (XRD) and scanning electron microscopy (SEM).

2. Experimental

The amorphous aluminum alloy powders used in the present investigation were prepared by gas atomization. A mixture of pure elemental Al(99.99%), Ni(99.9%) and La(99.9%) with the chemical composition of $Al_{85}Ni_{10}La_5$ (at.%) was induction melted under a high-purity Ar atmosphere and then atomized at a superheat level of approximately 250°C using high-purity He gas with an atomizing pressure of 6.2 MPa. The powders were mechanically sieved with the sizes of 500 mesh (25 μm) and 270 mesh (53μm). XRD analyis shows that particles between 25 μm and 53 μm are partially amorphous whereas particles less than 25 μm are essentially amorphous, as seen in Figure 1. The microstructure of the particles less than 25 μm was also examined by SEM under the back-scattering electron (BSE) mode, showing that tiny intermetallic compounds could be present in some particles (Figure 2) although they were not detected by XRD. As a result, the crystallization behaviors of the amorphous phase were studied using the powders less than 25 μm.

The XRD experiments were carried out with a Scintag XDS 2000 X-ray Diffractometer using CuK_α radiation. The SEM was performed using a XL30 FEG Scanning Electron Microscope. The continuous heating and isothermal annealing were carried out on a Perkin-Elmer DSC-7 using ultra-high purity N_2 as the purging gas.

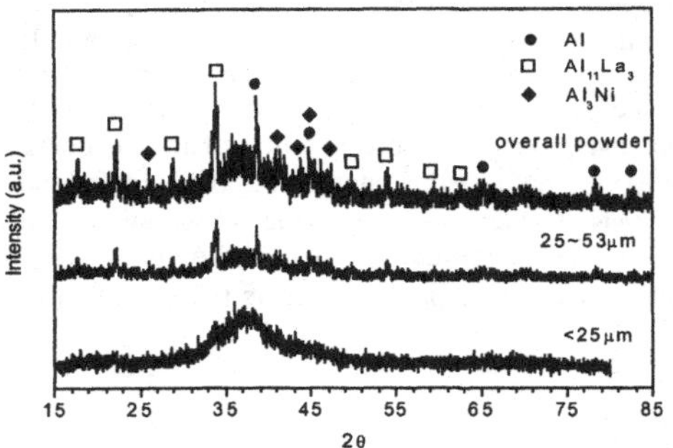

Figure 1 XRD patterns of gas atomized $Al_{85}Ni_{10}La_5$ powders.

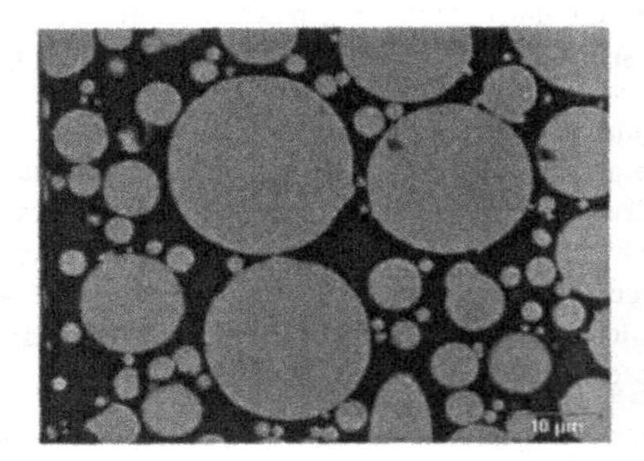

Figure 2 SEM back scattered electron pictures of the cross-sections of amorphous $Al_{85}Ni_{10}La_5$ powders (< 25 μm).

3. Results

Figure 3 shows the continuous heating DSC traces of the $Al_{85}Ni_{10}La_5$ amorphous powders (<25 μm) using heating rates from 2.5 °C/min to 40 °C/min. Crystallization processes occurred in the temperature range of 250°C to 390°C. Three distinct exothermic peaks were observed during the devitrification process. At high heating rates, a minor endothermic signal was observed prior to the first crystallization peak, which corresponded to the glass transition, while another minor exothermic peak was found after the third exothermic peak. In order to determine the crystallized phase corresponding to the first peak, the amorphous powders were heated to the peak temperature of 263°C at a heating rate of 10 °C/min and held for 5 minutes. The heat-treated powders were then analyzed using XRD. The XRD results showed the first exothermic peak corresponded to a eutectic-like reaction. The precipitated phase consisted of

Al, Al_3Ni, $Al_{11}La_3$ and a metastable phase of Al_3La. XRD analysis of amorphous powders that were heated to 500°C indicated that the fully crystallized phases consisted of Al, Al_3Ni and $Al_{11}La_3$; the metastable phase Al_3La had disappeared. In other words, the phase composition of the fully crystallized powder is the same as that of the coarse powders larger than 25 μm.

The crystallization onset temperatures and the peak temperatures for the first three peaks determined by DSC tracing are listed in Table 1. According to Kissinger [6], the apparent activation energy controlling the exothermic reactions is given by the following equation:

$$\ln\left(\frac{v}{T^2}\right) = -\frac{E_a}{RT} + C$$

where v is the heating rate, R is the gas constant, T is the characteristic temperature and can be frequently selected as the peak temperature T_p or the onset temperature T_{on}. The term C is a constant. As a result, the activation energy can be determined from the slope of the curves $\ln(v/T_p^2)$ vs. $1/T_p$. These plots were given in Figure 4. The activation energies corresponding the first peaks and the third peaks are linearly fitted to be 344 kJ/mol and 198 kJ/mol respectively. The second peak does not exhibit a linear relationship and the reasons for this behavior will be discussed subsequently.

Table 1 Onset and Peak Temperatures of the Exothermic Reactions at Various Heating Rates

Heating rate (°C/min)	First peak Onset (°C)	First peak Peak (°C)	Second peak Onset (°C)	Second peak Peak (°C)	Third peak Onset (°C)	Third peak Peak (°C)
2.5	249.8	253.6	269.3	286.2	--	295.2
5	253.1	259.0	276.6	293.8	298.3	305.3
10	258.8	263.4	276.7	297.0	304.7	312.9
20	264.6	267.6	276.8	286.6	306.4	323.6
40	268.5	272.8	282.9	290.0	310.9	333.8

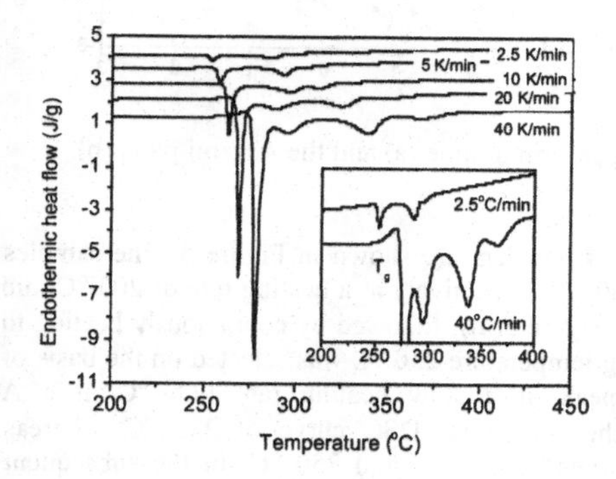

Figure 3 Continuous heating DSC traces of $Al_{85}Ni_{10}La_5$ amorphous powders (<25 μm). The inset picture shows the enlarged curves for the heating rate of 40 °C /min and 2.5 °C /min.

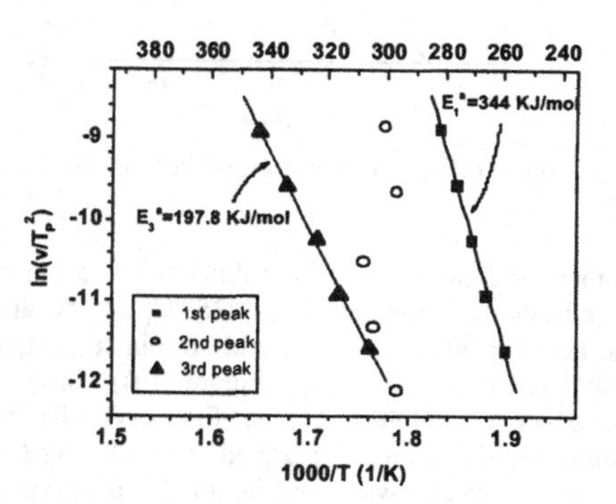

Figure 4 Kissinger plots of gas atomized $Al_{85}Ni_{10}La_5$ powders.

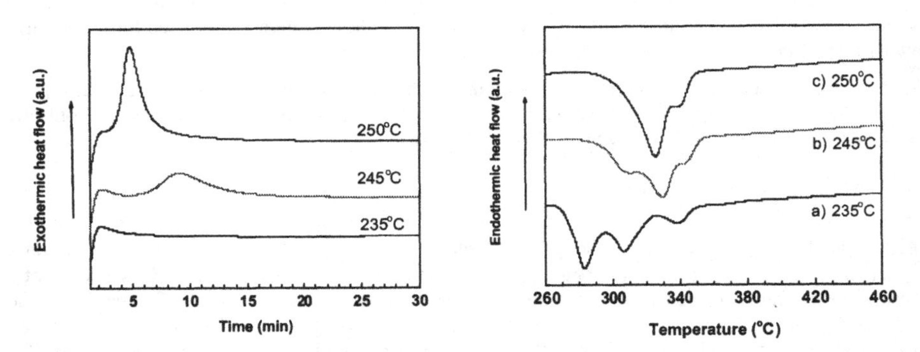

Figure 5 Isothermal annealing of $Al_{85}Ni_{10}La_5$ amorphous powders (< 25 μm) at different temperatures (a) and then continuous heating to 500 °C with a heating rate of 40 °C/min (b).

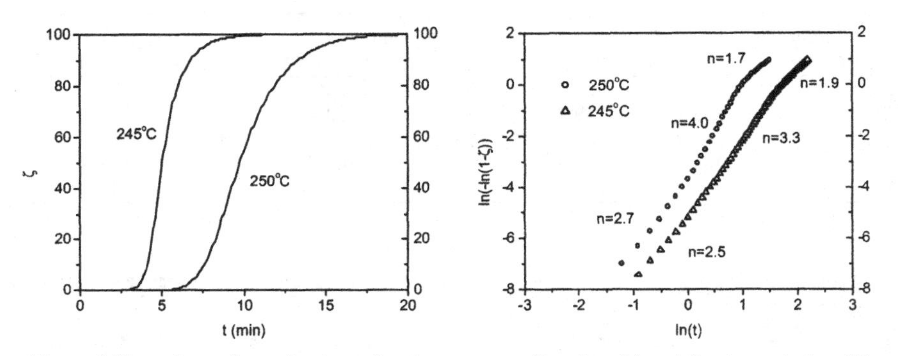

Figure 6 Plots of transformed volume fraction vs. annealing time (a) and the Avrami plots (b).

Isothermal DSC traces of the Al85Ni10La5 amorphous powders are shown in Figure 5. The samples were heated to temperatures of 235 °C, 245 °C and 250 °C, respectively at a heating rate of 200 °C/min and held for 60 min, 30 min and 30 min respectively (Figure 5(a)), followed by continuously heating to 500°C at a rate of 40°C/min (Figure 5(b)). The holding temperature 250 °C was selected on the basis of the onset temperature of the first crystallization peak at a slow heating rate (2.5 °C/min). A monotonically decreasing signal was observed in the isothermal DSC curve of 235 °C, whereas exothermic peaks were present in the isothermal curves of 245 °C and 250 °C. In the subsequent continuous DSC traces, three distinct exothermic peaks were found after annealing at 235 °C whereas the exothermic peaks became overlapped or absent in the curves for the samples annealed at 245 °C and 250 °C. It is therefore suggested that different crystallization mechanisms controlled the crystallization at 235 °C and above 245 °C.

By integrating the exothermic peaks in Figure 5(a), a sigmoidal relationship of the transformed volume fraction vs. the annealing time can be obtained, as seen in Figure 6(a). The sigmoid phase transformation curve is often described by the Johnson-Mehl-Avrami equation [7]:

$$\zeta = 1 - \exp\left(-k(t-\tau)^n\right)$$

where ζ is the transformed volume fraction, n is the Avrami exponent which depends on the transformation mechanism, τ is the incubation time at which a region nucleates. The term k is the reaction constant. The Avrami plots of $\ln(-\ln(1-\zeta))$ vs. $\ln(t)$ are given in Figure 6(b). The Avrami exponents at 245 °C are equal to 2.5 at the initial stage, then increase to 3.3 at the intermediate stage and decrease to 1.9 at the final stage. In the case of 250 °C, the Avrami exponent starts at 2.7 and increases to 4.0 before finally decreasing to 1.7. The non-linear nature of the Avrami plots is indicative of non-steady state transformation. The XRD patterns for the samples annealed at 235 °C, 245 °C and 250 °C are shown in Figure 7. It is observed that crystallization of Al phase occurred at 235 °C. Al phase and intermetallic compounds of $Al_{11}La_3$, Al_3Ni and Al_3La precipitated at 245 °C with a eutectic-like reaction. No apparent differences were found in terms of phase formations for crystallization at 245 °C and 250 °C. For comparison, the XRD pattern for the powders which were annealed at 283 °C for 5 min is also plotted on Figure 7. It can be seen that the metastable phase Al_3La has disappeared. The eutectic-like precipitation manner may contribute to the variation of the Avrami exponents during the annealing process.

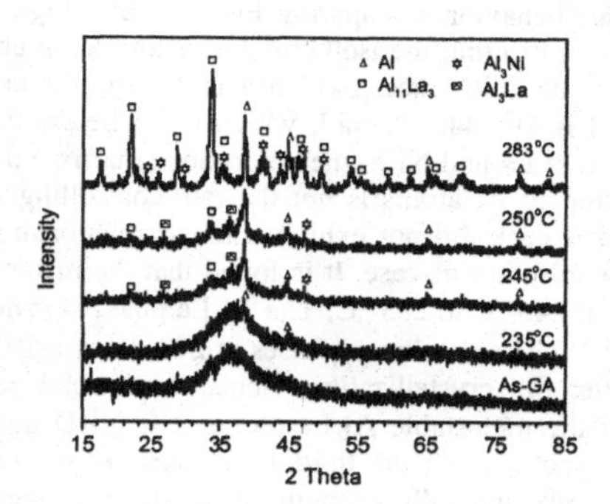

Figure 7 XRD pattern of $Al_{85}Ni_{10}La_5$ powders (<25μm) annealed at various temperatures.

4. Discussion

The present results show that an amorphous powder of the $Al_{85}Ni_{10}La_5$ alloy can be produced by gas atomization. The amorphous phase is stable up to an annealing temperature of 235 °C. The observation of a monotonically decreasing signal during the isothermal DSC tracing performed at 235 °C suggested a grain growth reaction [8], indicating the presence of quenched-in nuclei in the amorphous matrix. However, a heating rate of 200 °C to the holding temperatures was employed in Figure 5. In order to rule out that the signal resulted from an instrumental artifact due to heat flow overshoot, a cross-examination test was completed using a heating rate of 20 °C/min, followed by re-running the temperature program with the transformed sample after it had cooled to room temperature, as shown in

Figure 8. The baseline signal gave a perfect horizontal line indicating that the monotonically decreasing signal resulted from phase transformation in the sample. This confirms that the observation of a monotonically decreasing curve is a characteristic feature due to grain growth in the amorphous matrix. In contrast, observation of an exothermic peak suggested a nucleation and growth reaction during the isothermal annealing process at 245 °C and 250 °C. Such different crystallization mechanisms for 235 °C and above 245 °C were also supported by the subsequent continuous DSC scans. As shown in Figure 5(b), the DSC curve for the sample pre-annealed at 235 °C is obviously different from the other two curves.

At a temperature of 245 °C, the Avrami exponent is 2.5 at the initial stage, followed by an increase to 3.3 and then a decrease to 1.9. An Avrami exponent with a value of 2.5 is a typical value for diffusion-controlled growth with a constant nucleation rate [7]. The value of 3.3 corresponds to interface-controlled growth of nuclei with decreasing nucleation rate. The Avrami exponent in the range of 1.5 to 2.5 corresponds to diffusion-controlled growth with a decreasing nucleation rate. At the temperature of 250 °C, the Avrami exponent is 2.7 at the initial stage, followed by an increase to 4.0 and then a decreas to 1.7. The value of 2.7, slightly larger than 2.5, corresponds to the diffusion-controlled growth with increasing nucleation rate. The Avrami exponent of 4.0 is typical for interface-controlled growth with constant nucleation rate.

It should be noted that continuous DSC curves mainly exhibited a three-stage crystallization behavior (Figure 3) and the first crystallization peaks in these curves occurred above 250 °C. Figure 5(b) showed that the curve for the samples held at 235 °C exhibited a similar three-stage behavior, whereas those held above 245 °C did not display this behavior. Comparing Figure 3 and Figure 5(b), it is reasonable to assume that the activation energy controlling the isothermal transformation above 245 °C is identical to that for the first peaks of continuous DSC traces, which was determined to be 344 kJ/mol. The self-diffusion activation energy of Al is Q=144.2 kJ/mol, which is far below the value of 344 kJ/mol. It implies that presence of the elements La and Ni in the amorphous matrix either considerably influences the precipitation of Al or diffusion of Al atoms is not the rate-controlling limit in the crystallization process above 245 °C. The second peak did not exhibit a linear relationship, but the presence of the metastable phase Al_3La is noteworthy in this case. It is found that the metastable phase Al_3La could be present in the temperature range of 245 °C to 283 °C. The Al_3La phase was not observed by XRD in the powders held at 283 °C for 5 min. The crystallized phases at 283 °C consist of Al, $Al_{11}La_3$ and Al_3Ni, which are the same as those after full crystallization. Hence the second peak may be related to the formation and decomposition of the metastable Al_3La phase. The XRD traces do not allow a simple identification of the phases corresponding to the third DSC peak. However, in comparing the X-ray diffraction intensities for the partially and fully crystallized phases, it is reasonable to assume that the third peak should be related to the precipitation of Al_3Ni.

Precipitation of the elemental Al phase will imply that the remnant amorphous phase becomes enriched with alloying elements, especially near the aluminum phase nuclei. These increased concentrations stabilize the amorphous phase and impede crystallization [9, 10]. Therefore it can be deduced that the crystallization process should be affected by the pre-annealing procedure if the $Al_{85}Ni_{10}La_5$ amorphous powders are held at 235 °C. Figure 8 shows the influence of annealing time. $Al_{85}Ni_{10}La_5$ powders were first pre-annealed at 235 °C and then heated to 283 °C at a heating rate of 40 °C/min. It was found that increasing the pre-annealing time from 30 min to 60 min caused a peak shift from 277 °C to 283 °C. This suggests that the thermal stability of the amorphous phase has been increased.

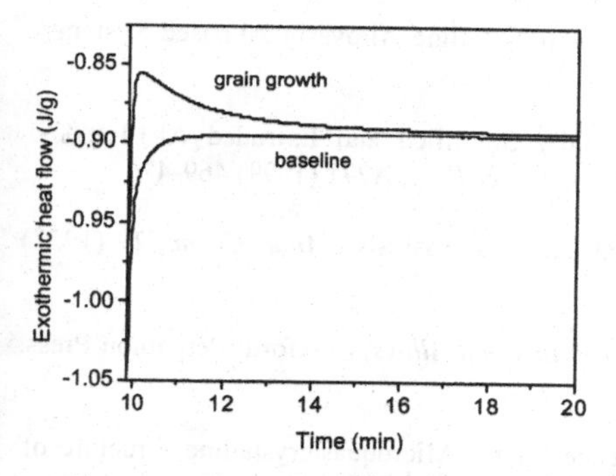

Figure 8 DSC isothermal traces of $Al_{85}Ni_{10}La_5$ powders (<25μm) held at 235°C.

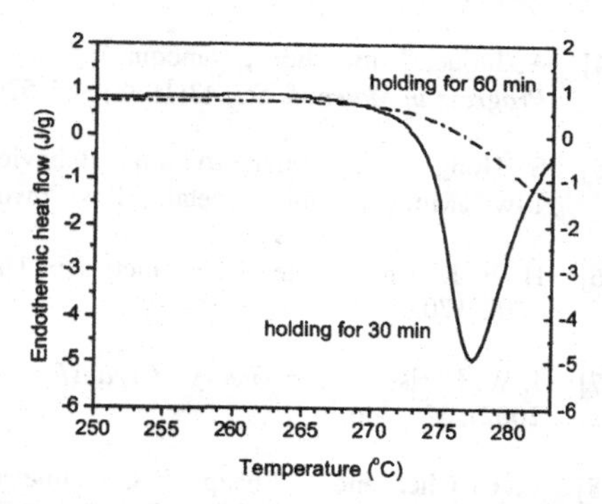

Figure 9 Continuous DSC traces showing that the thermal stability can be improved by a pre-annealing at 235°C.

5. Conclusions

(1) Amorphous $Al_{85}Ni_{10}La_5$ powders with a particle size less than 25 μm were produced by gas atomization. The amorphous phase is stable on heating up to a temperature of 235 °C. A pre-annealing at this temperature could improve the thermal stability of the amorphous phase.

(2) Isothermal DSC traces indicate that quenched-in Al nuclei existed in the amorphous matrix. Crystallization behavior at the temperature of 235 °C exhibited a manner of grain growth of aluminum phase. A nucleation and growth behavior of devitrification were suggested at the temperatures above 245 °C and the phase transformation exhibited a eutectic-like reaction process, during which Al phase was precipitated concomitantly with intermetallic compounds of $Al_{11}La_3$ and Al_3Ni as well as a metastable phase Al_3La.

Acknowledgements

The authors would like to acknowledge the Army Research Office (Grant No. DAAD 19-03-1-0020 & Grant No. DAAD19-01-1-0627) for financial support. Particular thanks also go to Dr. William Mullins for his support and assistance.

References

[1] J -P. Immarigeon et al., "Lightweight Materials for Aircraft Applications", *Materials Characterization*, 35 (1995) 41-67

[2] Y. H. Kim, A. Inoue, and T. Masumoto, "Ultrahigh Tensile Strength of $Al_{88}Y_2Ni_9Mn_1$ or $Al_{88}Y_2Ni_9Fe_1$ Amorphous-Alloys Containing Finely Dispersed Fcc-Al Particles", *Mater. Trans. J. I. M.*, 31 (1990) 747-749

[3] Y. He, S. J. Poon, and G. J. Shiflet, "Synthesis and Properties of Metallic Glasses that Contain Aluminum", *Science*, 241 (1988) 1640-1642

[4] A. Inoue, "Amorphous, Nanoquasicrystalline and Nanocrystalline Alloys in Al-based Systems", *Progress in Mater. & Sci*, 43(1998) 365-520

[5] S. Hong et al., "Microstructural Behavior of Rapidly Solidified and Extruded Al-14wt%Ni-14wt%Mm (Mm, misch metal) Alloy Powders", *Mater. Sci. & Eng.*, A271 (1999) 469-476

[6] H. E. Kissinger, "Reaction Kinetics in Differential Thermal Analysis", *Anal. Chem.*, 29 (1957) 1702-1705

[7] J. W. Christian, *The Theory of Transformations in Metals and Alloys*, (Oxford: Pergamon Press, 2002), 529-546

[8] L. C. Chen and F. Spaepen, "Calorimetric Evidence for the Microquasicrystalline Structure of 'Amorphous' Al /Transition Metal Alloys", *Nature*, 336 (1988) 366-368

[9] X. Y. Jiang, Z. C. Zhong, and A. L. Greer, "Particle Size effects in Primary Crystallization of Amorphous Al-Ni-Y Alloys", *Mater. Sci. & Eng.*, A226-228 (1997) 789-793

[10] D. R. Allen, J. C. Foley, and J. H. Perepezko, "Nanocrystal Development during Primary Crystallization of Amorphous Alloys", *Acta Mater.*, 46 (1998) 431-440

Processing and Properties of Structural Nanomaterials
Edited by Leon L. Shaw, C. Suryanarayana and Rajiv S. Mishra
TMS (The Minerals, Metals & Materials Society), 2003

Texture Evolution in Nanocrystalline Fe-36wt%Ni Alloy Foil

J. H. Seo, J. K. Kim and Y. B. Park

Department of Materials Science and Metallurgical Engineering,
Nano Materials Research Center, Sunchon National University,
Sunchon 540-742, Korea

Keywords: Electrodeposition, Grain Growth, Invar, Nanostructure, OIM, Texture.

Abstract

The texture evolution that takes place during annealing was investigated in a nanocrystalline Fe-36wt%Ni alloy fabricated by using an electrodeposition method. The as-deposited texture was characterized by a strong <100>//ND fibre and a weak <111>//ND fibre. Grain growth occurred during annealing beyond 375°C and resulted in the texture change that the <111>//ND fibre strongly developed at the expense of the <100>//ND fibre. It was observed using orientation image microscopy that the <111>//ND oriented grains abnormally grew in the early stages of grain growth. The relationship between the texture evolution and the microstructural change during annealing is interpreted and discussed in terms of the orientation dependency of grain growth.

Introduction

Fe-36wt%Ni referred to as Invar has been representative of alloys with low thermal expansion [1, 2]. Invar anomalies, namely its thermal expansion coefficient close to zero at room temperature, have been extensively studied, but there exist by far many different explanations proposed for these phenomena [3, 4]. Although Invar anomalies are not yet been fully understood, Fe-36wt%Ni and neighbor alloys have been widely used for industrial applications such as shadow masks for the cathode ray tubes, bi/tri-metals, electronic devices, etc.

Nanocrystalline materials consisting of nanometer-sized crystallites contain a large number of interfaces, and thus, a large volume fraction of the atoms are associated with the intercrystalline region [5]. Due to this morphological characteristic, in general, nanocrystalline materials are far away from thermodynamic equilibrium and as a result reveal quite different properties in comparison with conventional coarse-grained materials [6, 7]. In ferromagnetic materials, the soft magnetic properties can be significantly modified by reducing grain size to nanometer scale [8]. Since Invar anomalies are strongly related to a magnetism behavior [9], a particular attention has recently been paid to the synthesis of nanocrystalline Invar.

In the current work, a nanocrystalline Fe-36wt%Ni alloy foil was fabricated using an electrodeposition method. Such nanocrystalline electrodeposits are in a high non-equilibrium (metastable) state, i.e. a high energy state due to a large number of interfaces [5-7]. In nanocrystalline materials, therefore, grain growth can occur at relatively lower temperatures than in counterpart microcrystalline materials [10]. Grain growth that takes place at low temperature is related to thermal stability of nanocrystalline materials from the viewpoint of their technological applications. In materials, the occurrence of grain growth is accompanied not merely by change in dimension but also by a structural rearrangement, i.e. change of crystallographic textures. However, the texture phenomena due to grain growth have not been fully explained by using prevailing models [11]. Therefore, the present study is aimed at elucidating the relationship between the texture evolution and grain growth that take place during annealing in the nanocrystalline Fe-36wt%Ni alloy foil.

Experimental Procedure

A nanocrystalline Fe-36wt%Ni alloy foil was fabricated by using an electrodeposition method in a newly developed electrolytic bath containing nickel chloride, iron sulfate, boric acid and saccharine [12]. Deposition was carried out at a temperature of 45°C and at a current density of 100 mA/cm^2. The final thickness of the electrodeposited foils was approximately 20 μm.

In order to characterize the Fe-36wt%Ni alloy, synchrotron X-ray diffraction (XRD) was employed in the 8C1 POSCO beam line at the Pohang Light Source (PLS) in Korea. The incident X-rays were vertically focused by a mirror, and monochromatized to the wavelength of 1.7714 □ by a double bounce Si (111) monochromator. The monochromator also focused the X-rays horizontally. The momentum transfer resolution was controlled by two pairs of slits on the detector arm, and was set at 0.001 Å^{-1} in this experiment.

Thermal behavior of the specimens was examined using a DSC, where the samples were heated from room temperature up to 450°C at a heating rate of 10°C/min. X-ray pole figures were measured using Co K_{α} to determine the development of macro-textures. For the analysis of micro- textures, the samples were observed in a field emission gun scanning electron microscope (FEGSEM) with an automated orientation imaging microscopy (OIM) system.

Results

Figure 1 shows the distribution of grain size calculated on the basis of XRD peak broadening in terms of {111} and {200} in the nanocrystalline Fe-Ni electrodeposits fabricated under the current experimental conditions. The average grain size increased with Ni content, but it approached approximately 5 nanometers for Invar and its neighbor compositions. In DSC curve obtained during heating the nanocrystalline Fe-36wt%Ni electrodeposit, a heat release peak appeared, as shown in Fig. 2, starting at 375.7°C, i.e. the onset temperature and reaching the peak maximum at 392.3°C.

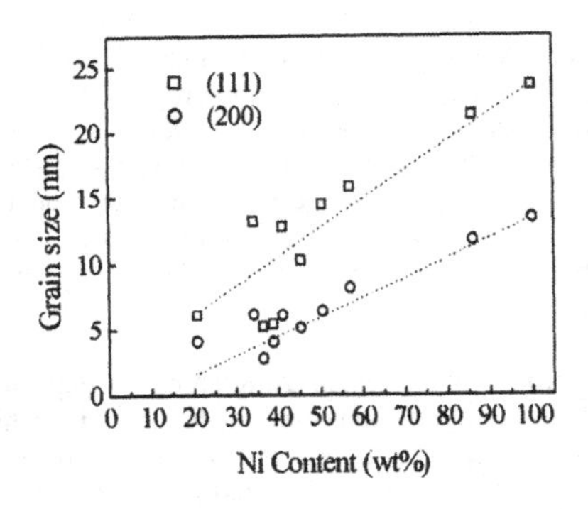

Figure 1: Change of grain size as a function of Ni content in nanocrystalline Fe-Ni electrodeposits.

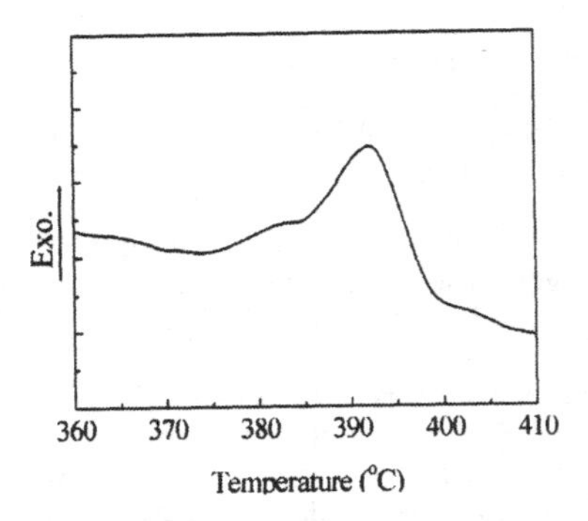

Figure 2: DSC curve obtained during heating the nanocrystalline Fe-36wt%Ni alloy electrodeposit.

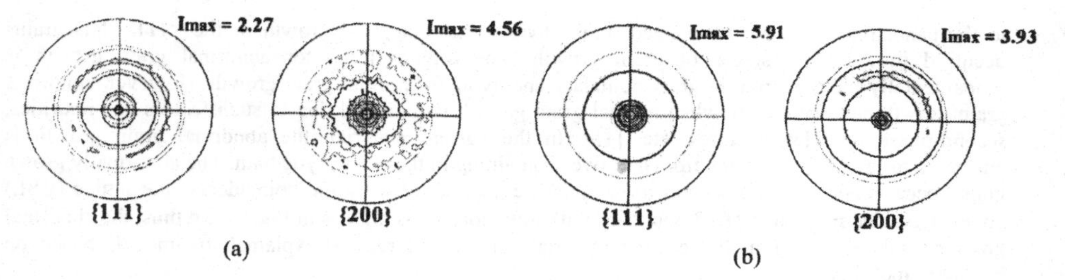

Figure 3: {111} and {200} pole figures measured in the Fe-36wt%Ni alloy electrodeposit. (a) As-deposited. (b) After annealing at 390°C for 30 min.

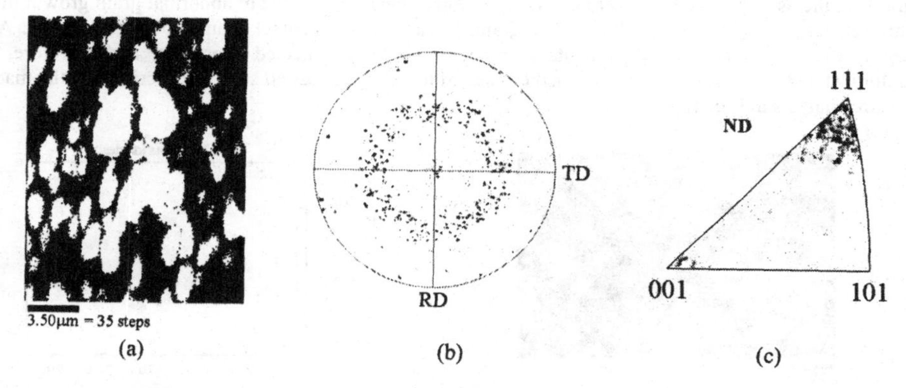

Figure 4: Early stage of grain growth in the Fe-36wt%Ni alloy electrodeposit after annealing at 390 °C for 10 sec. (a) OIM map (white; <111>, grey; <100> and black; nanostructured state). (b) {200} pole figure from the OIM map. (c) ND inverse pole figure from the OIM map.

This exothermal process is attributed to the occurrence of grain growth in the sample. Grain growth occurs in polycrystalline materials to decrease the interfacial energy and hence the total energy of the system. Since nanocrystalline materials are in a high energy state due to a large volume fraction of the interfacial component, the driving force for grain growth is so large that the grain boundaries can move far below temperatures at which grain growth is expected to occur in microcrystalline materials [7].

Grain growth that took place in the specimen resulted in change of crystallographic textures as shown in Fig. 3. The as-deposited texture is characterized by a mixture of a strong <100>//ND and a weak <111>//ND fibre components (Fig. 3(a)). This textural state was completely reversed after annealing. As shown in Fig. 3(b), in the specimen annealed at 390 °C for 30 min, the development of the <111>//ND fibre texture became predominant while the intensity of the <100> fibre texture decreased to some extent. Therefore, it is obvious that during annealing, the <111>//ND oriented grains grew faster than <100>//ND and other oriented grains.

The OIM map obtained in the sample annealed at 390°C for 10 sec shows that the <111>//ND grains were rapidly grown to be much coarser than other oriented grains as in Fig. 4. In this OIM map, white and grey colors indicate <111>//ND and <100>//ND grains, respectively, and the dark area corresponds to the nanostructured state unchanged. It is obvious that abnormal growth of the <111>//ND grains occurred in the early stages of grain growth. The driving force for abnormal grain growth is fundamentally the reduction in grain boundary energy as for normal grain growth. However, abnormal grain growth can occur only when normal grain growth is inhibited due to such factors as impurities, second-phase particles, textures, etc. [11]. In the current material, the abnormal grain growth is attributed to impurities resulted from additives containing in the electrolytic bath. The abnormally grown grains occasionally contain several misoirented regions divided by the coincidence site lattice (CSL) boundaries. The frequency of Σ3 boundary with low mobility is high as in Fig. 5, and thus, the abnormal grain growth occurring in the current material can not be readily explained by models based on boundary mobility.

Figure 6 shows the OIM map in the specimen annealed at 390 °C for 30 min. The nanometer-sized grains in the as-deposited state were grown up to micrometer scale. Due to abnormal grain growth in the early stages of grain growth, <111>//ND grains became much coarser than <100>//ND grains. As a result, <111>//ND texture components dominated the fully annealed specimen, and therefore, the textural state became completely reversed compared to the as-deposited state as described by the macro-texture analysis in Fig. 3.

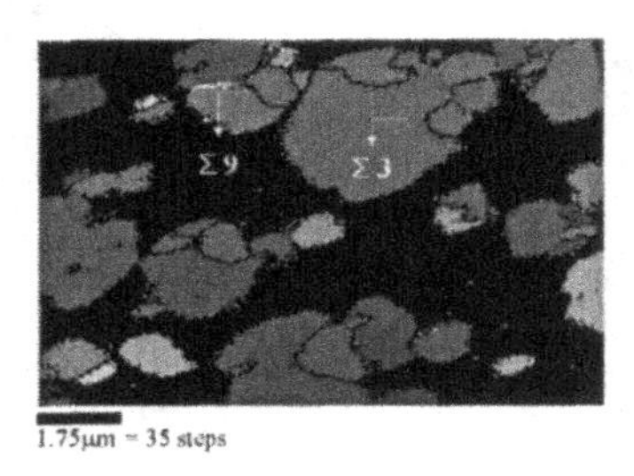

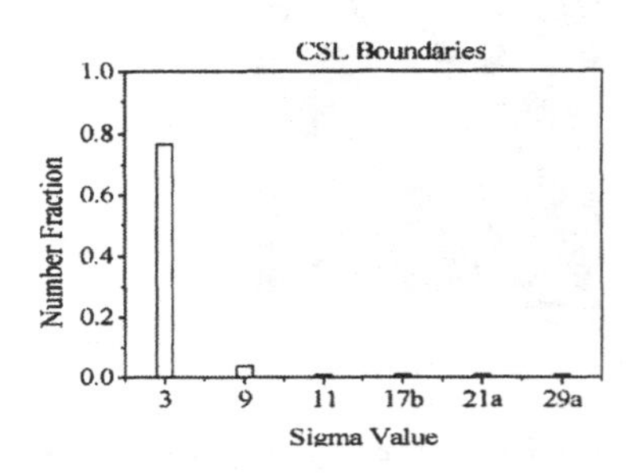

Figure 5: Boundary characterization on the basis of the OIM map in the Fe-36wt%Ni alloy electrodeposit after annealing at 390 °C for 5 min.

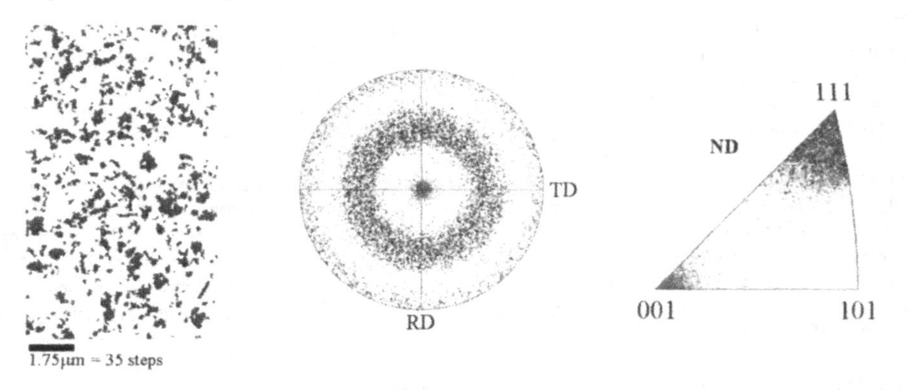

Figure 6: Orientation dependency of grain size in the Fe-36wt%Ni alloy after annealing at 390 °C for 30 min. White and dark areas in the OIM map indicate <111>//ND and <100>//ND grains, respectively.

Discussion

The texture evolution accompanied by grain growth as described above in the nanocrystalline Invar can be explained by a model which Park et al. [14] proposed for nanocrystalline Permalloy as follows. Recently, Li, Czerwinski and Szpunar [13] calculated the orientation dependency of surface energy by using atomistic interactions by Lennard-Jones in electrodeposited Ni-Fe alloys. According to their calculation, the surface energy is the lowest for the {111} plane and increases in the order of {100}, {110}, {311}, {210}, etc. It follows that the <111>//ND texture is expected to strongly develop when the electrodeposition is carried out under a condition close to equilibrium. As the deposition condition becomes farther from equilibrium, crystallographic planes having higher surface energies, e.g. {100}, would increase rather than {111}.

Suppose two nanocrystalline electrodeposits consisting of only the <111>//ND and the <100>//ND grains, respectively. It is apparent that the free energy of the <111>//ND-grained material is lower than that of the <100>//ND-grained material since the former may be electrodeposited under a condition closer to equilibrium. There can exist two possibilities that the energy difference between the two materials results in microstructural difference: first, in case that the two materials have the same grain size, energy per unit volume of the <100>//ND grains would be higher than that of the <111>//ND grains; second, in case that grains with different orientations have the same energy density, the size of the <111>//ND grains would be larger than that of the <100>//ND grains.

These two possibilities will be now employed to explain the texture change that took place during the annealing of the current material. The as-deposited specimen consisted of the <100>//ND grains as the major component and the <111>//ND grains as the minor component. Therefore, our material lies somewhere between the two extremities above, i.e., only-<100>//ND-grained and only-<111>//ND-grained materials. Under assumption that energy density is independent of crystallographic orientation, the <111>//ND grains whose size might be the largest in the material would grow faster due to the size advantage and thus dominate the annealing texture. However, from OIM observation of the as-deposited specimen, there is no evidence that the size of the <111>//ND grains is larger than that of other oriented grains. On the other hand, if grain size is the same regardless of the orientations of grains, the <111>//ND grains having the lowest energy density may grow into the different oriented grains to decrease the total energy of the system. According to the calculation of Park et al. [10], in nanocrystalline Ni electrodeposit the energy density difference between <111>//ND and <100> ND grains is of the order of 1 J/mole, which can affect the 'direction' of grain growth.

Conclusions

Grain growth that takes place during the annealing of electrodeposited nanocrystalline Fe-36wt%Ni alloy foil was investigated and discussed in terms of its possible effect on the texture evolution. The texture of the as-deposited specimen was characterized by a mixture of a strong <100>//ND and a weak <111>//ND components. Abnormal growth of the <111>//ND grains in the early stages of grain growth occurred on annealing beyond 375°C and resulted in the texture change that the <111>//ND texture components strongly developed at the expense of the <100>//ND texture components. This texture evolution was interpreted by means of the orientation dependency of grain growth.

Acknowledgment

This research was supported by a grant from the Center for Advanced Materials Processing (CAMP) of the 21st Century Frontier R&D Program funded by the Ministry of Science and Technology, Korea.

References

1. D. Wenschhof, ed., "Low Expansion Alloys," Metals Handbook 9th ed. vol. 3 (American Society for Metals, 1980), 792-798.
2. S.G. Steinemann, "Invar-Type Alloys - Magnetomechanical Effects," *J. Mag. Mag. Mat.*, 7 (1978), 84-100.
3. S. Chikazumi, "Invar Anomalies," *J. Mag. Mag. Mat.*, 10 (1979), 113-119.
4. D.G. Rancourt, "Invar Problem," *Physics in Canada*, 45 (1989), 3-10.
5. G. Palumbo, S.J. Thorpe and K. Aust, "On the Contribution of Triple Junctions to the Structure and Properties of Nanocrystalline Materials," *Scripta Metall. Mater.*, 24 (1990), 1347-1350.
6. H. Gleiter, "Nanocrystalline Materials," *Progress in Materials Science*, 33 (1989), 223-315.
7. C. Suryanarayana, "Nanocrystalline Materials," *Int. Mater. Rev.*, 40 (1995), 41-64.
8. U. Erb, "Electrodeposited Nanocrystals: Synthesis, Properties and Industrial Applications," NanoStructured Materials, 6 (1995), 533-538.
9. D.G. Rancourt, S. Chehab and G. Lamarche, "Reentrant Magnetism, Antiferromagnetism, and Domain Wall Pinning in Norminally Ferromagnetic Fe-Ni Invar," *J. Mag. Mag. Mat.*, 78 (1989), 129-152.
10. Y.B. Park, S.-H. Hong, C.S. Ha, H.Y. Lee and T.H. Yim, "Orientation Dependency of Grain Growth in a Nanocrystalline Ni Foil," *Materials Science Forum*, 408-412 (2002), 931-936.
11. F.J. Humphreys and M. Hatherly, *Recrystallization and Related Annealing Phenomena*, (Oxford, Elsevier Science Ltd., 1995), 281.
12. Y.B. Park, Nano Invar Alloy and the Process of Producing the Same," Korean Patent No. 10-2003-0026108, (2003).
13. H. Li, F. Czerwinski and J.A. Szpunar, "Monte-Carlo Simulation of Texture and Microstructure Development in Nanocrystalline Electrodeposists," *Nanostruct. Mater.* 9 (1997), 673-676.
14. Y.B. Park, J. Park, C.S. Ha and T.H. Yim, "Texture Evolution during Annealing of Nanocrystalline Permalloy," *Materials Science Forum*, 408-412 (2002), 919-924.

AUTHOR INDEX

SUBJECT INDEX

9 780873 395588